Suzuki
GS, GN, GZ & DR125
Service and Repair Manual

by Jeremy Churchill

with an additional Chapter on the DR125, SF, SH and SJ 'Raider', GN125 and GZ125 'Marauder' models

by Pete Shoemark & Louise Brown

Models covered
GS125. 124cc. 1982 to 1987
GS125ES. 124cc. 1982 to 1999
GN125. 124cc. 1993 to 1999
GZ125 Marauder. 124cc. 1998 to 1999
DR125S. 124cc. 1982 to 1986
DR125S Raider. 124cc. 1985 to 1993

(888 - 208 - 11Y7)

ABCD

A book in the **Haynes Service and Repair Manual Series**

ISBN **1 85960 563 X**

British Library Cataloguing in Publication Data
A catalogue record of this book is available from the British Library

Printed in the USA

Haynes Publishing
Sparkford, Nr Yeovil, Somerset BA22 7JJ, England

Haynes North America, Inc
861 Lawrence Drive, Newbury Park, California 91320, USA

Editions Haynes S.A.
Tour Aurore - IBC, 18 Place des Reflets,
92975 Paris La Défense 2, Cedex, France

Haynes Publishing Nordiska AB
Box 1504, 751 45 UPPSALA, Sweden

Acknowledgements

Thanks are due to Paul Branson Motorcycles of Yeovil, APS Motorcycles of Wells and GT Motorcycles of Yeovil for providing the machines featured in this manual.

We should also like to thank Heron Suzuki (GB) for permission to use their line illustrations, and members of the Technical Service Department of that company for their help and advice.

Finally, we would like to thank the Avon Rubber Company, who kindly supplied information and technical assistance on tyre fitting; NGK Spark Plugs (UK) Ltd for information on spark plug maintenance and electrode conditions, and Renold Ltd for advice on chain care and renewal.

About this manual

The purpose of this manual is to present the owner with a concise and graphic guide which will enable him to tackle any operation from basic routine maintenance to a major overhaul. It has been assumed that any work would be undertaken without the luxury of a well-equipped workshop and a range of manufacturer's service tools.

To this end, the machine featured in the manual was stripped and rebuilt in our own workshop, by a team comprising a mechanic, a photographer and the author. The resulting photographic sequence depicts events as they took place, the hands shown being those of the author and the mechanic.

The use of specialised, and expensive, service tools was avoided unless their use was considered to be essential due to risk of breakage or injury. There is usually some way of improvising a method of removing a stubborn component, providing that a suitable degree of care is exercised.

The author learnt his motorcycle mechanics over a number of years, faced with the same difficulties and using similar facilities to those encountered by most owners. It is hoped that this practical experience can be passed on through the pages of this manual.

Where possible, a well-used example of the machine is chosen for the workshop project, as this highlights any areas which might be particularly prone to giving rise to problems. In this way, any such difficulties are encountered and resolved before the text is written, and the techniques used to deal with them can be incorporated in the relevant section. Armed with a working knowledge of the machine, the author undertakes a considerable amount of research in order that the maximum amount of data can be included in the manual.

A comprehensive section, preceding the main part of the manual, describes procedures for carrying out the routine maintenance of the machine at intervals of time and mileage. This section is included particularly for those owners who wish to ensure the efficient day-to-day running of their motorcycle, but who choose not to undertake overhaul or renovation work.

Each Chapter is divided into numbered sections. Within these sections are numbered paragraphs. Cross reference throughout the manual is quite straightforward and logical. When reference is made 'See Section 6.10' it means Section 6, paragraph 10 in the same Chapter. If another Chapter were intended, the reference would read, for example, 'See Chapter 2, Section 6.10'. All the photographs are captioned with a section/paragraph number to which they refer and are relevant to the Chapter text adjacent.

Figures (usually line illustrations) appear in a logical but numerical order, within a given Chapter. Fig. 1.1 therefore refers to the first figure in Chapter 1.

Left-hand and right-hand descriptions of the machines and their components refer to the left and right of a given machine when the rider is seated normally.

Motorcycle manufacturers continually make changes to specifications and recommendations, and these, when notified, are incorporated into our manuals at the earliest opportunity.

We take great pride in the accuracy of information given in this manual, but motorcycle manufacturers make alterations and design changes during the production run of a particular motorcycle of which they do not inform us. No liability can be accepted by the authors or publishers for loss, damage or injury caused by any errors in, or omissions from, the information given.

Contents

Left-hand view of the DR125 SD

Engine/gearbox unit of the GS125 ESZ

Introduction to the Suzuki GS125 ES, GS125, and DR125 S models

Before 1982 the 125cc capacity class of motorcycles was not particularly popular in the UK, since the machines were too slow to attract in large numbers the younger riders looking for performance above all else, and yet were too large to be attractive to the commuter, or non-enthusiast rider. With a few notable exceptions, principally in the trial bike class, 125s were not good sellers and their relatively dull specifications reflected the manufacturer's natural lack of interest.

This situation was altered radically by the legislation that came into force during 1982 and early 1983; all learner motorcyclists were to be restricted to machines of a maximum engine size of 125cc, the power outputs being restricted to 9kW (12.2 bhp). Almost immediately the four Japanese manufacturers responded with a confusing mass of new models designed to attract the new captive market. Features formerly considered worthwhile only on machines of much larger engine capacity were included in the specification of the new models, many of which incorporated the latest ideas in suspension and styling.

Suzuki's new models were introduced early in 1982, becoming available from June onwards. The basic road model is the GS125 which, apart from its styling, is in the mould of the traditional 125. Thanks to the restrictions imposed, however, it is the equal in performance of its more attractive partner, the GS125 ES, and its lower price might well make it the more popular machine in the long term. The GS125 ES, with its electric start, disc front brake, alloy wheels and sports fairing is intended to be the star of the range, however, and can match the specification and performance of any of the opposition. The DR125 S variant is aleady acknowledged as a first-class trailbike since its performance and agility are the equal of the opposition and it is fitted with Suzuki's 'Full Floater' rear suspension system which is considered by many to be the best available at the moment.

All three machines share the same basic engine/gearbox unit which follows the now standard Japanese practice of using a single overhead camshaft for greater reliability at high engine speeds. The engine is entirely traditional in design and is therefore likely to prove as reliable as its forerunners; the only feature of interest being Suzuki's Twin Dome Combustion Chamber. This was developed as a result of experience with the larger capacity two- and four-cylinder machines in Suzuki's range and is claimed to improve greatly engine efficiency; this would seem to be supported by the excellent fuel consumption figures obtained by most road testers.

The GS125 ES and GS125 employ 'Katana'-inspired styling which makes them at least as attractive as their competitors, while the DR125 S follows the current trend towards 'enduro' styling favoured by all the Japanese manufacturers; all three, however, retain a good measure of practicality which should make them a pleasure to own and maintain in the future.

Modified versions of each model have appeared over the years, each being indicated by a different suffix letter which is Suzuki's own code for each factory production year. It is essential to identify as exactly as possible the machine to be serviced, since while servicing procedures and specifications may not always vary from version to version, in many cases modified components have been fitted which will require care when ordering replacement parts. Given below are the model designation for each version, the initial frame number with which each version's production run commenced and the approximate date of import into the UK. Note that the latter need not necessarily coincide with a machine's date of sale or registration.

Model	Frame number	Date
GS125 Z	NF41B-100001 on	Jan '82 to Mar '84
GS125 D	NF41B-106767 on	Mar '84 to Apr '87
GS125 ESZ	NF41B-100001 on	Jan '82 to Jun '83
GS125 ESD	NF41B-106767 on	Jun '83 to Feb '87
GS125 ESF	NF41B-116269 on	Feb '87 to Sep '89
GS125 ESK	NF41B-174982 on	Sep '89 to Apr '91
GS125 ESM	NF41B-216035 on	Apr '91 to Dec '95
GS125 ESR	NF41B-457356 on	Jan '96 to Nov '98
GS125 ESX	NF41B-554896 on	Nov '98 on
DR125 SZ	SF41A-100001 on	Jan '82 to Jun '83
DR125 SD	SF41A-108134 on	Jun '83 to Jun '85
DR125 SE	SF41A-112155 on	Jun '85 to May '86

Refer to Chapter 7 for DR125 SF, SH and SJ 'Raider' models

Model dimensions and weights

	GS125	GS125 ES	DR125 S
Overall length	1945 mm (76.6 in)	1945 mm (76.6 in)	2095 mm (82.5 in)
Overall width	710 mm (28.0 in)	710 mm (28.0 in)	850 mm (33.5 in)
Overall height	1040 mm (41.0 in)	1122 mm (44.2 in)	1165 mm (46.0 in)
Wheelbase	1270 mm (50.0 in)	1270 mm (50.0 in)	1345 mm (53.0 in)
Ground clearance	170 mm (6.7 in)	170 mm (6.7 in)	265 mm (10.4 in)
Seat height	745 mm (29.3 in)	745 mm (29.3 in)	820 mm (32.3 in)
Dry weight	99 kg (218 lbs)	103 kg (227 lbs)	102 kg (225 lbs)

Ordering spare parts

When ordering spare parts for any Suzuki, it is advisable to deal direct with an official Suzuki agent who should be able to supply most of the parts ex-stock. Parts cannot be obtained from Suzuki direct, even if the parts required are not held in stock. Always quote the engine and frame numbers in full, especially if parts are required for earlier models. The engine number is stamped on the top surface of the left-hand crankcase half. The frame number is stamped on the left-hand side of the steering head.

Use only genuine Suzuki spares. Some pattern parts are available that are made in Japan and may be packed in similar looking packages. They should only be used if genuine parts are hard to obtain or in an emergency, for they do not normally last as long as genuine parts, even though there may be a price advantage.

Some of the more expendable parts such as spark plugs, bulbs, oils and greases etc, can be obtained from accessory shops and motor factors, who have convenient opening hours, and can be found not far from home. It is also possible to obtain parts on a Mail Order basis from a number of specialists who advertise regularly in the motorcycle magazines.

Location of frame number

Location of engine number

Safety first!

Professional motor mechanics are trained in safe working procedures. However enthusiastic you may be about getting on with the job in hand, do take the time to ensure that your safety is not put at risk. A moment's lack of attention can result in an accident, as can failure to observe certain elementary precautions.

There will always be new ways of having accidents, and the following points do not pretend to be a comprehensive list of all dangers; they are intended rather to make you aware of the risks and to encourage a safety-conscious approach to all work you carry out on your vehicle.

Essential DOs and DON'Ts

DON'T start the engine without first ascertaining that the transmission is in neutral.
DON'T suddenly remove the filler cap from a hot cooling system – cover it with a cloth and release the pressure gradually first, or you may get scalded by escaping coolant.
DON'T attempt to drain oil until you are sure it has cooled sufficiently to avoid scalding you.
DON'T grasp any part of the engine, exhaust or silencer without first ascertaining that it is sufficiently cool to avoid burning you.
DON'T allow brake fluid or antifreeze to contact the machine's paintwork or plastic components.
DON'T syphon toxic liquids such as fuel, brake fluid or antifreeze by mouth, or allow them to remain on your skin.
DON'T inhale dust – it may be injurious to health (see *Asbestos* heading).
DON'T allow any spilt oil or grease to remain on the floor – wipe it up straight away, before someone slips on it.
DON'T use ill-fitting spanners or other tools which may slip and cause injury.
DON'T attempt to lift a heavy component which may be beyond your capability – get assistance.
DON'T rush to finish a job, or take unverified short cuts.
DON'T allow children or animals in or around an unattended vehicle.
DON'T inflate a tyre to a pressure above the recommended maximum. Apart from overstressing the carcase and wheel rim, in extreme cases the tyre may blow off forcibly.
DO ensure that the machine is supported securely at all times. This is especially important when the machine is blocked up to aid wheel or fork removal.
DO take care when attempting to slacken a stubborn nut or bolt. It is generally better to pull on a spanner, rather than push, so that if slippage occurs you fall away from the machine rather than on to it.
DO wear eye protection when using power tools such as drill, sander, bench grinder etc.
DO use a barrier cream on your hands prior to undertaking dirty jobs – it will protect your skin from infection as well as making the dirt easier to remove afterwards; but make sure your hands aren't left slippery. Note that long-term contact with used engine oil can be a health hazard.
DO keep loose clothing (cuffs, tie etc) and long hair well out of the way of moving mechanical parts.
DO remove rings, wristwatch etc, before working on the vehicle – especially the electrical system.
DO keep your work area tidy – it is only too easy to fall over articles left lying around.
DO exercise caution when compressing springs for removal or installation. Ensure that the tension is applied and released in a controlled manner, using suitable tools which preclude the possibility of the spring escaping violently.
DO ensure that any lifting tackle used has a safe working load rating adequate for the job.
DO get someone to check periodically that all is well, when working alone on the vehicle.
DO carry out work in a logical sequence and check that everything is correctly assembled and tightened afterwards.
DO remember that your vehicle's safety affects that of yourself and others. If in doubt on any point, get specialist advice.
IF, in spite of following these precautions, you are unfortunate enough to injure yourself, seek medical attention as soon as possible.

Asbestos

Certain friction, insulating, sealing, and other products – such as brake linings, clutch linings, gaskets, etc – contain asbestos. *Extreme care must be taken to avoid inhalation of dust from such products since it is hazardous to health.* If in doubt, assume that they *do* contain asbestos.

Fire

Remember at all times that petrol (gasoline) is highly flammable. Never smoke, or have any kind of naked flame around, when working on the vehicle. But the risk does not end there – a spark caused by an electrical short-circuit, by two metal surfaces contacting each other, by careless use of tools, or even by static electricity built up in your body under certain conditions, can ignite petrol vapour, which in a confined space is highly explosive.

Always disconnect the battery earth (ground) terminal before working on any part of the fuel or electrical system, and never risk spilling fuel on to a hot engine or exhaust.

It is recommended that a fire extinguisher of a type suitable for fuel and electrical fires is kept handy in the garage or workplace at all times. Never try to extinguish a fuel or electrical fire with water.

Note: *Any reference to a 'torch' appearing in this manual should always be taken to mean a hand-held battery-operated electric lamp or flashlight. It does* **not** *mean a welding/gas torch or blowlamp.*

Fumes

Certain fumes are highly toxic and can quickly cause unconsciousness and even death if inhaled to any extent. Petrol (gasoline) vapour comes into this category, as do the vapours from certain solvents such as trichloroethylene. Any draining or pouring of such volatile fluids should be done in a well ventilated area.

When using cleaning fluids and solvents, read the instructions carefully. Never use materials from unmarked containers – they may give off poisonous vapours.

Never run the engine of a motor vehicle in an enclosed space such as a garage. Exhaust fumes contain carbon monoxide which is extremely poisonous; if you need to run the engine, always do so in the open air or at least have the rear of the vehicle outside the workplace.

The battery

Never cause a spark, or allow a naked light, near the vehicle's battery. It will normally be giving off a certain amount of hydrogen gas, which is highly explosive.

Always disconnect the battery earth (ground) terminal before working on the fuel or electrical systems.

If possible, loosen the filler plugs or cover when charging the battery from an external source. Do not charge at an excessive rate or the battery may burst.

Take care when topping up and when carrying the battery. The acid electrolyte, even when diluted, is very corrosive and should not be allowed to contact the eyes or skin.

If you ever need to prepare electrolyte yourself, always add the acid slowly to the water, and never the other way round. Protect against splashes by wearing rubber gloves and goggles.

Mains electricity and electrical equipment

When using an electric power tool, inspection light etc, always ensure that the appliance is correctly connected to its plug and that, where necessary, it is properly earthed (grounded). Do not use such appliances in damp conditions and, again, beware of creating a spark or applying excessive heat in the vicinity of fuel or fuel vapour. Also ensure that the appliances meet the relevant national safety standards.

Ignition HT voltage

A severe electric shock can result from touching certain parts of the ignition system, such as the HT leads, when the engine is running or being cranked, particularly if components are damp or the insulation is defective. Where an electronic ignition system is fitted, the HT voltage is much higher and could prove fatal.

Tools and working facilities

The first priority when undertaking maintenance or repair work of any sort on a motorcycle is to have a clean, dry, well-lit working area. Work carried out in peace and quiet in the well-ordered atmosphere of a good workshop will give more satisfaction and much better results than can usually be achieved in poor working conditions. A good workshop must have a clean flat workbench or a solidly constructed table of convenient working height. The workbench or table should be equipped with a vice which has a jaw opening of at least 4 in (100 mm). A set of jaw covers should be made from soft metal such as aluminium alloy or copper, or from wood. These covers will minimise the marking or damaging of soft or delicate components which may be clamped in the vice. Some clean, dry, storage space will be required for tools, lubricants and dismantled components. It will be necessary during a major overhaul to lay out engine/gearbox components for examination and to keep them where they will remain undisturbed for as long as is necessary. To this end it is recommended that a supply of metal or plastic containers of suitable size is collected. A supply of clean, lint-free, rags for cleaning purposes and some newspapers, other rags, or paper towels for mopping up spillages should also be kept. If working on a hard concrete floor note that both the floor and one's knees can be protected from oil spillages and wear by cutting open a large cardboard box and spreading it flat on the floor under the machine or workbench. This also helps to provide some warmth in winter and to prevent the loss of nuts, washers, and other tiny components which have a tendency to disappear when dropped on anything other than a perfectly clean, flat, surface.

Unfortunately, such working conditions are not always available to the home mechanic. When working in poor conditions it is essential to take extra time and care to ensure that the components being worked on are kept scrupulously clean and to ensure that no components or tools are lost or damaged.

A selection of good tools is a fundamental requirement for anyone contemplating the maintenance and repair of a motor vehicle. For the owner who does not possess any, their purchase will prove a considerable expense, offsetting some of the savings made by doing-it-yourself. However, provided that the tools purchased meet the relevant national safety standards and are of good quality, they will last for many years and prove an extremely worthwhile investment.

To help the average owner to decide which tools are needed to carry out the various tasks detailed in this manual, we have compiled three lists of tools under the following headings: *Maintenance and minor repair, Repair and overhaul,* and *Specialized.* The newcomer to practical mechanics should start off with the simpler jobs around the vehicle. Then, as his confidence and experience grow, he can undertake more difficult tasks, buying extra tools as and when they are needed. In this way, a *Maintenance and minor repair* tool kit can be built-up into a *Repair and overhaul* tool kit over a considerable period of time without any major cash outlays. The experienced home mechanic will have a tool kit good enough for most repair and overhaul procedures and will add tools from the specialized category when he feels the expense is justified by the amount of use these tools will be put to.

It is obviously not possible to cover the subject of tools fully here. For those who wish to learn more about tools and their use there is a book entitled *Motorcycle Workshop Practice Manual* available from the publishers of this manual.

As a general rule, it is better to buy the more expensive, good quality tools. Given reasonable use, such tools will last for a very long time, whereas the cheaper, poor quality, item will wear out faster and need to be renewed more often, thus nullifying the original saving. There is also the risk of a poor quality tool breaking while in use, causing personal injury or expensive damage to the component being worked on.

For practically all tools, a tool factor is the best source since he will have a very comprehensive range compared with the average garage or accessory shop. Having said that, accessory shops often offer excellent quality tools at discount prices, so it pays to shop around. There are plenty of tools around at reasonable prices, but always aim to purchase items which meet the relevant national safety standards. If in doubt, seek the advice of the shop proprietor or manager before making a purchase.

The basis of any toolkit is a set of spanners. While open-ended spanners with their slim jaws, are useful for working on awkwardly-positioned nuts, ring spanners have advantages in that they grip the nut far more positively. There is less risk of the spanner slipping off the nut and damaging it, for this reason alone ring spanners are to be preferred. Ideally, the home mechanic should acquire a set of each, but if expense rules this out a set of combination spanners (open-ended at one end and with a ring of the same size at the other) will provide a good compromise. Another item which is so useful it should be considered an essential requirement for any home mechanic is a set of socket spanners. These are available in a variety of drive sizes. It is recommended that the $\frac{1}{2}$-inch drive type is purchased to begin with as although bulkier and more expensive than the $\frac{3}{8}$-inch type, the larger size is far more common and will accept a greater variety of torque wrenches, extension pieces and socket sizes. The socket set should comprise sockets of sizes between 8 and 24 mm, a reversible ratchet drive, an extension bar of about 10 inches in length, a spark plug socket with a rubber insert, and a universal joint. Other attachments can be added to the set at a later date.

Maintenance and minor repair tool kit

Set of spanners 8 – 24 mm
Set of sockets and attachments
Spark plug spanner with rubber insert – 10, 12, or 14 mm as appropriate
Adjustable spanner
C-spanner/pin spanner
Torque wrench (same size drive as sockets)
Set of screwdrivers (flat blade)
Set of screwdrivers (cross-head)
Set of Allen keys 4 – 10 mm
Impact screwdriver and bits
Ball pein hammer – 2 lb
Hacksaw (junior)
Self-locking pliers – Mole grips or vice grips
Pliers – combination
Pliers – needle nose
Wire brush (small)
Soft-bristled brush
Tyre pump
Tyre pressure gauge
Tyre tread depth gauge
Oil can
Fine emery cloth
Funnel (medium size)
Drip tray
Grease gun
Set of feeler gauges
Brake bleeding kit
Strobe timing light
Continuity tester (dry battery and bulb)
Soldering iron and solder
Wire stripper or craft knife
PVC insulating tape
Assortment of split pins, nuts, bolts, and washers

Repair and overhaul toolkit

The tools in this list are virtually essential for anyone undertaking major repairs to a motorcycle and are additional to the tools listed above. Concerning Torx driver bits, Torx screws are encountered on some of the more modern machines where their use is restricted to fastening certain components inside the engine/gearbox unit. It is therefore recommended that if Torx bits cannot be borrowed from a local dealer, they are purchased individually as the need arises. They are not in regular use in the motor trade and will therefore only be available in specialist tool shops.

Plastic or rubber soft-faced mallet
Torx driver bits
Pliers – electrician's side cutters
Circlip pliers – internal (straight or right-angled tips are available)
Circlip pliers – external
Cold chisel
Centre punch
Pin punch
Scriber
Scraper (made from soft metal such as aluminium or copper)
Soft metal drift
Steel rule/straight edge
Assortment of files
Electric drill and bits
Wire brush (large)
Soft wire brush (similar to those used for cleaning suede shoes)
Sheet of plate glass
Hacksaw (large)
Valve grinding tool
Valve grinding compound (coarse and fine)
Stud extractor set (E-Z out)

Specialized tools

This is not a list of the tools made by the machine's manufacturer to carry out a specific task on a limited range of models. Occasional references are made to such tools in the text of this manual and, in general, an alternative method of carrying out the task without the manufacturer's tool is given where possible. The tools mentioned in this list are those which are not used regularly and are expensive to buy in view of their infrequent use. Where this is the case it may be possible to hire or borrow the tools against a deposit from a local dealer or tool hire shop. An alternative is for a group of friends or a motorcycle club to join in the purchase.

Valve spring compressor
Piston ring compressor
Universal bearing puller
Cylinder bore honing attachment (for electric drill)
Micrometer set
Vernier calipers
Dial gauge set
Cylinder compression gauge
Vacuum gauge set
Multimeter
Dwell meter/tachometer

Care and maintenance of tools

Whatever the quality of the tools purchased, they will last much longer if cared for. This means in practice ensuring that a tool is used for its intended purpose; for example screwdrivers should not be used as a substitute for a centre punch, or as chisels. Always remove dirt or grease and any metal particles but remember that a light film of oil will prevent rusting if the tools are infrequently used. The common tools can be kept together in a large box or tray but the more delicate, and more expensive, items should be stored separately where they cannot be damaged. When a tool is damaged or worn out, be sure to renew it immediately. It is false economy to continue to use a worn spanner or screwdriver which may slip and cause expensive damage to the component being worked on.

Fastening systems

Fasteners, basically, are nuts, bolts and screws used to hold two or more parts together. There are a few things to keep in mind when working with fasteners. Almost all of them use a locking device of some type; either a lock washer, lock nut, locking tab or thread adhesive. All threaded fasteners should be clean, straight, have undamaged threads and undamaged corners on the hexagon head where the spanner fits. Develop the habit of replacing all damaged nuts and bolts with new ones.

Rusted nuts and bolts should be treated with a rust penetrating fluid to ease removal and prevent breakage. After applying the rust penetrant, let it 'work' for a few minutes before trying to loosen the nut or bolt. Badly rusted fasteners may have to be chiseled off or removed with a special nut breaker, available at tool shops.

Flat washers and lock washers, when removed from an assembly should always be replaced exactly as removed. Replace any damaged washers with new ones. Always use a flat washer between a lock washer and any soft metal surface (such as aluminium), thin sheet metal or plastic. Special lock nuts can only be used once or twice before they lose their locking ability and must be renewed.

If a bolt or stud breaks off in an assembly, it can be drilled out and removed with a special tool called an E-Z out. Most dealer service departments and motorcycle repair shops can perform this task, as well as others (such as the repair of threaded holes that have been stripped out).

Spanner size comparison

Jaw gap (in)	Spanner size	Jaw gap (in)	Spanner size
0.250	$\frac{1}{4}$ in AF	0.945	24 mm
0.276	7 mm	1.000	1 in AF
0.313	$\frac{5}{16}$ in AF	1.010	$\frac{9}{16}$ in Whitworth; $\frac{5}{8}$ in BSF
0.315	8 mm	1.024	26 mm
0.344	$\frac{11}{32}$ in AF; $\frac{1}{8}$ in Whitworth	1.063	1$\frac{1}{16}$ in AF; 27 mm
0.354	9 mm	1.100	$\frac{5}{8}$ in Whitworth; $\frac{11}{16}$ in BSF
0.375	$\frac{3}{8}$ in AF	1.125	1$\frac{1}{8}$ in AF
0.394	10 mm	1.181	30 mm
0.433	11 mm	1.200	$\frac{11}{16}$ in Whitworth; $\frac{3}{4}$ in BSF
0.438	$\frac{7}{16}$ in AF	1.250	1$\frac{1}{4}$ in AF
0.445	$\frac{3}{16}$ in Whitworth; $\frac{1}{4}$ in BSF	1.260	32 mm
0.472	12 mm	1.300	$\frac{3}{4}$ in Whitworth; $\frac{7}{8}$ in BSF
0.500	$\frac{1}{2}$ in AF	1.313	1$\frac{5}{16}$ in AF
0.512	13 mm	1.390	$\frac{13}{16}$ in Whitworth; $\frac{15}{16}$ in BSF
0.525	$\frac{1}{4}$ in Whitworth; $\frac{5}{16}$ in BSF	1.417	36 mm
0.551	14 mm	1.438	1$\frac{7}{16}$ in AF
0.563	$\frac{9}{16}$ in AF	1.480	$\frac{7}{8}$ in Whitworth; 1 in BSF
0.591	15 mm	1.500	1$\frac{1}{2}$ in AF
0.600	$\frac{5}{16}$ in Whitworth; $\frac{3}{8}$ in BSF	1.575	40 mm; $\frac{15}{16}$ in Whitworth
0.625	$\frac{5}{8}$ in AF	1.614	41 mm
0.630	16 mm	1.625	1$\frac{5}{8}$ in AF
0.669	17 mm	1.670	1 in Whitworth; 1$\frac{1}{8}$ in BSF
0.686	$\frac{11}{16}$ in AF	1.688	1$\frac{11}{16}$ in AF
0.709	18 mm	1.811	46 mm
0.710	$\frac{3}{8}$ in Whitworth; $\frac{7}{16}$ in BSF	1.813	1$\frac{13}{16}$ in AF
0.748	19 mm	1.860	1$\frac{1}{8}$ in Whitworth; 1$\frac{1}{4}$ in BSF
0.750	$\frac{3}{4}$ in AF	1.875	1$\frac{7}{8}$ in AF
0.813	$\frac{13}{16}$ in AF	1.969	50 mm
0.820	$\frac{7}{16}$ in Whitworth; $\frac{1}{2}$ in BSF	2.000	2 in AF
0.866	22 mm	2.050	1$\frac{1}{4}$ in Whitworth; 1$\frac{3}{8}$ in BSF
0.875	$\frac{7}{8}$ in AF	2.165	55 mm
0.920	$\frac{1}{2}$ in Whitworth; $\frac{9}{16}$ in BSF	2.362	60 mm
0.938	$\frac{15}{16}$ in AF		

Standard torque settings

Specific torque settings will be found at the end of the specifications section of each chapter. Where no figure is given, bolts should be secured according to the table below.

Fastener type (thread diameter)	**kgf m**	**lbf ft**
5mm bolt or nut	0.45 – 0.6	3.5 – 4.5
6 mm bolt or nut	0.8 – 1.2	6 – 9
8 mm bolt or nut	1.8 – 2.5	13 – 18
10 mm bolt or nut	3.0 – 4.0	22 – 29
12 mm bolt or nut	5.0 – 6.0	36 – 43
5 mm screw	0.35 – 0.5	2.5 – 3.6
6 mm screw	0.7 – 1.1	5 – 8
6 mm flange bolt	1.0 – 1.4	7 – 10
8 mm flange bolt	2.4 – 3.0	17 – 22
10 mm flange bolt	3.0 – 4.0	22 – 29

Choosing and fitting accessories

The range of accessories available to the modern motorcyclist is almost as varied and bewildering as the range of motorcycles. This Section is intended to help the owner in choosing the correct equipment for his needs and to avoid some of the mistakes made by many riders when adding accessories to their machines. It will be evident that the Section can only cover the subject in the most general terms and so it is recommended that the owner, having decided that he wants to fit, for example, a luggage rack or carrier, seeks the advice of several local dealers and the owners of similar machines. This will give a good idea of what makes of carrier are easily available, and at what price. Talking to other owners will give some insight into the drawbacks or good points of any one make. A walk round the motorcycles in car parks or outside a dealer will often reveal the same sort of information.

The first priority when choosing accessories is to assess exactly what one needs. It is, for example, pointless to buy a large heavy-duty carrier which is designed to take the weight of fully laden panniers and topbox when all you need is a place to strap on a set of waterproofs and a lunchbox when going to work. Many accessory manufacturers have ranges of equipment to cater for the individual needs of different riders and this point should be borne in mind when looking through a dealer's catalogues. Having decided exactly what is required and the use to which the accessories are going to be put, the owner will need a few hints on what to look for when making the final choice. To this end the Section is now sub-divided to cover the more popular accessories fitted. Note that it is in no way a customizing guide, but merely seeks to outline the practical considerations to be taken into account when adding aftermarket equipment to a motorcycle.

Fairings and windscreens

A fairing is possibly the single, most expensive, aftermarket item to be fitted to any motorcycle and, therefore, requires the most thought before purchase. Fairings can be divided into two main groups: front fork mounted handlebar fairings and windscreens, and frame mounted fairings.

The first group, the front fork mounted fairings, are becoming far more popular than was once the case, as they offer several advantages over the second group. Front fork mounted fairings generally are much easier and quicker to fit, involve less modification to the motorcycle, do not as a rule restrict the steering lock, permit a wider selection of handlebar styles to be used, and offer adequate protection for much less money than the frame mounted type. They are also lighter, can be swapped easily between different motorcycles, and are available in a much greater variety of styles. Their main disadvantages are that they do not offer as much weather protection as the frame mounted types, rarely offer any storage space, and, if poorly fitted or naturally incompatible, can have an adverse effect on the stability of the motorcycle.

The second group, the frame mounted fairings, are secured so rigidly to the main frame of the motorcycle that they can offer a substantial amount of protection to motorcycle and rider in the event of a crash. They offer almost complete protection from the weather and, if double-skinned in construction, can provide a great deal of useful storage space. The feeling of peace, quiet and complete relaxation encountered when riding behind a good full fairing has to be experienced to be believed. For this reason full fairings are considered essential by most touring motorcyclists and by many people who ride all year round. The main disadvantages of this type are that fitting can take a long time, often involving removal or modification of standard motorcycle components, they restrict the steering lock and they can add up to about 40 lb to the weight of the machine. They do not usually affect the stability of the machine to any great extent once the front tyre pressure and suspension have been adjusted to compensate for the extra weight, but can be affected by sidewinds.

The first thing to look for when purchasing a fairing is the quality of the fittings. A good fairing will have strong, substantial brackets constructed from heavy-gauge tubing; the brackets must be shaped to fit the frame or forks evenly so that the minimum of stress is imposed on the assembly when it is bolted down. The brackets should be properly painted or finished – a nylon coating being the favourite of the better manufacturers – the nuts and bolts provided should be of the same thread and size standard as is used on the motorcycle and be properly plated. Look also for shakeproof locking nuts or locking washers to ensure that everything remains securely tightened down. The fairing shell is generally made from one of two materials: fibreglass or ABS plastic. Both have their advantages and disadvantages, but the main consideration for the owner is that fibreglass is much easier to repair in the event of damage occurring to the fairing. Whichever material is used, check that it is properly finished inside as well as out, that the edges are protected by beading and that the fairing shell is insulated from vibration by the use of rubber grommets at all mounting points. Also be careful to check that the windscreen is retained by plastic bolts which will snap on impact so that the windscreen will break away and not cause personal injury in the event of an accident.

Having purchased your fairing or windscreen, read the manufacturer's fitting instructions very carefully and check that you have all the necessary brackets and fittings. Ensure that the mounting brackets are located correctly and bolted down securely. Note that some manufacturers use hose clamps to retain the mounting brackets; these should be discarded as they are convenient to use but not strong enough for the task. Stronger clamps should be substituted; car exhaust pipe clamps of suitable size would be a good alternative. Ensure that the front forks can turn through the full steering lock available without fouling the fairing. With many types of frame-mounted fairing the handlebars will have to be altered or a different type fitted and the steering lock will be restricted by stops provided with the fittings. Also check that the fairing does not foul the front wheel or mudguard, in any steering position, under full fork compression. Re-route any cables, brake pipes or electrical wiring which may snag on the fairing and take great care to protect all electrical connections, using insulating tape. If the manufacturer's instructions are followed carefully at every stage no serious problems should be encountered. Remember that hydraulic pipes that have been disconnected must be carefully re-tightened and the hydraulic system purged of air bubbles by bleeding.

Two things will become immediately apparent when taking a motorcycle on the road for the first time with a fairing – the first i s the tendency to underestimate the road speed because of the lack of wind pressure on the body. This must be very carefully watched until one has grown accustomed to riding behind the fairing. The second thing is the alarming increase in engine noise which is an unfortunate but inevitable by-product of fitting any type of fairing or windscreen, and is caused by normal engine noise being reflected, and in some cases amplified, by the flat surface of the fairing.

Luggage racks or carriers

Carriers are possibly the commonest item to be fitted to modern motorcycles. They vary enormously in size, carrying capacity, and durability. When selecting a carrier, always look for one which is made specifically for your machine and which is bolted on with as few separate brackets as possible. The universal-type carrier, with its mass of brackets and adaptor pieces, will generally prove too weak to be of any real use. A good carrier should bolt to the main frame, generally using the two suspension unit top mountings and a mudguard mounting bolt as attachment points, and have its luggage platform as low and as far forward as possible to minimise the effect of any load on the machine's stability. Look for good quality, heavy gauge tubing, good welding and good finish. Also ensure that the carrier does not prevent opening of the seat, sidepanels or tail compartment, as appropriate. When using a carrier, be very careful not to overload it. Excessive weight placed so high and so far to the rear of any motorcycle will have an adverse effect on the machine's steering and stability.

Luggage

Motorcycle luggage can be grouped under two headings: soft and hard. Both types are available in many sizes and styles and have advantages and disadvantages in use.

Soft luggage is now becoming very popular because of its lower cost and its versatility. Whether in the form of tankbags, panniers, or strap-on bags, soft luggage requires in general no brackets and no modification to the motorcycle. Equipment can be swapped easily from one motorcycle to another and can be fitted and removed in seconds. Awkwardly shaped loads can easily be carried. The disadvantages of soft luggage are that the contents cannot be secure against the casual thief, very little protection is afforded in the event of a crash, and waterproofing is generally poor. Also, in the case of panniers, carrying capacity is restricted to approximately 10 lb, although this amount will vary considerably depending on the manufacturer's recommendation. When purchasing soft luggage, look for good quality material, generally vinyl or nylon, with strong, well-stitched attachment points. It is always useful to have separate pockets, especially on tank bags, for items which will be needed on the journey. When purchasing a tank bag, look for one which has a separate, well-padded, base. This will protect the tank's paintwork and permit easy access to the filler cap at petrol stations.

Hard luggage is confined to two types: panniers, and top boxes or tail trunks. Most hard luggage manufacturers produce matching sets of these items, the basis of which is generally that manufacturer's own heavy-duty luggage rack. Variations on this theme occur in the form of separate frames for the better quality panniers, fixed or quickly-detachable luggage, and in size and carrying capacity. Hard luggage offers a reasonable degree of security against theft and good protection against weather and accident damage. Carrying capacity is greater than that of soft luggage, around 15 – 20 lb in the case of panniers, although top boxes should never be loaded as much as their apparent capacity might imply. A top box should only be used for lightweight items, because one that is heavily laden can have a serious effect on the stability of the machine. When purchasing hard luggage look for the same good points as mentioned under fairings and windscreens, ie good quality mounting brackets and fittings, and well-finished fibreglass or ABS plastic cases. Again as with fairings, always purchase luggage made specifically for your motorcycle, using as few separate brackets as possible, to ensure that everything remains securely bolted in place. When fitting hard luggage, be careful to check that the rear suspension and brake operation will not be impaired in any way and remember that many pannier kits require re-siting of the indicators. Remember also that a non-standard exhaust system may make fitting extremely difficult.

Handlebars

The occupation of fitting alternative types of handlebar is extremely popular with modern motorcyclists, whose motives may vary from the purely practical, wishing to improve the comfort of their machines, to the purely aesthetic, where form is more important than function. Whatever the reason, there are several considerations to be borne in mind when changing the handlebars of your machine. If fitting lower bars, check carefully that the switches and cables do not foul the petrol tank on full lock and that the surplus length of cable, brake pipe, and electrical wiring are smoothly and tidily disposed of. Avoid tight kinks in cable or brake pipes which will produce stiff controls or the premature and disastrous failure of an overstressed component. If necessary, remove the petrol tank and re-route the cable from the engine/gearbox unit upwards, ensuring smooth gentle curves are produced. In extreme cases, it will be necessary to purchase a shorter brake pipe to overcome this problem. In the case of higher handlebars than standard it will almost certainly be necessary to purchase extended cables and brake pipes. Fortunately, many standard motorcycles have a custom version which will be equipped with higher handlebars and, therefore, factory-built extended components will be available from your local dealer. It is not usually necessary to extend electrical wiring, as switch clusters may be used on several different motorcycles, some being custom versions. This point should be borne in mind however when fitting extremely high or wide handlebars.

When fitting different types of handlebar, ensure that the mounting clamps are correctly tightened to the manufacturer's specifications and that cables and wiring, as previously mentioned, have smooth easy runs and do not snag on any part of the motorcycle throughout the full steering lock. Ensure that the fluid level in the front brake master cylinder remains level to avoid any chance of air entering the hydraulic system. Also check that the cables are adjusted correctly and that all handlebar controls operate correctly and can be easily reached when riding.

Crashbars

Crashbars, also known as engine protector bars, engine guards, or case savers, are extremely useful items of equipment which can contribute protection to the machine's structure if a crash occurs. They do not, as has been inferred in the US, prevent the rider from crashing, or necessarily prevent rider injury should a crash occur.

It is recommended that only the smaller, neater, engine protector type of crashbar is considered. This type will offer protection while restricting, as little as is possible, access to the engine and the machine's ground clearance. The crashbars should be designed for use specifically on your machine, and should be constructed of heavy-gauge tubing with strong, integral mounting brackets. Where possible, they should bolt to a strong lug on the frame, usually at the engine mounting bolts.

The alternative type of crashbar is the larger cage type. This type is not recommended in spite of their appearance which promises some protection to the rider as well as to the machine. The larger amount of leverage imposed by the size of this type of crashbar increases the risk of severe frame damage in the event of an accident. This type also decreases the machine's ground clearance and restricts access to the engine. The amount of protection afforded the rider is open to some doubt as the design is based on the premise that the rider will stay in the normally seated position during an accident, and the crash bar structure will not itself fail. Neither result can in any way be guaranteed.

As a general rule, always purchase the best, ie usually the most expensive, set of crashbars you an afford. The investment will be repaid by minimising the amount of damage incurred, should the machine be involved in an accident. Finally, avoid the universal type of crashbar. This should be regarded only as a last resort to be used if no alternative exists. With its usual multitude of separate brackets and spacers, the universal crashbar is far too weak in design and construction to be of any practical value.

Exhaust systems

The fitting of aftermarket exhaust systems is another extremely popular pastime amongst motorcyclists. The usual motive is to gain more performance from the engine but other considerations are to gain more ground clearance, to lose weight from the motorcycle, to obtain a more distinctive exhaust note or to find a cheaper alternative to the manufacturer's original equipment exhaust system. Original equipment exhaust systems often cost more and may well have a relatively short life. It should be noted that it is rare for an aftermarket exhaust system alone to give a noticeable increase in the engine's power output. Modern motorcycles are designed to give the highest power output possible allowing for factors such as quietness, fuel economy, spread of power, and long-term reliability. If there were a magic formula which allowed the exhaust system to produce more power without affecting these other considerations you can be sure that the manufacturers, with their large research and development facilities, would have found it and made use of it. Performance increases of a worthwhile and noticeable nature only come from well-tried and properly matched modifications to the entire engine, from the air filter, through the carburettors, port timing or camshaft and valve design, combustion chamber shape, compression ratio, and the exhaust system. Such modifications are well outside the scope of this manual but interested owners might refer to specialist books produced by the publisher of this manual which go into the whole subject in great detail.

Whatever your motive for wishing to fit an alternative exhaust system, be sure to seek expert advice before doing so. Changes to the carburettor jetting will almost certainly be required for which you must consult the exhaust system manufacturer. If he cannot supply adequately specific information it is reasonable to assume that insufficient development work has been carried out, and that particular make should be avoided. Other factors to be borne in mind are whether the exhaust system allows the use of both centre and side stands, whether it allows sufficient access to permit oil and filter changing and whether modifications are necessary to the standard exhaust system. Many two-stroke expansion chamber systems require the use of the standard exhaust pipe; this is all very well if the standard exhaust pipe and silencer are separate units but can cause problems if the two, as with so many modern two-strokes, are a one-piece unit. While the exhaust pipe can be removed easily by means of a hacksaw it is not so easy to refit the original silencer should you at any time wish to return the machine to standard trim. The same applies to several four-stroke systems.

On the subject of the finish of aftermarket exhausts, avoid black-painted systems unless you enjoy painting. As any trail-bike owner will tell you, rust has a great affinity for black exhausts and re-painting or rust removal becomes a task which must be carried out with monotonous regularity. A bright chrome finish is, as a general rule, a far better proposition as it is much easier to keep clean and to prevent rusting. Although the general finish of aftermarket exhaust systems is not always up to the standard of the original equipment the lower cost of such systems does at least reflect this fact.

When fitting an alternative system always purchase a full set of new exhaust gaskets, to prevent leaks. Fit the exhaust first to the cylinder head or barrel, as appropriate, tightening the retaining nuts or bolts by hand only and then line up the exhaust rear mountings. If the new system is a one-piece unit and the rear mountings do not line up exactly, spacers must be fabricated to take up the difference. Do not force the system into place as the stress thus imposed will rapidly cause cracks and splits to appear. Once all the mountings are loosely fixed, tighten the retaining nuts or bolts securely, being careful not to overtighten them. Where the motorcycle manufacturer's torque settings are available, these should be used. Do not forget to carry out any carburation changes recommended by the exhaust system's manufacturer.

Electrical equipment

The vast range of electrical equipment available to motorcyclists is so large and so diverse that only the most general outline can be given here. Electrical accessories vary from electronic ignition kits fitted to replace contact breaker points, to additional lighting at the front and rear, more powerful horns, various instruments and gauges, clocks, anti-theft systems, heated clothing, CB radios, radio-cassette players, and intercom systems, to name but a few of the more popular items of equipment.

As will be evident, it would require a separate manual to cover this subject alone and this section is therefore restricted to outlining a few basic rules which must be borne in mind when fitting electrical equipment. The first consideration is whether your machine's electrical system has enough reserve capacity to cope with the added demand of the accessories you wish to fit. The motorcycle's manufacturer or importer should be able to furnish this sort of information and may also be able to offer advice on uprating the electrical system. Failing this, a good dealer or the accessory manufacturer may be able to help. In some cases, more powerful generator components may be available, perhaps from another motorcycle in the manufacturer's range. The second consideration is the legal requirements in force in your area. The local police may be prepared to help with this point. In the UK for example, there are strict regulations governing the position and use of auxiliary riding lamps and fog lamps.

When fitting electrical equipment always disconnect the battery first to prevent the risk of a short-circuit, and be careful to ensure that all connections are properly made and that they are waterproof. Remember that many electrical accesories are designed primarily for use in cars and that they cannot easily withstand the exposure to vibration and to the weather. Delicate components must be rubber-mounted to insulate them from vibration, and sealed carefully to prevent the entry of rainwater and dirt. Be careful to follow exactly the accessory manufacturer's instructions in conjunction with the wiring diagram at the back of this manual.

Accessories – general

Accessories fitted to your motorcycle will rapidly deteriorate if not cared for. Regular washing and polishing will maintain the finish and will provide an opportunity to check that all mounting bolts and nuts are securely fastened. Any signs of chafing or wear should be watched for, and the cause cured as soon as possible before serious damage occurs.

As a general rule, do not expect the re-sale value of your motorcycle to increase by an amount proportional to the amount of money and effort put into fitting accessories. It is usually the case that an absolutely standard motorcycle will sell more easily at a better price than one that has been modified. If you are in the habit of exchanging your machine for another at frequent intervals, this factor should be borne in mind to avoid loss of money.

Fault diagnosis

Contents

1 Introduction

This Section provides an easy reference-guide to the more common ailments that are likely to afflict your machine. Obviously, the opportunities are almost limitless for faults to occur as a result of obscure failures, and to try and cover all eventualities would require a book. Indeed, a number have been written on the subject.

Successful fault diagnosis is not a mysterious 'black art' but the application of a bit of knowledge combined with a systematic and logical approach to the problem. Approach any fault diagnosis by first accurately identifying the symptom and then checking through the list of possible causes, starting with the simplest or most obvious and progressing in stages to the most complex. Take nothing for granted, but above all apply liberal quantities of common sense.

The main symptom of a fault is given in the text as a major heading below which are listed, as Section headings, the various systems or areas which may contain the fault. Details of each possible cause for a fault and the remedial action to be taken are given, in brief, in the paragraphs below each Section heading. Further information should be sought in the relevant Chapter.

Starter motor problems

2 Starter motor not rotating

Engine stop switch off.

Fuse blown. Check the main fuse located behind the battery side cover.

Battery voltage low. Switching on the headlamp and operating the horn will give a good indication of the charge level. If necessary recharge the battery from an external source.

Ignition switch defective. Check switch for continuity and connections for security.

Engine stop switch defective. Check switch for continuity in 'Run' position. Fault will be caused by broken, wet or corroded switch contacts. Clean or renew as necessary.

Starter button switch faulty. Check continuity of switch. Faults as for engine stop switch.

Starter relay (solenoid) faulty. If the switch is functioning correctly a pronounced click should be heard when the starter button is depressed. This presupposes that current is flowing to the solenoid when the button is depressed.

Wiring open or shorted. Check first that the battery terminal connections are tight and corrosion free. Follow this by checking that all wiring connections are dry, tight and corrosion free. Check also for frayed or broken wiring. Occasionally a wire may become trapped between two moving components, particularly in the vicinity of the steering head, leading to breakage of the internal core but leaving the softer but more resilient outer cover intact. This can cause mysterious intermittent or total power loss.

Starter motor defective. A badly worn starter motor may cause high current drain from a battery without the motor rotating. If current is found to be reaching the motor, after checking the starter button and starter relay, suspect a damaged motor. The motor should be removed for inspection.

3 Starter motor rotates but engine does not turn over

Starter motor clutch defective. Suspect jammed or worn engagement rollers, plungers and springs.

Damaged starter motor drive train. Inspect and renew component where necessary. Failure in this area is unlikely.

4 Starter motor and clutch function but engine will not turn over

Engine seized. Seizure of the engine is always a result of damage to internal components due to lubrication failure, or component breakage resulting from abuse, neglect or old age. A seizing or partially seized component may go un-noticed until the engine has cooled down and an attempt is made to restart the engine. Suspect first seizure of the valves, valve gear and the piston. Instantaneous seizure whilst the engine is running indicates component breakage. In either case major dismantling and inspection will be required.

Engine does not start when turned over

5 No fuel flow to carburettor

No fuel or insufficient fuel in tank.

Fuel tap lever position incorrectly selected.

Tank filler cap air vent obstructed. Usually caused by dirt or water. Clean the vent orifice.

Fuel tap or filter blocked. Blockage may be due to accumulation of rust or paint flakes from the tank's inner surface or of foreign matter from contaminated fuel. Remove the tap and clean it and the filter. Look also for water droplets in the fuel.

Fuel line blocked. Blockage of the fuel line is more likely to result from a kink in the line rather than the accumulation of debris.

6 Fuel not reaching cylinder

Float chamber not filling. Caused by float needle or floats sticking in up position. This may occur after the machine has been left standing for an extended length of time allowing the fuel to evaporate. When this occurs a gummy residue is often left which hardens to a varnish-like substance. This condition may be worsened by corrosion and crystaline deposits produced prior to the total evaporation of contaminated fuel. Sticking of the float needle may also be caused by wear. In any case removal of the float chamber will be necessary for inspection and cleaning.

Blockage in starting circuit, slow running circuit or jets. Blockage of these items may be attributable to debris from the fuel tank bypassing the filter system or to gumming up as described in paragraph 1. Water droplets in the fuel will also block jets and passages. The carburettor should be dismantled for cleaning.

Fuel level too low. The fuel level in the float chamber is controlled by float height. The float height may increase with wear or damage but will never reduce, thus a low float height is an inherent rather than developing condition. Check the float height and make any necessary adjustment.

7 Engine flooding

Float valve needle worn or stuck open. A piece of rust or other debris can prevent correct seating of the needle against the valve seat thereby permitting an uncontrolled flow of fuel. Similarly, a worn needle or needle seat will prevent valve closure. Dismantle the carburettor float bowl for cleaning and, if necessary, renewal of the worn components.

Fuel level too high. The fuel level is controlled by the float height which may increase due to wear of the float needle, pivot pin or operating tang. Check the float height, and make any necessary adjustment. A leaking float will cause an increase in fuel level, and thus should be renewed.

Cold starting mechanism. Check the choke (starter mechanism) for correct operation. If the mechanism jams in the 'On' position subsequent starting of a hot engine will be difficult.

Blocked air filter. A badly restricted air filter will cause flooding. Check the filter and clean or renew as required. A collapsed inlet hose will have a similar effect.

8 No spark at plug

Ignition switch not on.

Engine stop switch off.

Fuse blown. Check fuse for ignition circuit. See wiring diagram.

Battery voltage low. The current draw required by a starter motor is sufficiently high that an under-charged battery may not have enough spare capacity to provide power for the ignition circuit during starting.

Starter motor inefficient. A starter motor with worn brushes and a worn or dirty commutator will draw excessive amounts of current causing power starvation in the ignition system. See the preceding paragraph. Starter motor overhaul will be required.

Spark plug failure. Clean the spark plug thoroughly and reset the electrode gap. Refer to the spark plug section and the colour condition guide in Chapter 3. If the spark plug shorts internally or has sustained visible damage to the electrodes, core or ceramic insulator it should be renewed. On rare occasions a plug that appears to spark vigorously will fail to do so when refitted to the engine and subjected to the compression pressure in the cylinder.

Spark plug cap or high tension (HT) lead faulty. Check condition and security. Replace if deterioration is evident.

Spark plug cap loose. Check that the spark plug cap fits securely over the plug and, where fitted, the screwed terminal on the plug end is secure.

Shorting due to moisture. Certain parts of the ignition system are susceptible to shorting when the machine is ridden or parked in wet weather. Check particularly the area from the spark plug cap back to the ignition coil. A water dispersant spray may be used to dry out waterlogged components. Recurrence of the problem can be prevented by using an ignition sealant spray after drying out and cleaning.

Ignition or stop switch shorted. May be caused by water, corrosion or wear. Water dispersant and contact cleaning sprays may be used. If this fails to overcome the problem dismantling and visual inspection of the switches will be required.

Shorting or open circuit in wiring. Failure in any wire connecting any of the ignition components will cause ignition malfunction. Check also that all connections are clean, dry and tight.

Ignition coil failure. Check the coil, referring to Chapter 3.

Electronic ignition component failure, see Chapter 3.

9 Weak spark at plug

Feeble sparking at the plug may be caused by any of the faults mentioned in the preceding Section other than those items in paragraphs 1 and 2. Check first the spark plug, this being the most likely culprit.

10 Compression low

Spark plug loose. This will be self-evident on inspection, and may be accompanied by a hissing noise when the engine is turned over. Remove the plug and check that the threads in the cylinder head are not damaged. Check also that the plug sealing washer is in good condition.

Cylinder head gasket leaking. This condition is often accompanied by a high pitched squeak from around the cylinder head and oil loss, and may be caused by insufficiently tightened cylinder head fasteners, a warped cylinder head or mechanical failure of the gasket material. Re-torqueing the fasteners to the correct specification may seal the leak in some instances but if damage has occurred this course of action will provide, at best, only a temporary cure.

Valve not seating correctly. The failure of a valve to seat may be caused by insufficient valve clearance, pitting of the valve seat or face, carbon deposits on the valve seat or seizure of the valve stem or valve gear components. Valve spring breakage will also prevent correct valve closure. The valve clearances should be checked first and then, if these are found to be in order, further dismantling will be required to inspect the relevant components for failure.

Cylinder, piston and ring wear. Compression pressure will be lost if any of these components are badly worn. Wear in one component is invariably accompanied by wear in another. A top end overhaul will be required.

Piston rings sticking or broken. Sticking of the piston rings may be caused by seizure due to lack of lubrication or heating as a result of poor carburation or incorrect fuel type. Gumming of the rings may result from lack of use, or carbon deposits in the ring grooves. Broken rings result from over-revving, overheating or general wear. In either case a top-end overhaul will be required.

Engine stalls after starting

11 General causes

Improper cold start mechanism operation. Check that the operating controls function smoothly and, where applicable, are correctly adjusted. A cold engine may not require application of an enriched mixture to start initially but may baulk without choke once firing. Likewise a hot engine may start with an enriched mixture but will stop almost immediately if the choke is inadvertently in operation.

Ignition malfunction. See Section 9, 'Weak spark at plug'.

Carburettor incorrectly adjusted. Maladjustment of the mixture strength or idle speed may cause the engine to stop immediately after starting. See Chapter 2.

Fuel contamination. Check for filter blockage by debris or water which reduces, but does not completely stop, fuel flow or blockage of the slow speed circuit in the carburettor by the same agents. If water is present it can often be seen as droplets in the bottom of the float bowl. Clean the filter and, where water is in evidence, drain and flush the fuel tank and float bowl.

Intake air leak. Check for security of the carburettor mounting and hose connections, and for cracks or splits in the hoses. Check also that the carburettor top is secure.

Air filter blocked or omitted. A blocked filter will cause an over-rich mixture; the omission of a filter will cause an excessively weak mixture. Both conditions will have a detrimental affect on carburation. Clean or renew the filter as necessary.

Fuel filler cap air vent blocked. Usually caused by dirt or water. Clean the vent orifice.

Poor running at idle and low speed

12 Weak spark at plug or erratic firing

Battery voltage low. In certain conditions low battery charge, especially when coupled with a badly sulphated battery, may result in misfiring. If the battery is in good general condition it should be recharged; an old battery suffering from sulphated plates should be renewed.

Spark plug fouled, faulty or incorrectly adjusted. See Section 8 or refer to Chapter 3.

Spark plug cap or high tension lead shorting. Check the condition of both these items ensuring that they are in good condition and dry and that the cap is fitted correctly.

Spark plug type incorrect. Fit plug of correct type and heat range as given in Specifications. In certain conditions a plug of hotter or colder type may be required for normal running.

Igniting timing incorrect. Check the ignition timing statically and dynamically, ensuring that the advance is functioning correctly.

Faulty ignition coil. Partial failure of the coil internal insulation will diminish the performance of the coil. No repair is possible, a new component must be fitted.

Ignition system failure, see Chapter 3.

13 Fuel/air mixture incorrect

Intake air leak. See Section 11.

Mixture strength incorrect. Adjust slow running mixture strength using pilot adjustment screw.

Pilot jet or slow running circuit blocked. The carburettor should be removed and dismantled for thorough cleaning. Blow through all jets and air passages with compressed air to clear obstructions.

Air cleaner clogged or omitted. Clean or fit air cleaner element as necessary. Check also that the element and air filter cover are correctly seated.

Cold start mechanism in operation. Check that the choke has not been left on inadvertently and the operation is correct.

Fuel level too high or too low. Check the float height and adjust as necessary. See Section 7.

Fuel tank air vent obstructed. Obstruction usually caused by dirt or water. Clean vent orifice.

Valve clearance incorrect. Check, and if necessary, adjust, the clearances.

14 Compression low

See Section 10.

Acceleration poor

15 General causes

All items as for previous Section.
Timing not advancing.
Brakes binding. Usually caused by maladajustment or partial seizure of the operating mechanism due to poor maintenance. Check brake adjustment (where applicable). A bent wheel spindle or warped brake disc can produce similar symptoms.

Poor running or lack of power at high speeds

16 Weak spark at plug or erratic firing

All items as for Section 12.
HT lead insulation failure. Insulation failure of the HT lead and spark plug cap due to old age or damage can cause shorting when the engine is driven hard. This condition may be less noticeable, or not noticeable at all at lower engine speeds.

17 Fuel/air mixture incorrect

All items as for Section 13, with the exception of items 2 and 3.
Main jet blocked. Debris from contaminated fuel, or from the fuel tank, and water in the fuel can block the main jet. Clean the fuel filter, the float bowl area, and if water is present, flush and refill the fuel tank.
Main jet is the wrong size. The standard carburettor jetting is for sea level atmospheric pressure. For high altitudes, usually above 5000 ft, a smaller main jet will be required.
Jet needle and needle jet worn. These can be renewed individually but should be renewed as a pair. Renewal of both items requires partial dismantling of the carburettor.
Air bleed holes blocked. Dismantle carburettor and use compressed air to blow out all air passages.
Reduced fuel flow. A reduction in the maximum fuel flow from the fuel tank to the carburettor will cause fuel starvation, proportionate to the engine speed. Check for blockages through debris or a kinked fuel line.

18 Compression low

See Section 10.

Knocking or pinking

19 General causes

Carbon build-up in combustion chamber. After high mileages have been covered large accumulation of carbon may occur. This may glow red hot and cause premature ignition of the fuel/air mixture, in advance of normal firing by the spark plug. Cylinder head removal will be required to allow inspection and cleaning.
Fuel incorrect. A low grade fuel, or one of poor quality may result in compression induced detonation of the fuel resulting in knocking and pinking noises. Old fuel can cause similar problems. A too highly leaded fuel will reduce detonation but will accelerate deposit formation in the combustion chamber and may lead to early pre-ignition as described in item 1.
Spark plug heat range incorrect. Uncontrolled pre-ignition can result from the use of a spark plug the heat range of which is too hot.
Weak mixture. Overheating of the engine due to a weak mixture can result in pre-ignition occurring where it would not occur when engine temperature was within normal limits. Maladjustment, blocked jets or passages and air leaks can cause this condition.

Overheating

20 Firing incorrect

Spark plug fouled, defective or maladjusted. See Section 6.
Spark plug type incorrect. Refer to the Specifications and ensure that the correct plug type is fitted.
Incorrect ignition timing. Timing that is far too much advanced or far too much retarded will cause overheating. Check the ignition timing is correct and that the advance circuit is functioning.

21 Fuel/air mixture incorrect

Slow speed mixture strength incorrect. Adjust pilot air screw.
Main jet wrong size. The carburettor is jetted for sea level atmospheric conditions. For high altitudes, usually above 5000 ft, a smaller main jet will be required.
Air filter badly fitted or omitted. Check that the filter element is in place and that it and the air filter box cover are sealing correctly. Any leaks will cause a weak mixture.
Induction air leaks. Check the security of the carburettor mountings and hose connections, and for cracks and splits in the hoses. Check also that the carburettor top is secure.
Fuel level too low. See Section 6.
Fuel tank filler cap air vent obstructed. Clear blockage.

22 Lubrication inadequate

Engine oil too low. Not only does the oil serve as a lubricant by preventing friction between moving components, but it also acts as a coolant. Check the oil level and replenish.
Engine oil overworked. The lubricating properties of oil are lost slowly during use as a result of changes resulting from heat and also contamination. Always change the oil at the recommended interval.
Engine oil of incorrect viscosity or poor quality. Always use the recommended viscosity and type of oil.
Oil filter and filter by-pass valve blocked. Renew filter.

23 Miscellaneous causes

Engine fins clogged. A build-up of mud in the cylinder head and cylinder barrel cooling fins will decrease the cooling capabilities of the fins. Clean the fins as required.

Clutch operating problems

24 Clutch slip

No clutch lever play. Adjust clutch lever end play according to the procedure in Chapter 1
Friction plates worn or warped. Overhaul clutch assembly, replacing plates out of specification (Chapter 1).
Steel plates worn or warped. Overhaul clutch assembly, replacing plates out of specification (Chapter 1).
Clutch springs broken or worn. Old or heat-damaged (from slipping clutch) springs should be replaced with new ones (Chapter 1).
Clutch release not adjusted properly. See Chapter 1.
Clutch inner cable snagging. Caused by a frayed cable or kinked outer cable. Replace the cable with a new one. Repair of a frayed cable is not advised.
Clutch release mechanism defective. Worn or damaged parts in the clutch release mechanism could include the shaft, cam, actuating arm or pilot. Replace parts as necessary (Chapter 1).
Clutch hub and outer drum worn. Severe indentation by the clutch plate tangs of the channels in the hub and drum will cause snagging of the plates preventing correct engagement. If this damage occurs, renewal of the worn components is required.
Lubricant incorrect. Use of a transmission lubricant other than that specified may allow the plates to slip.

25 Clutch drag

Clutch lever play excessive. Adjust lever at bars or at cable end if necessary (Chapter 1).

Clutch plates warped or damaged. This will cause a drag on the clutch, causing the machine to creep. Overhaul clutch assembly (Chapter 1).

Clutch spring tension uneven. Usually caused by a sagged or broken spring. Check and replace springs (Chapter 1).

Engine oil deteriorated. Badly contaminated engine oil and a heavy deposit of oil sludge and carbon on the plates will cause plate sticking. The oil recommended for this machine is of the detergent type, therefore it is unlikely that this problem will arise unless regular oil changes are neglected.

Engine oil viscosity too high. Drag in the plates will result from the use of an oil with too high a viscosity. In very cold weather clutch drag may occur until the engine has reached operating temperature.

Clutch hub and outer drum worn. Indentation by the clutch plate tangs of the channels in the hub and drum will prevent easy plate disengagement. If the damage is light the affected areas may be dressed with a fine file. More pronounced damage will necessitate renewal of the components.

Clutch outer drum seized to shaft. Lack of lubrication, severe wear or damage can cause the drum to seize to the shaft. Overhaul of the clutch, and perhaps the transmission, may be necessary to repair damage (Chapter 1).

Clutch release mechanism defective. Worn or damaged release mechanism parts can stick and fail to provide leverage. Overhaul clutch release components (Chapter 1).

Loose clutch centre nut. Causes drum and centre misalignment, putting a drag on the engine. Engagement adjustment continually varies. Overhaul clutch assembly (Chapter).

Gear selection problems

26 Gear lever does not return

Weak or broken centraliser spring. Renew the spring.

Gearchange shaft bent or seized. Distortion of the gearchange shaft often occurs if the machine is dropped heavily on the gear lever. Provided that damage is not severe straightening of the shaft is permissible.

27 Gear selection difficult or impossible

Clutch not disengaging fully. See Section 25.

Gearchange shaft bent. This often occurs if the machine is dropped heavily on the gear lever. Straightening of the shaft is permissible if the damage is not too great.

Gearchange arms, pawls or pins worn or damaged. Wear or breakage of any of these items may cause difficulty in selecting one or more gears. Overhaul the selector mechanism.

Gearchange arm spring broken. Renew spring.

Gearchange drum stopper cam or detent plunger damage. Failure, rather than wear, of these items may jam the drum thereby preventing gearchanging. The damaged items must be renewed.

Selector forks bent or seized. This can be caused by dropping the machine heavily on the gearchange lever or as a result of lack of lubrication. Though rare, bending of a shaft can result from a missed gearchange or false selection at high speed.

Selector fork end and pin wear. Pronounced wear of these items and the grooves in the gearchange drum can lead to imprecise selection and, eventually, no selection. Renewal of the worn components will be required.

Structural failure. Failure of any one component of the selector rod and change mechanism will result in improper or fouled gear selection.

28 Jumping out of gear

Detent plunger assembly worn or damaged. Wear of the plunger and the cam with which it locates and breakage of the detent spring can cause imprecise gear selection resulting in jumping out of gear. Renew the damaged components.

Gear pinion dogs worn or damaged. Rounding off the dog edges and the mating recesses in adjacent pinion can lead to jumping out of gear when under load. The gears should be inspected and renewed. Attempting to reprofile the dogs is not recommended.

Selector forks, gearchange drum and pinion grooves worn. Extreme wear of these interconnected items can occur after high mileages especially when lubrication has been neglected. The worn components must be renewed.

Gear pinions, bushes and shafts worn. Renew the worn components.

Bent gearchange shaft. Often caused by dropping the machine on the gear lever.

Gear pinion tooth broken. Chipped teeth are unlikely to cause jumping out of gear once the gear has been selected fully; a tooth which is completely broken off, however, may cause problems in this respect and in any event will cause transmission noise.

29 Overselection

Pawl spring weak or broken. Renew the spring.

Detent plunger worn or broken. Renew the damaged items.

Stopper arm spring worn or broken. Renew the spring.

Gearchange arm stop pads worn. Repairs can be made by welding and reprofiling with a file.

Abnormal engine noise

30 Knocking or pinking

See Section 19.

31 Piston slap or rattling from cylinder

Cylinder bore/piston clearance excessive. Resulting from wear, partial seizure or improper boring during overhaul. This condition can often be heard as a high, rapid tapping noise when the engine is under little or no load, particularly when power is just beginning to be applied. Reboring to the next correct oversize should be carried out and a new oversize piston fitted.

Connecting rod bent. This can be caused by over-revving, trying to start a very badly flooded engine (resulting in a hydraulic lock in the cylinder) or by earlier mechanical failure such as a dropped valve. Attempts at straightening a bent connecting rod from a high performance engine are not recommended. Careful inspection of the crankshaft should be made before renewing the damaged connecting rod.

Gudgeon pin, piston boss bore or small-end bearing wear or seizure. Excess clearance or partial seizure between normal moving parts of these items can cause continuous or intermittent tapping noises. Rapid wear or seizure is caused by lubrication starvation resulting from an insufficient engine oil level or oilway blockage.

Piston rings worn, broken or sticking. Renew the rings after careful inspection of the piston and bore.

32 Valve noise or tapping from the cylinder head

Valve clearance incorrect. Adjust the clearances with the engine cold.

Valve spring broken or weak. Renew the spring set.

Camshaft or cylinder head worn or damaged. The camshaft lobes are the most highly stressed of all components in the engine and are subject to high wear if lubrication becomes inadequate. The bearing surfaces on the camshaft and cylinder head are also sensitive to a lack of lubrication. Lubrication failure due to blocked oilways can occur, but over-enthusiastic revving before engine warm-up is complete is the usual cause.

Rocker arm or spindle wear. Rapid wear of a rocker arm, and the resulting need for frequent valve clearance adjustment, indicates breakthrough or failure of the surface hardening on the rocker arm tips.

Similar wear in the cam lobes can be expected. Renew the worn components after checking for lubrication failure.

Worn camshaft drive components. A rustling noise or light tapping which is not improved by correct re-adjustment of the cam chain tension can be emitted by a worn cam chain or worn sprockets and chain. If uncorrected, subsequent cam chain breakage may cause extensive damage. The worn components must be renewed before wear becomes too far advanced.

33 Other noises

Big-end bearing wear. A pronounced knock from within the crankcase which worstens rapidly is indicative of big-end bearing failure as a result of extreme normal wear or lubrication failure. Remedial action in the form of a bottom end overhaul should be taken; continuing to run the engine will lead to further damage including the possibility of connecting rod breakage.

Main bearing failure. Extreme normal wear or failure of the main bearings is characteristically accompanied by a rumble from the crankcase and vibration felt through the frame and footrests. Renew the worn bearings and carry out a very careful examination of the crankshaft.

Crankshaft excessively out of true. A bent crank may result from over-revving or damage from an upper cylinder component or gearbox failure. Damage can also result from dropping the machine on either crankshaft end. Straightening of the crankshaft is not possible in normal circumstances; a replacement item should be fitted.

Engine mounting loose. Tighten all the engine mounting nuts and bolts.

Cylinder head gasket leaking. The noise most often associated with a leaking head gasket is a high pitched squeaking, although any other noise consistent with gas being forced out under pressure from a small orifice can also be emitted. Gasket leakage is often accompanied by oil seepage from around the mating joint or from the cylinder head holding down bolts and nuts. Leakage into the cam chain tunnel or oil return passages will increase crankcase pressure and may cause oil leakage at joints and oil seals. Also, oil contamination will be accelerated. Leakage results from insufficient or uneven tightening of the cylinder head fasteners, or from random mechanical failure. Retightening to the correct torque figure will, at best, only provide a temporary cure. The gasket should be renewed at the earliest opportunity.

Exhaust system leakage. Popping or crackling in the exhaust system, particularly when it occurs with the engine on the overrun, indicates a poor joint either at the cylinder port or at the exhaust pipe/silencer connection. Failure of the gasket or looseness of the clamp should be looked for.

Abnormal transmission noise

34 Clutch noise

Clutch outer drum/friction plate tang clearance excessive.
Clutch outer drum/spacer clearance excessive.
Clutch outer drum/thrust washer clearance excessive.
Primary drive gear teeth worn or damaged.
Clutch shock absorber assembly worn or damaged.

35 Transmission noise

Bearing or bushes worn or damaged. Renew the affected components.

Gear pinions worn or chipped. Renew the gear pinions.

Metal chips jams in gear teeth.This can occur when pieces of metal from any failed component are picked up by a meshing pinion. The condition will lead to rapid bearing wear or early gear failure.

Engine/transmission oil level too low. Top up immediately to prevent damage to gearbox and engine.

Gearchange mechanism worn or damaged. Wear or failure of certain items in the selection and change components can induce mis-selection of gears (see Section 27) where incipient engagement of more than one gear set is promoted. Remedial action, by the overhaul of the gearbox, should be taken without delay.

Loose gearbox sprocket. Remove the sprocket and check for impact damage to the splines of the sprocket and shaft. Excessive slack between the splines will promote loosening of the securing nut; renewal of the worn components is required. When retightening the nut ensure that it is tightened fully and that, where fitted, the lock washer is bent up against one flat of the nut.

Chain snagging on cases or cycle parts. A badly worn chain or one that is excessively loose may snag or smack against adjacent components.

Exhaust smokes excessively

36 White/blue smoke (caused by oil burning)

Piston rings worn or broken. Breakage or wear of any ring, but particularly the oil control ring, will allow engine oil past the piston into the combustion chamber. Overhaul the cylinder barrel and piston.

Cylinder cracked, worn or scored. These conditions may be caused by overheating, lack of lubrication, component failure or advanced normal wear. The cylinder barrel should be renewed or rebored and the next oversize piston fitted.

Valve oil seal damages or worn. This can occur as a result of valve guide failure or old age. The emission of smoke is likely to occur when the throttle is closed rapidly after acceleration, for instance, when changing gear. Renew the valve oil seals and, if necessary, the valve guides.

Valve guides worn. See the preceding paragraph.

Engine oil level too high. This increases the crankcase pressure and allows oil to be forced pass the piston rings. Often accompanied by seepage of oil at joints and oil seals.

Cylinder head gasket blown between cam chain tunnel or oil return passage. Renew the cylinder head gasket.

Abnormal crankcase pressure. This may be caused by blocked breather passages or hoses causing back-pressure at high engine revolutions.

37 Black smoke (caused by over-rich mixture)

Air filter element clogged. Clean or renew the element.

Main jet loose or too large. Remove the float chamber to check for tightness of the jet. If the machine is used at high altitudes rejetting will be required to compensate for the lower atmospheric pressure.

Cold start mechanism jammed on. Check that the mechanism works smoothly and correctly and that, where fitted, the operating cable is lubricated and not snagged.

Fuel level too high. The fuel level is controlled by the float height which can increase as a result of wear or damage. Remove the float bowl and check the float height. Check also that floats have not punctured; a punctured float will loose buoyancy and allow an increased fuel level.

Float valve needle stuck open. Caused by dirt or a worn valve. Clean the float chamber or renew the needle and, if necessary, the valve seat.

Lubrication problems

38 Engine lubrication system failure

Engine oil defective. Oil pump shaft or locating pin sheared off from ingesting debris or seizing from lack of lubrication (low oil level) (Chapter 1).

Engine oil screen clogged. Change oil and filter and service pickup screen (Chapter 1).

Engine oil level too low. Inspect for leak or other problem causing low oil level and add recommended lubricant (Chapter 1).

Engine oil viscosity too low. Very old, thin oil, or an improper weight of oil used in engine. Change to correct lubricant (Chapter 1).

Camshaft or journals worn. High wear causing drop in oil pressure. Replace cam and/or head. Abnormal wear could be caused by oil starvation at high rpm from low oil level, improper oil grade or type.

Crankshaft and/or bearings worn. Same problems as paragraph 5. Overhaul lower end (Chapter 1).

Poor handling or roadholding

39 Directional instability

Steering head bearing adjustment too tight. This will cause rolling or weaving at low speeds. Re-adjust the bearings.

Steering head bearing worn or damaged. Correct adjustment of the bearing will prove impossible to achieve if wear or damage has occurred. Inconsistent handling will occur including rolling or weaving at low speed and poor directional control at indeterminate higher speeds. The steering head bearing should be dismantled for inspection and renewed if required. Lubrication should also be carried out.

Bearing races pitted or dented. Impact damage caused, perhaps, by an accident or riding over a pot-hole can cause indentation of the bearing, usually in one position. This should be noted as notchiness when the handlebars are turned. Renew and lubricate the bearings.

Steering stem bent. This will occur only if the machine is subjected to a high impact such as hitting a curb or a pot-hole. The lower yoke/stem should be renewed; do not attempt to straighten the stem.

Front or rear tyre pressures too low.

Front or rear tyre worn. General instability, high speed wobbles and skipping over white lines indicates that tyre renewal may be required. Tyre induced problems, in some machine/tyre combinations, can occur even when the tyre in question is by no means fully worn.

Swinging arm bearings worn. Difficulties in holding line, particularly when cornering or when changing power settings indicates wear in the swinging arm bearings. The swinging arm should be removed from the machine and the bearings renewed.

Swinging arm flexing. The symptoms given in the preceding paragraph will also occur if the swinging arm fork flexes badly. This can be caused by structural weakness as a result of corrosion, fatigue or impact damage, or because the rear wheel spindle is slack.

Wheel bearings worn. Renew the worn bearings.

Loose wheel spokes. The spokes should be tightened evenly to maintain tension and trueness of the rim.

Tyres unsuitable for machine. Not all available tyres will suit the characteristics of the frame and suspension, indeed, some tyres or tyre combinations may cause a transformation in the handling characteristics. If handling problems occur immediately after changing to a new tyre type or make, revert to the original tyres to see whether an improvement can be noted. In some instances a change to what are, in fact, suitable tyres may give rise to handling deficiences. In this case a thorough check should be made of all frame and suspension items which affect stability.

40 Steering bias to left or right

Rear wheel out of alignment. Caused by uneven adjustment of chain tensioner adjusters allowing the wheel to be askew in the fork ends. A bent rear wheel spindle will also misalign the wheel in the swinging arm.

Wheels out of alignment. This can be caused by impact damage to the frame, swinging arm, wheel spindles or front forks. Although occasionally a result of material failure or corrosion it is usually as a result of a crash.

Front forks twisted in the steering yokes. A light impact, for instance with a pot-hole or low curb, can twist the fork legs in the steering yokes without causing structural damage to the fork legs or the yokes themselves. Re-alignment can be made by loosening the yoke pinch bolts, wheel spindle and mudguard bolts. Re-align the wheel with the handlebars and tighten the bolts working upwards from the wheel spindle. This action should be carried out only when there is no chance that structural damage has occurred.

41 Handlebar vibrates or oscillates

Tyres worn or out of balance. Either condition, particularly in the front tyre, will promote shaking of the fork assembly and thus the handlebars. A sudden onset of shaking can result if a balance weight is displaced during use.

Tyres badly positioned on the wheel rims. A moulded line on each wall of a tyre is provided to allow visual verification that the tyre is correctly positioned on the rim. A check can be made by rotating the tyre; any misalignment will be immediately obvious.

Wheels rims warped or damaged. Inspect the wheels for runout as described in Chapter 5.

Swinging arm bearings worn. Renew the bearings.

Wheel bearings worn. Renew the bearings.

Steering head bearings incorrectly adjusted. Vibration is more likely to result from bearings which are too loose rather than too tight. Re-adjust the bearings.

Loosen fork component fasteners. Loose nuts and bolts holding the fork legs, wheel spindle, mudguards or steering stem can promote shaking at the handlebars. Fasteners on running gear such as the forks and suspension should be check tightened occasionally to prevent dangerous looseness of components occurring.

Engine mounting bolts loose. Tighten all fasteners.

42 Poor front fork performance

Damping fluid level incorrect. If the fluid level is too low poor suspension control will occur resulting in a general impairment of roadholding and early loss of tyre adhesion when cornering and braking. Too much oil is unlikely to change the fork characteristics unless severe overfilling occurs when the fork action will become stiffer and oil seal failure may occur.

Damping oil viscosity incorrect. The damping action of the fork is directly related to the viscosity of the damping oil. The lighter the oil used, the less will be the damping action imparted. For general use, use the recommended viscosity of oil, changing to a slightly higher or heavier oil only when a change in damping characteristic is required. Overworked oil, or oil contaminated with water which has found its way past the seals, should be renewed to restore the correct damping performance and to prevent bottoming of the forks.

Damping components worn or corroded. Advanced normal wear of the fork internals is unlikely to ocur until a very high mileage has been covered. Continual use of the machine with damaged oil seals which allows the ingress of water, or neglect, will lead to rapid corrosion and wear. Dismantle the forks for inspection and overhaul. See Chapter 4.

Weak fork springs. Progressive fatigue of the fork springs, resulting in a reduced spring free length, will occur after extensive use. This condition will promote excessive fork dive under braking, and in its advanced form will reduce the at-rest extended length of the forks and thus the fork geometry. Renewal of the springs as a pair is the only satisfactory course of action.

Bent stanchions or corroded stanchions. Both conditions will prevent correct telescoping of the fork legs, and in an advanced state can cause sticking of the fork in one position. In a mild form corrosion will cause stiction of the fork thereby increasing the time the suspension takes to react to an uneven road surface. Bent fork stanchions should be attended to immediately because they indicate that impact damage has occurred, and there is a danger that the forks will fail with disastrous consequences.

43 Front fork judder when braking (see also Section 55)

Wear between the fork stanchions and the fork legs. Renewal of the affected components is required.

Slack steering head bearings. Re-adjust the bearings.

Warped brake disc or drum. If irregular braking action occurs fork judder can be induced in what are normally serviceable forks. Renew the damaged brake components.

44 Poor rear suspension performances

Rear suspension unit damper worn out or leaking. The damping performance of most rear suspension units falls off with age. This is a gradual process, and thus may not be immediately obvious. Indications of poor damping include hopping of the rear end when cornering or braking, and a general loss of positive stability. See Chapter 4.

Weak rear springs. If the suspension unit springs fatigue they will promote excessive pitching of the machine and reduce the ground clearance when cornering. Suspension linkage bearings worn.

Swinging arm flexing or bearings worn. See Sections 39 and 40.

Bent suspension unit damper rod. This is likely to occur only if the machine is dropped or if seizure of the piston occurs. If either happens the suspension units should be renewed as a pair.

Abnormal frame and suspension noise

45 Front end noise

Oil level low or too thin. This can cause a 'spurting' sound and is usually accompanied by irregular fork action (Chapter 4).

Spring weak or broken. Makes a clicking or scraping sound. Fork oil will have a lot of metal particles in it (Chapter 4).

Steering head bearings loose or damaged. Clicks when braking. Check, adjust or replace (Chapter 4).

Fork clamps loose. Make sure all fork clamp pinch bolts are tight (Chapter 4).

Fork stanchion bent. Good possibility if machine has been dropped. Repair or replace tube (Chapter 4).

46 Rear suspension noise

Fluid level too low. Leakage of a suspension unit, usually evident by oil on the outer surfaces, can cause a spurting noise. The suspension units should be renewed as a pair.

Defective rear suspension unit with internal damage. Renew the suspension units as a pair.

Worn suspension linkage bearings. Renew.

Brake problems

47 Brakes are spongy or ineffective – disc brakes

Air in brake circuit. This is only likely to happen in service due to neglect in checking the fluid level or because a leak has developed. The problem should be identified and the brake system bled of air.

Pad worn. Check the pad wear against the wear lines provided and renew the pads if necessary.

Contaminated pads. Cleaning pads which have been contaminated with oil, grease or brake fluid is unlikely to prove successful; the pads should be renewed.

Pads glazed. This is usually caused by overheating. The surface of the pads may be roughened using glass-paper or a fine file.

Brake fluid deterioration. A brake which on initial operation is firm but rapidly becomes spongy in use may be failing due to water contamination of the fluid. The fluid should be drained and then the system refilled and bled.

Master cylinder seal failure. Wear or damage of master cylinder internal parts will prevent pressurisation of the brake fluid. Overhaul the master cylinder unit.

Caliper seal failure. This will almost certainly be obvious by loss of fluid, a lowering of fluid in the master cylinder reservoir and contamination of the brake pads and caliper. Overhaul the caliper assembly.

48 Brakes drag – disc brakes

Disc warped. The disc must be renewed.

Caliper piston, caliper or pads corroded. The brake caliper assembly is vulnerable to corrosion due to water and dirt, and unless cleaned at regular intervals and lubricated in the recommended manner, will become sticky in operation.

Piston seal deteriorated. The seal is designed to return the piston in the caliper to the retracted position when the brake is released. Wear or old age can affect this function. The caliper should be overhauled if this occurs.

Brake pad damaged. Pad material separating from the backing plate due to wear or faulty manufacture. Renew the pads. Faulty installation of a pad also will cause dragging.

Wheel spindle bent. The spindle may be straightened if no structural damage has occurred.

Brake lever not returning. Check that the lever works smoothly throughout its operating range and does not snag on any adjacent cycle parts. Lubricate the pivot if necessary.

Twisted caliper support bracket. This is likely to occur only after impact in an accident. No attempt should be made to re-align the caliper; the bracket should be renewed.

49 Brake lever or pedal pulsates in operation – disc brakes

Disc warped or irregularly worn. The disc must be renewed.

Wheel spindle bent. The spindle may be straightened provided no structural damage has occurred.

50 Disc brake noise

Brake squeal. This can be caused by the omission or incorrect installation of the anti-squeal shim fitted to the rear of one pad. The arrow on the shim should face the direction of wheel normal rotation. Squealing can also be caused by dust on the pads, usually in combination with glazed pads, or other contamination from oil, grease, brake fluid or corrosion. Persistent squealing which cannot be traced to any of the normal causes can often be cured by applying a thin layer of high temperature silicone grease to the rear of the pads. Make absolutely certain that no grease is allowed to contaminate the braking surface of the pads.

Glazed pads. This is usually caused by high temperatures or contamination. The pad surfaces may be roughened using glass-paper or a fine file. If this approach does not effect a cure the pads should be renewed.

Disc warped. This can cause a chattering, clicking or intermittent squeal and is usually accompanied by a pulsating brake lever or pedal or uneven braking. The disc must be renewed.

Brake pads fitted incorrectly or undersize. Longitudinal play in the pads due to omission of the locating springs (where fitted) or because pads of the wrong size have been fitted will cause a single tapping noise every time the brake is operated. Inspect the pads for correct installation and security.

51 Brakes are spongy or ineffective – drum brakes

Brake cable deterioration. Damage to the outer cable by stretching or being trapped will give a spongy feel to the brake lever. The cable should be renewed. A cable which has become corroded due to old age or neglect of lubrication will partially seize making operation very heavy. Lubrication at this stage may overcome the problem but the fitting of a new cable is recommended.

Worn brake linings. Determine lining wear using the external brake wear indicator on the brake backplate, or by removing the wheel and withdrawing the brake backplate. Renew the shoe/lining units as a pair if the linings are worn below the recommended limit.

Worn brake camshaft. Wear between the camshaft and the bearing surface will reduce brake feel and reduce operating efficiency. Renewal of one or both items will be required to rectify the fault.

Worn brake cam and shoe ends. Renew the worn components.

Linings contaminated with dust or grease. Any accumulations of dust should be cleaned from the brake assembly and drum using a petrol dampened cloth. Do not blow or brush off the dust because it is asbestos based and thus harmful if inhaled. Light contamination from grease can be removed from the surface of the brake linings using a solvent; attempts at removing heavier contamination are less likely to be successful because some of the lubricant will have been absorbed by the lining material which will severely reduce the braking performance.

52 Brake drag – drum brakes

Incorrect adjustment. Re-adjust the brake operating mechanism.

Drum warped or oval. This can result from overheating, impact or uneven tension of the wheel spokes. The condition is difficult to correct, although if slight ovality only occurs, skimming the surface of the brake drum can provide a cure. This is work for a specialist engineer. Renewal of the complete wheel hub is normally the only satisfactory solution.

Weak brake shoe return springs. This will prevent the brake lining/shoe units from pulling away from the drum surface once the brake is released. The springs should be renewed.

Brake camshaft, lever pivot or cable poorly lubricated. Failure to attend to regular lubrication of these areas will increase operating resistance which, when compounded, may cause tardy operation and poor release movement.

53 Brake lever or pedal pulsates in operation – drum brakes

Drums warped or oval. This can result from overheating, impact or uneven spoke tension. This condition is difficult to correct, although if slight ovality only occurs skimming the surface of the drum can provide a cure. This is work for a specialist engineer. Renewal of the hub is normally the only satisfactory solution.

54 Drum brake noise

Drum warped or oval. This can cause intermittent rubbing of the brake linings against the drum. See the preceding Section.

Brake linings glazed. This condition, usually accompanied by heavy lining dust contamination, often induces brake squeal. The surface of the linings may be roughened using glass-paper or a fine file.

55 Brake induced fork judder

Worn front fork stanchions and legs, or worn or badly adjusted steering head bearings. These conditions, combined with uneven or pulsating braking as described in Sections 49 and 53 will induce more or less judder when the brakes are applied, dependent on the degree of wear and poor brake operation. Attention should be given to both areas of malfunction. See the relevant Sections.

Electrical problems

56 Battery dead or weak

Battery faulty. Battery life should not be expected to exceed 3 to 4 years, particularly where a starter motor is used regularly. Gradual sulphation of the plates and sediment deposits will reduce the battery performance. Plate and insulator damage can often occur as a result of vibration. Complete power failure, or intermittent failure, may be due to a broken battery terminal. Lack of electrolyte will prevent the battery maintaining charge.

Battery leads making poor contact. Remove the battery leads and clean them and the terminals, removing all traces of corrosion and tarnish. Reconnect the leads and apply a coating of petroleum jelly to the terminals.

Load excessive. If additional items such as spot lamps, are fitted, which increase the total electrical load above the maximum alternator output, the battery will fail to maintain full charge. Reduce the electrical load to suit the electrical capacity.

Regulator/rectifier failure.

Alternator generating coils open-circuit or shorted.

Charging circuit shorting or open circuit. This may be caused by frayed or broken wiring, dirty connectors or a faulty ignition switch. The system should be tested in a logical manner. See Section 59.

57 Battery overcharged

Rectifier/regulator faulty. Overcharging is indicated if the battery becomes hot or it is noticed that the electrolyte level falls repeatedly between checks. In extreme cases the battery will boil causing corrosive gases and electrolyte to be emitted through the vent pipes.

Battery wrongly matched to the electrical circuit. Ensure that the specified battery is fitted to the machine.

58 Total electrical failure

Fuse blown. Check the main fuse. If a fault has occurred, it must be rectified before a new fuse is fitted.

Battery faulty. See Section 56.

Earth failure. Check that the frame main earth strap from the battery is securely affixed to the frame and is making a good contact.

Ignition switch or power circuit failure. Check for current flow through the battery positive lead (red) to the ignition switch. Check the ignition switch for continuity.

59 Circuit failure

Cable failure. Refer to the machine's wiring diagram and check the circuit for continuity. Open circuits are a result of loose or corroded connections, either at terminals or in-line connectors, or because of broken wires. Occasionally, the core of a wire will break without there being any apparent damage to the outer plastic cover.

Switch failure. All switches may be checked for continuity in each switch position, after referring to the switch position boxes incorporated in the wiring diagram for the machine. Switch failure may be a result of mechanical breakage, corrosion or water.

Fuse blown. Refer to the wiring diagram to check whether or not a circuit fuse is fitted. Replace the fuse, if blown, only after the fault has been identified and rectified.

60 Bulbs blowing repeatedly

Vibration failure. This is often an inherent fault related to the natural vibration characteristics of the engine and frame and is, thus, difficult to resolve. Modifications of the lamp mounting, to change the damping characteristics may help.

Intermittent earth. Repeated failure of one bulb, particularly where the bulb is fed directly from the generator, indicates that a poor earth exists somewhere in the circuit. Check that a good contact is available at each earthing point in the circuit.

Reduced voltage. Where a quartz-halogen bulb is fitted the voltage to the bulb should be maintained or early failure of the bulb will occur. Do not overload the system with additional electrical equipment in excess of the system's power capacity and ensure that all circuit connections are maintained clean and tight.

SUZUKI GS/DR 125 SINGLES

Check list

Daily (pre-ride) checks

1 Check the engine/transmission oil level
2 Check the level of fuel in the tank
3 Check the hydraulic fluid level in the front brake reservoir (where applicable). Check the correct operation of both brakes
4 Inspect the tyres for damage and check the pressures
5 Check the final drive chain adjustment and lubricate the chain
6 Check that all the controls function smoothly and correctly. Check that the suspension and steering are in good working order
7 Check that the lights and electrical system functions correctly. Also check that the speedometer is in working order

Monthly or every 600 miles (1000 km)

1 Check that the final drive chain is correctly adjusted and lubricated

Three monthly or every 1500 miles (2500 km)

1 Change the engine/transmission oil

Four monthly or every 2000 miles (3000 km)

1 Adjust the cam chain tension
2 Clean the air filter element

Six monthly or every 3000 miles (5000 km)

1 Check the compression pressure and the oil pressure
2 Change the engine/transmission oil and the oil filter element
3 Check and adjust the valve clearances
4 Clean the spark plug and reset the electrode gap
5 Check the carburettor for smooth operation and make any necessary adjustment. Lubricate and adjust the throttle cable, and check the fuel line
6 Check and adjust the clutch cable free play
7 Check the battery electrolyte level
8 Check brake pad and shoe wear, and adjust the brakes
9 Check the steering head bearing adjustment
10 Inspect carefully the suspension and the wheels
11 Grease the rear suspension linkage – DR125 S model
12 Make a general check of all controls and fasteners for correct operation and tightness. Lubricate all controls and pivot points.

Annually or every 6000 miles (10 000 km)

1 Clean the oil pump filter gauze
2 Renew the spark plug
3 Clean the fuel filter
4 Change the fork oil
5 Grease the drum brake camshaft
6 Grease the wheel bearings and instrument drive cables

Additional maintenance items – *see text for frequency*

1 Grease the steering head and swinging arm pivot bearings
2 Renew the brake fluid
3 Renew the fuel feed pipe
4 Renew the hydraulic brake hose
5 Cleaning the machine

Adjustment data

	GS125	DR125
Tyre pressures		
Front:		
Solo	24 psi	22 psi
Pillion	24 psi	22 psi
Rear:		
Solo	28 psi	24 psi
Pillion	32 psi	28 psi
Compression pressure		
Minimum	114 psi (8 kg/cm²)	142 psi (10 kg/cm²)
Valve clearances (cold)	0.08 – 0.13 mm (0.003 – 0.005 in)	
Spark plug gap	0.6 – 0.7 mm (0.024 – 0.028 in)	
Spark plug type	NGK DR8ES-L or ND X24ESR-U	
Idle speed	1450 ± 50 rpm	

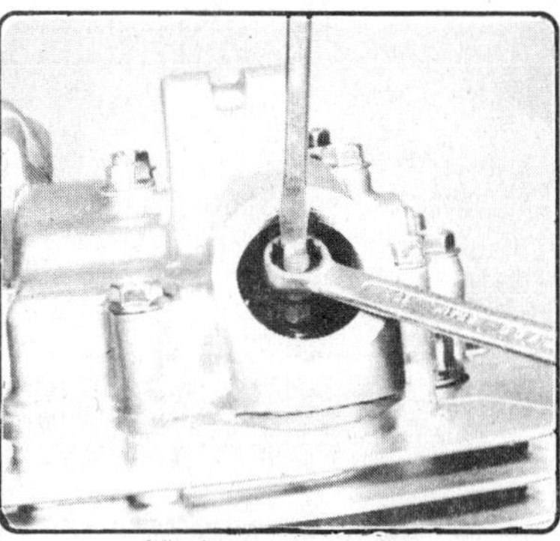

Adjusting the valve clearances

Recommended lubricants

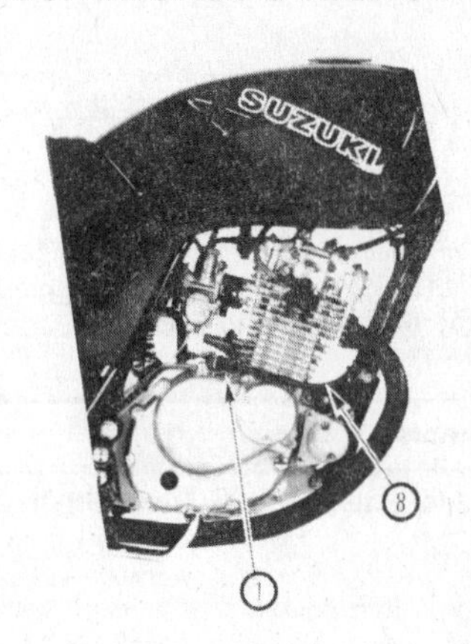

Component	Quantity	Type/viscosity
① Engine/transmission:		
At oil change	850 cc (1.5 pint)	SAE 10W/40 SE or SF engine oil
At oil and filter change	950 cc (1.7 pint)	SAE 10W/40 SE or SF engine oil
② Front forks (per leg:)		
GS 125	136 cc (4.79 fl oz)	SAE 15 fork oil
DR125	177 cc (6.23 fl oz)	SAE 10 fork oil
③ Final drive chain	As required	Aerosol chain lubricant or special chain grease
④ Hydraulic front brake	As required	SAE J1703 or DOT 3 or 4 hydraulic fluid
⑤ Wheel bearings	As required	High melting-point grease
⑥ Steering head bearings	As required	High melting-point grease
⑦ Pivot points	As required	High melting-point grease, molybdenum disulphide-based
⑧ Control cables	As required	Multi-purpose aerosol lubricant or light machine oil

ROUTINE MAINTENANCE GUIDE

Refer to Chapter 7 for information relating to DR125 SF, SH, SJ 'Raider', GN125 and GZ125 'Marauder' models

Routine maintenance

Refer to Chapter 7 for information relating to DR125 SF, SH, SJ 'Raider', GN125 and GZ125 'Marauder' models

Periodic routine maintenance is a continuous process which should commence immediately the machine is used. The object is to maintain all adjustments and to diagnose and rectify minor defects before they develop into more extensive, and often more expensive, problems.

It follows that if the machine is maintained properly, it will both run and perform with optimum efficiency, and be less prone to unexpected breakdowns. Regular inspection of the machine will show up any parts which are wearing, and with a little experience, it is possible to obtain the maximum life from any one component, renewing it when it becomes so worn that it is liable to fail.

Regular cleaning can be considered as important as mechanical maintenance. This will ensure that all the cycle parts are inspected regularly and are kept free from accumulations of road dirt and grime.

The various maintenance tasks are described under their respective mileage and calendar headings, and are accompanied by diagrams and photographs where pertinent.

It should be noted that the intervals between each maintenance task serve only as a guide. As the machine gets older, or if it is used under particularly arduous conditions, it is advisable to reduce the period between each check.

For ease of reference, most service operations are described in detail under the relevant heading. However, if further general information is required, this can be found under the pertinent Section heading and Chapter in the main text.

Although no special tools are required for routine maintenance, a good selection of general workshop tools is essential. Included in the tools must be a range of metric ring or combination spanners, a selection of crosshead screwdrivers, and two pairs of circlip pliers, one external opening and the other internal opening. Additionally, owing to the extreme tightness of most casing screws on Japanese machines, an impact screwdriver, together with a choice of large or small crosshead screw bits, is absolutely indispensable. This is particularly so if the engine has not been dismantled since leaving the factory. Another tool which is essential for DR125 S owners is a stand of some sort. A strong wooden box is normally recommended, but this is not always convenient for routine maintenance tasks, and so the purchase or fabrication of a metal stand, which will support the machine securely upright with enough height for either wheel to be removed, is advised. 'Paddock' type stands are frequently advertised in the national motorcycle press and are, in the author's view, well worth the money spent. DR125 S owners should note that where a centre stand is mentioned in the following instructions, they must substitute their own stand.

Daily (pre-riding checks)

Before taking the machine out on the road, there are certain checks which should be completed to ensure that it is in a safe and legal condition to be used.

1 *Engine/transmission oil level*

With the machine standing upright on its wheels on level ground, check that the oil level visible through the sight glass set in the crankcase right-hand cover is between the 'L' (Low) and 'F' (Full) lines stamped on the crankcase cover. If topping-up is necessary, add oil slowly via the filler plug orifice in the crankcase cover. Use only a good quality SAE 10W/40 SE or SF engine oil, and refit securely the filler plug. If the engine has just been run, allow one or two minutes for the oil to drain back into the crankcase before checking the level.

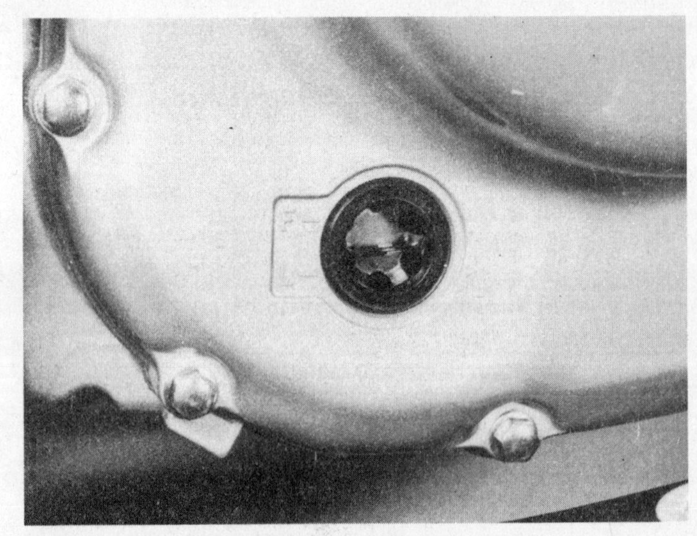

Oil level must be between 'F' and 'L' marks

Use only good quality engine oil of correct grade when topping up

2 *Petrol level*

Checking the petrol level may seem obvious, but it is all too easy to forget. Ensure that you have enough petrol to complete your journey, or at least to get you to the nearest petrol station.

3 *Brakes*

Check that the front and rear brakes work effectively and without binding. Ensure that the rod linkages and the cables, as applicable, are lubricated and properly adjusted. Check the fluid level in the master cylinder reservoir of GS125 ES models, and ensure that there are no fluid leaks. Should topping-up be required, use only the recommended hydraulic fluid to specification SAE J1703 or DOT 3 or 4.

4 *Tyres*

Check the tyre pressures with a gauge that is known to be accurate. It is worthwhile puchasing a pocket gauge for this purpose because the gauges on garage forecourt airlines are notoriously inaccurate. The pressures should be checked with the tyres cold. Even a few miles travelled will warm up the tyres to a point where pressures increase and an inaccurate reading will result. Tyre pressures for these models are:

	GS125 models	**DR125 S**
Front - solo	24 psi	22 psi
Rear - solo	28 psi	24 psi
Front - pillion	24 psi	22 psi
Rear - pillion	32 psi	28 psi

At the same time as the tyre pressures are checked, examine the tyres themselves. Check them for damage, especially splitting of the sidewalls. Remove any small stones or other road debris caught between the treads. This is particularly important on the rear tyre, where rapid deflation due to penetration of the inner tube will almost certainly cause total loss of control. When checking the tyres for damage, they should be examined for tread depth in view of both the legal and safety aspects. It is vital to keep the tread depth within the UK legal limits of 1 mm of depth over three-quarters of the tread breadth around the entire circumference, with no bald patches. Many riders, however, consider nearer 2 mm to be the limit for secure roadholding, traction, and braking, especially in adverse weather conditions.

5 *Final drive chain*

Check that the final drive chain is correctly adjusted and well lubricated. Remember that if the machine is used in adverse conditions the chain will require frequent, even daily, lubrication. Refer to the 1-monthly/600 mile service interval.

6 *Controls and steering*

Check throttle, clutch, gear lever and footrests to ensure that they are securely fastened and working properly. If a bolt is going to work loose, or a cable snap, it is better that it does so with the machine at a standstill than when riding. Check also that the steering and suspension are working correctly.

7 *Lights and speedometer*

Check that all lights, flashing indicators, horn and speedometer are working correctly to make sure that the machine complies with all legal requirements in this respect.

Monthly, or every 600 miles (1000 km)

This is where the proper procedure of routine maintenance begins. The daily checks serve to ensure that the machine is safe and legal to use, but contribute little to maintenance other than to give the owner an accurate picture of what item needs attention. However, if done conscientiously, they will give early warning, as has been stated, of any faults which are about to appear. When performing the following maintenance tasks, therefore, carry out the daily checks first.

1 *Check, adjust and lubricate the final drive chain*

It would appear that some GS125 models have been fitted with an O-ring chain; these chains are easily recognised by the O-rings set between the chain sideplates. The O-rings are employed to seal grease into the chain bushes and pins, hence the chain requires no maintenance other than checks for wear and the frequent application of SAE 80 or 90 gear oil to lubricate the chain roller/sprocket bearing surfaces. If it is necessary to clean the chain, use only paraffin; petrol or commercial solvents will rot the O-rings and allow the grease to escape. By the same token, use only aerosol spray lubricants that are designed for use with such chains; the solvent or propellant in most aerosol sprays will also damage the O-rings. Since O-ring chains do not have connecting links, it will be necessary to remove the swinging arm as described in Section 8 of Chapter 4 to allow the chain to be removed and refitted.

If a conventional chain is fitted, it must be checked for wear, cleaned and lubricated at regular intervals, as described below. A simple check for wear is as follows. With the chain fully lubricated and correctly adjusted as described below, attempt to pull the chain backwards off the rear sprocket. If the chain can be pulled clear of the sprocket teeth it must be considered worn out and renewed; chains should be renewed always in conjunction with the sprockets since the running together of new and part-worn components will greatly increase the rate of wear of both, necessitating renewal much sooner than would otherwise be the case.

A more accurate measurement of chain wear can be applied only to conventional chains, since it involves the removal of the chain from the machine and its thorough cleaning. Disconnect the chain at its split connecting link and pull the entire length of the chain clear of the spockets. Note that refitting the chain is greatly simplified if a worn-out length is temporarily connected to it. As the original chain is pulled off the sprockets, the worn-out chain will follow it and remain in place while the task of cleaning and examination is carried out. On reassembly, the process is repeated, pulling the worn-out chain over the sprockets so that the new chain, or the freshly cleaned and lubricated chain, is pulled easily into place.

To clean the chain, immerse it in a bath containing a mixture of petrol and paraffin and use a stiff-bristled brush to scrub away all the traces of road dirt and old lubricant. Take the necessary fire precautions when using this flammable solvent. Swill the chain around to ensure that the solvent penetrates fully into the bushes and rollers and can remove any lubricant which may still be present. When the chain is completely clean, remove it from the bath and hang it up to dry.

To assess accurately the amount of wear present in the chain, it must be cleaned and dried as described above, then laid out on a flat surface. Compress the chain fully and measure its length from end to end. Anchor one end of the chain and pull on the other end, drawing the chain out to its fullest extent. Measure the stretched length. If the stretched measurement exceeds the compressed measurement by more than $\frac{1}{4}$ in per foot, the chain must be considered worn out and be renewed. Suzuki's own recommendation is that a 20 pitch length, when cleaned and dried, must not exceed 259 mm (10.2 in). This is accomplished by marking any pin along the length of the chain, counting off 21 pins, and by measuring the distance between the two.

Chain lubrication is best carried out by immersing the chain in a molten lubricant such as Chainguard or Linklyfe. Lubrication carried out in this manner must be preceded by removing the chain from the machine, cleaning it, and drying it as described above. Follow the manufacturer's instructions carefully when using Chainguard or Linklyfe, and take great care to swill the chain gently in the molten lubricant to ensure that all bearing surfaces are fully greased.

Refitting a new, or freshly-lubricated, chain is a potentially messy affair which is greatly simplified by the substitution of a worn-out length of chain during removal. The new chain can then be connected to the worn-out length and pulled easily around the sprockets. Refit the connecting link, ensuring that the spring clip is fitted with its closed end facing the normal direction of travel of the chain.

For the purpose of daily or weekly lubrication, one of the many proprietary aerosol-applied chain lubricants is a far better proposition since this can be applied very quickly, while the chain is in place on the machine, and makes very little mess. It should be applied at least once a week, and daily if the machine is used in wet weather conditions. If the roller surfaces look dry, then they need lubrication. Engine oil can be used for this task, but remember that it is flung off the chain far more easily than grease, thus making the rear end of the machine unnecessarily dirty, and requires more frequent application if it is to perform its task adequately. Also remember that surplus oil will eventually find its way on to the tyre, with quite disastrous consequences.

Chain adjustment is necessary to take up wear in the multitude of bearing surfaces present in the chain. As this wear does not take place evenly along the length of the chain, tight spots will appear which must be compensated for when adjusting the chain. Place the machine securely on its side or centre stand with the transmission in neutral. Find the tightest spot in the chain by revolving the rear wheel and pushing upwards on the bottom run of the chain, midway between the front and rear sprockets, testing along the entire length of the chain. When the tightest spot has been found, measure the total amount of up and down movement available. This should be between 25-35 mm (1.0-1.4 in) on all GS125 models, and 30-40 mm (1.2-1.6 in) on DR125 S models.

To adjust the chain, withdraw the securing split pins from the wheel spindle retaining nut (not GS125 ESX) and from the brake torque arm securing nut, then slacken both nuts by just enough to permit the rear wheel assembly to be moved. Slacken the chain adjuster locknuts (where fitted) and tighten the adjuster nuts or bolts to draw the spindle backwards to the point where the chain is correctly tensioned. Note that the swinging arm fork ends are marked with a series of vertical lines and that each adjuster (GS125 models only) has a notch cut in it to provide a reference point. These marks are provided to assist in preserving accurate wheel alignment and are used by ensuring that the notch in each adjuster is aligned exactly with the same index mark stamped in each fork end. On DR125 S models align the forward edge

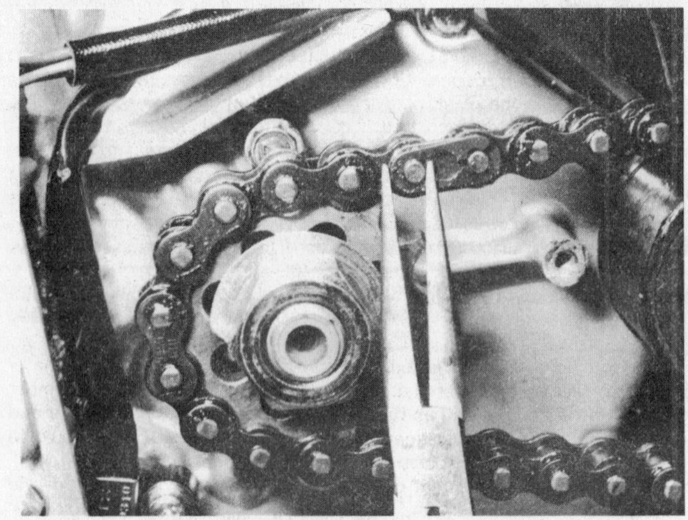
Spring clip is refitted with closed end facing direction of chain travel

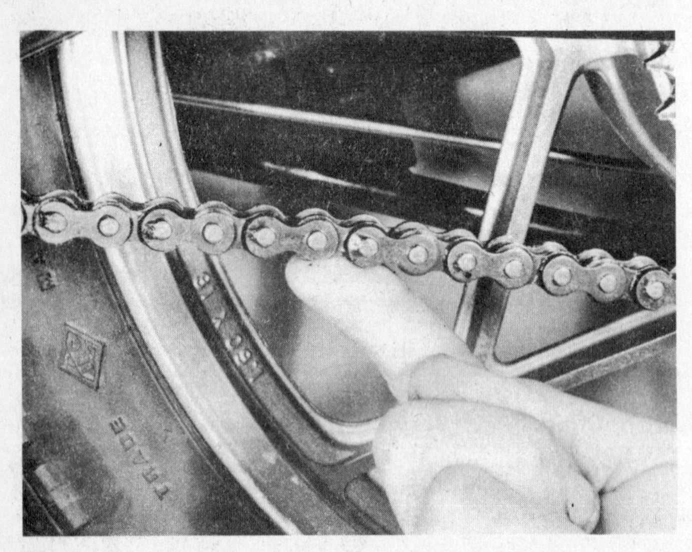
Check tension in position shown – find tightest spot in chain

Chain adjustment – all GS125 models – note alignment reference marks

Chain adjustment – DR125 S model

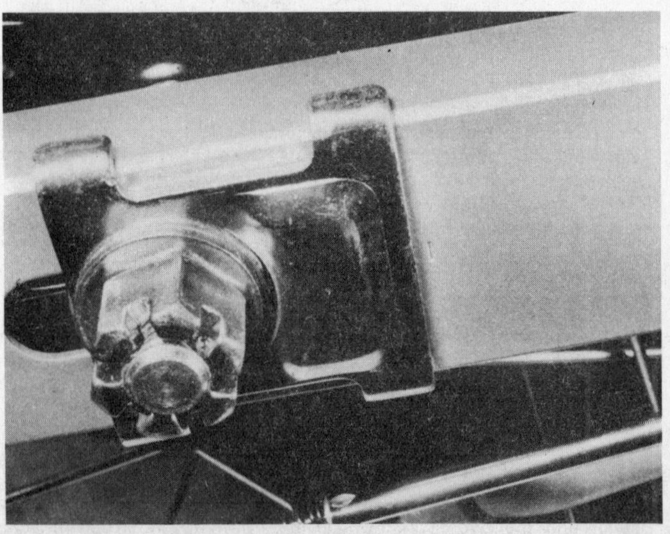
Index mark must be aligned with forward edge of spindle guide plate – DR125 S

of each spindle guide plate with the same index mark stamped in each fork end. A final check of accurate wheel alignment can be made by laying a plank of wood or drawing a length of string parallel to the machine so that it touches both walls of the rear tyre. Wheel alignment is correct when the plank or string is equidistant from both walls of the front tyre when tested on both sides of the machine. Note that if the front tyre is of smaller section that the rear, the plank or string will not touch the walls of the front tyre, as shown in the accompanying illustration. The task of preserving correct rear wheel alignment is made easier if care is taken to draw the spindle back in small stages, turning each adjuster nut or bolt by exactly the same amount.

When the chain is correctly tensioned, apply the rear brake to centralise the shoes on the drum, and tighten the spindle retaining nut to a torque setting of 5.0 – 8.0 kgf m (36 – 58 lbf ft), then on all models except the GS125 ESX (which uses a plain nut) secure it with a new split pin. Tighten the brake torque arm retaining nut to a torque setting of 1.0 – 1.5 kgf m (7 – 11 lbf ft) and secure it with a new split pin. Remember that if the chain tension has been altered significantly, the rear brake and stop lamp rear switch adjustment will also require resetting. These should be checked as a matter of course before taking the machine out on the road.

Note that replacement chains are now available in standard metric sizes from Renold Limited, the British chain manufacturer. When ordering a new chain, always quote the size, the number of chain links and the type of machine to which the chain is to be fitted. All the machines featured in this Manual use a 428 ($^{1}/_{2}$ x $^{5}/_{16}$ in) size chain, the number of links varying between 116-122, depending on for which model the chain is being ordered, and depending on what size of sprocket is fitted.

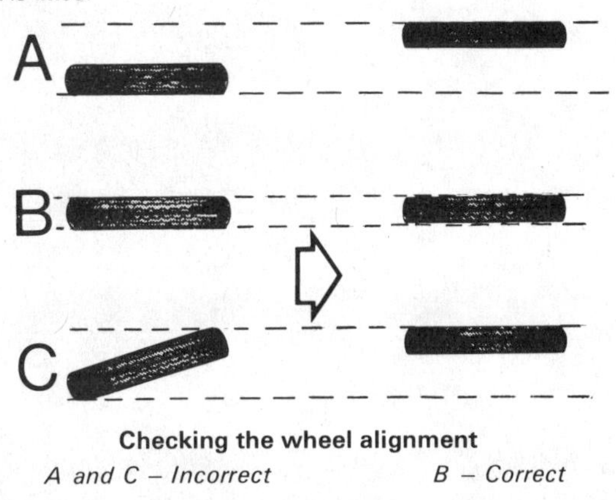

Checking the wheel alignment

A and C – Incorrect *B – Correct*

Three monthly, or every 1500 miles (2500 km)

Complete the tasks listed the previous mileage/time headings, and then carry out the following if necessary:

1 Additional engine/transmission oil change

Since the engine relies so heavily on the quantity and quality of its oil, and since the oil in any motorcycle engine is worked far harder than in other vehicles, it is recommended that the engine oil is changed at more frequent intervals than those specified by the manufacturer. This is particularly important if the machine is used at very high speeds for long periods of time, and even more important if the machine is used only at very slow speed or for very short journeys.

Follow the instructions given under the six monthly/3000 mile interval, but note that there will be no need to change the oil filter, and that therefore, only 850 cc (1.5 pint) of oil will be necessary, to refill the crankcase.

Four monthly, or every 2000 miles (3000 km)

Complete the tasks listed the previous mileage/time headings, where applicable, and then carry out the following:

1 Adjust the cam chain

Remove the two valve inspection caps from the cylinder head cover and the two inspection covers from the crankcase left-hand cover. Rotate the crankshaft until the engine is in the TDC position on the compression stroke; this is found by aligning the timing index mark stamped on the rotor rim with the arrow mark on the crankcase cover (all GS125 models) or with the centre of the inspection aperture (DR125 S models), when there is free play at both valve rocker arms. Slacken the cam chain tensioner locknut and unscrew the adjuster screw by one full turn. The pushrod should be heard to move out under spring pressure. Tighten the adjuster screw by just enough to retain the pushrod, then tighten the locknut. **Do not** overtighten either the screw or the locknut. Refit the inspection covers or plugs.

If adjustment fails to quieten the cam chain, remove the tensioner assembly and check it for faults as described in Section 24 of Chapter 1; removal and refitting are described in Sections 6 and 41 of the same Chapter.

Slacken locknut and adjuster screw to adjust cam chain tension – do not overtighten

2 Clean the air filter element

It is vitally important that the air filter element is kept clean and in good condition if the engine is to function properly. If the element becomes choked with dust it follows that the airflow to the engine will be impaired, leading to poor performance and high fuel consumption. Conversely, a damaged air filter will allow excessive amounts of unfiltered air to enter the engine, which can result in an increased rate of wear and possible damage due to the weak nature of the mixture. The interval specified above indicates the maximum time limit between each cleaning operation. Where the machine is used in particularly adverse conditions it is advised that cleaning takes place on a much more frequent basis.

Remove the left-hand sidepanel. On all GS125 models, remove the three screws and withdraw the filter element cover, then separate the element from its supporting frame. On DR125 S models, remove the single screw and withdraw the filter element cover, then remove the element retaining screw and pull out the metal supporting frame. Unscrew the butterfly nut, remove the plastic washer, and separate the element from its supporting frame.

Examine the element carefully. It must be renewed if there are any splits or tears, or if the foam appears to be hardened through age. To clean the element, immerse it in a bath of a non-flammable solvent such as white spirit. Petrol may be used, but be careful to take suitable precautions against the risk of fire. When the foam is clean, gently squeeze out the surplus solvent and allow the remainder to evaporate. Do not wring the element out, as this will damage the foam. Soak the cleaned, dry, element in engine oil and gently squeeze out the surplus to leave the element slightly oily to the touch. Filter reassembly is a straightforward reversal of the dismantling procedure, being careful to ensure that both the element and the filter cover are correctly seated so that no unfiltered air can bypass the element and enter the carburettor. A light application of grease to the sealing surfaces of the filter cover will help to achieve good sealing.

On no account should the air filter element be omitted while the engine is running in view of the increased noise level and of the high risk of severe damage to the engine due to overheating caused by the resultant weak mixture.

All GS125 models – remove three larger screws to release filter cover ...

... so that filter element can be separated from supporting frame

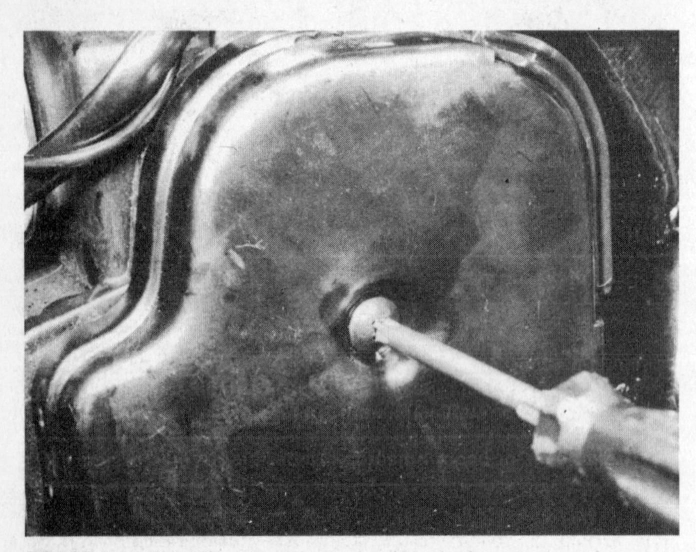
DR125 S model – filter cover is retained by single screw ...

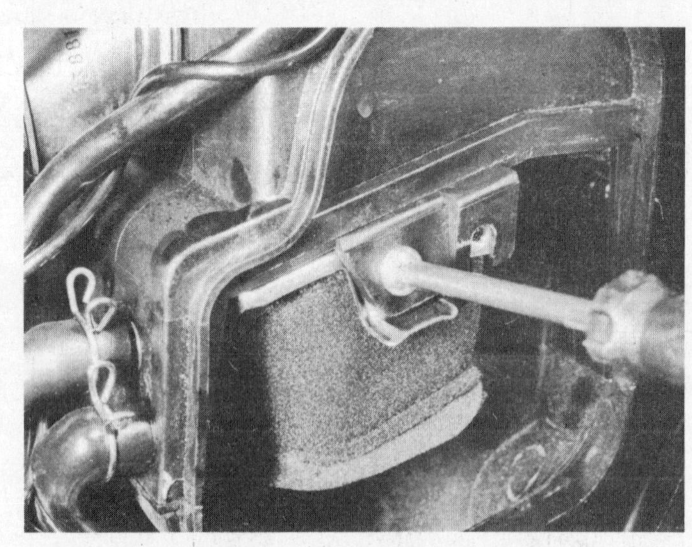
... remove element frame by releasing retaining screw ...

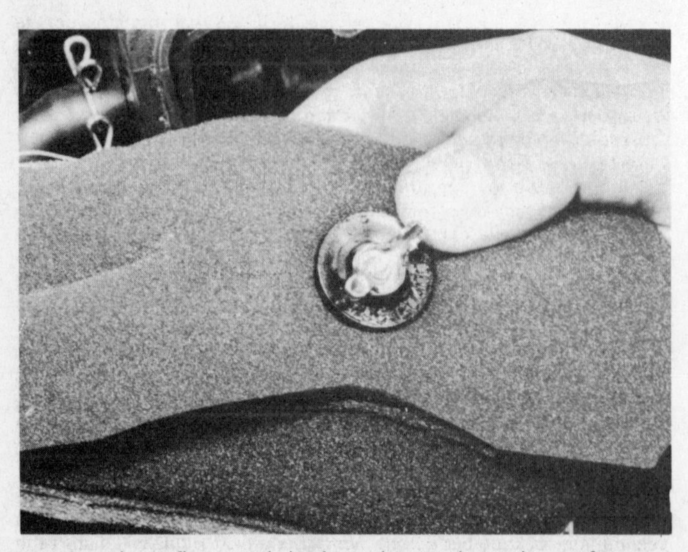
... remove butterfly nut and plastic washer to release element from frame

Six monthly, or every 3000 miles (5000 km)

Complete, where applicable, the tasks listed under the previous mileage/time headings, then carry out the following:

1 Check the compression pressure and oil pressure

The manufacturer recommends that the engine compression pressure and oil pressure be checked at this interval to give some idea of the degree of wear that has taken place in the upper cylinder and lubrication system components. Since this is a counsel of perfection, and since both tasks require expensive special tools, it is recommended that they are carried out only when the need arises.

Engine compression pressure is tested using a compression gauge with an adaptor suitable for a 12 mm spark plug thread (Suzuki service tools Part Number 09915-64510 and 09915-63210). The engine must be fully warmed up, with the valve clearances accurately set and all cylinder head retaining nuts tightened to the correct torque settings. Open fully the throttle twistgrip and turn the engine over several times with the starter motor or kickstart, noting the values recorded by the gauge. If the highest gauge reading is 114 psi (8 kg/cm²) on GS125 models or 142 psi (10 kg/cm²) on DR125 S models, or less, the engine is excessively worn and must be stripped for repair. The areas of wear are the piston/cylinder group, the head gasket, or the valves. The piston/cylinder group can be checked by removing the gauge, pouring a small amount of oil into the cylinder bore, then repeating the test. If the pressure recorded is significantly increased, the piston, piston rings or cylinder barrel are at fault; if the pressure remains unchanged, the head gasket or valves are faulty.

Oil pressure testing is described in Section 12 of Chapter 2. If either test reveals excessive wear, follow the instructions given in the relevant Sections of Chapter 1 to rectify the problem.

2 Change the engine/transmission oil and oil filter element

Oil changing is much easier and more efficient if the engine is fully warmed up so that the oil is thin and flows freely. Place the machine on its stand so that it is upright on level ground. A maximum of 1300 cc (2.3 pint) of oil is contained in the engine; place a suitable container under the drain plug, then remove both drain and filler plugs and allow the oil to drain. Note that the drain plug is situated in the middle of a separate triangular plate bolted to the crankcase underside; do not disturb the smaller hexagon-headed plug beside it (see accompanying photograph).

While the oil is draining, remove the three screws or nuts which fasten the oil filter chamber cap to the crankcase right-hand cover, then withdraw the chamber cap with its sealing O-ring, the coil spring, the filter element, and the second O-ring. Discard the used filter element. Examine the two O-rings; ideally these should be renewed whenever they are disturbed, and should be renewed always if damaged or worn. In practice, if undamaged, they can be re-used a maximum of three times (three filter changes) before renewal is essential. Wipe away all surplus oil from the filter chamber, then refit the small O-ring around the boss at the rear of the chamber. Fit the new filter element, ensuring that the aperture in one end fits over the chamber boss so that the filter blank end faces outwards. Refit the larger O-ring to the chamber cap, using a smear of grease to stick it in place, position the coil spring over the boss in the centre of the cap, then refit the cap, tightening securely the three retaining screws or nuts.

When the oil has finished draining, examine the condition of the drain plug sealing washer, renewing it if necessary, then refit the drain plug, tightening it to a torque setting of 1.8-2.0 kgf m (13-14.5 lbf ft). Refill the crankcase with 950 cc (1.7 pint) of good quality SAE 10W/40 SE or SF engine oil and refit the filler plug.

Start the engine and allow it to idle for one or two minutes, then stop it and wait a further few minutes while the oil level settles. With the machine standing upright on its wheels on level ground, the oil level visible through the sight glass set in the crankcase right-hand cover must be between the 'F' and 'L' marks. Add (or remove) oil as necessary; the engine should never be run with the level above the 'F' mark or below the 'L' mark. Check that the filler plug is securely tightened, remove all traces of surplus oil, and check for any oil leaks which may appear subsequently.

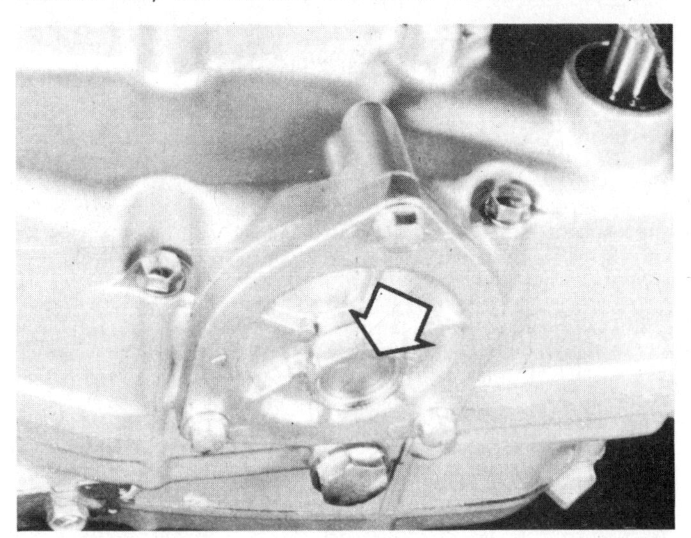

Engine oil drain plug (arrowed) – do not disturb the smaller plug

Small O-ring fits around filter chamber boss – do not omit

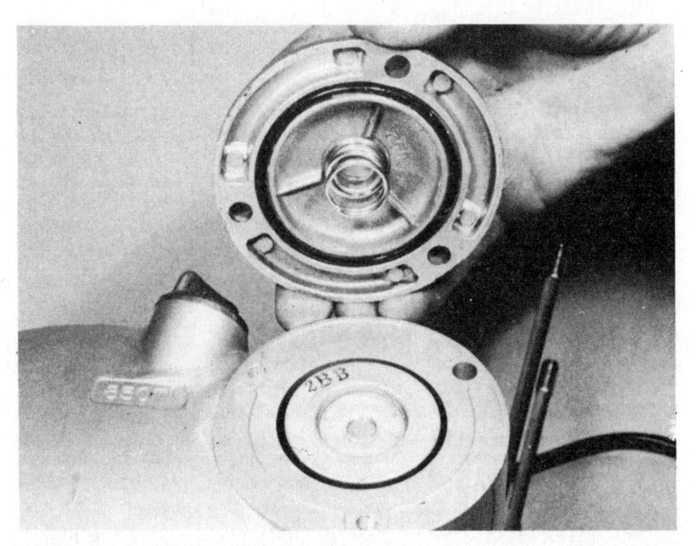

Refit filter element as shown – note position of coil spring and large O-ring

3 Check and adjust valve clearances

The valve clearances must be checked with the engine cold. Remove the sidepanels, the seat, the fuel tank, the spark plug, the inspection caps from the cylinder head cover and from the top of the crankcase left-hand cover, and the circular cap from the side of the crankcase left-hand cover. Apply a spanner to the alternator rotor retaining nut via the aperture in the side of the crankcase cover and rotate the crankshaft anti-clockwise until the piston is at top dead centre (TDC) on the compression stroke. This is achieved when the timing index mark (a straight line with the letter 'T' or 'O' adjacent) stamped on the rotor rim and visible via the aperture in the top of the crankcase cover is aligned exactly with the cast arrow on the crankcase (all GS125 models) or is exactly in the centre of the aperture (DR125 S model), and there is free play at the adjuster ends of both valve rockers.

The valve clearance is measured by sliding feeler gauges between the top of the valve stem and the tip of the adjuster. If the clearance is correct, a feeler gauge of 0.08-0.13 mm (0.003-0.005 in) will be a tight sliding fit between the two, for both inlet and exhaust valves.

To adjust the clearances, slacken the adjuster locknut and use a suitable screwdriver to rotate as necessary the adjuster screw. Tighten securely the locknut, but do not overtighten it since this will distort the threads and make future adjustment very difficult. Recheck the clearance and repeat the operation on the other valve.

Refit all inspection caps, checking that their sealing washers or O-rings are in good condition. Use a close-fitting ring spanner only on the valve inspection caps, and do not overtighten them since they are easily damaged.

4 Clean the spark plug

Detach the spark plug cap, and using the correct spanner remove the spark plug. Clean the electrodes using a wire brush followed by a strip of fine emery cloth or paper. Check the plug gap with a feeler gauge, adjusting it if necessary to within the range of 0.6-0.7 mm (0.024-0.028 in). Make adjustments by bending the outer electrode, never the inner (central) electrode. Before fitting the spark plug smear the threads with a graphited grease; this will aid subsequent removal. Refit the spark plug by hand only, screwing it down until the sealing washer is firmly seated, then tighten it by a further $\frac{1}{4}$ of a turn with the plug spanner.

5 Check the carburettor, throttle cable and fuel line

If rough running of the engine has developed, some adjustment of the carburettor pilot setting and tick-over speed may be required. If this is the case refer to Chapter 2, Section 8 for details. Do not make these adjustments unless they are obviously required, there is little to be gained by unwarranted attention to the carburettor.

The throttle cable must have 0.5-1.0 mm (0.02-0.04 in) of free play, measured at the in-line cable adjuster just under the twistgrip. Adjustment should be made firstly with the cable adjuster on the carburettor top (slide carburettor) or throttle pulley (CV carburettor), then fine adjustment can be made with the in-line adjuster at the cable's upper end. Check that the cable is smooth in operation and apply a few drops of oil to the inner cable at both adjusters, pulling the outer cable carefully away from the adjuster to gain access.

Complete carburettor maintenance by slackening the drain screw on the float chamber, turning the petrol on, and allowing a small amount of fuel to drain through, thus flushing any water or dirt from the carburettor. Tighten the drain screw again.

Give the pipe which connects the fuel tap and carburettor a close visual examination, checking for cracks or any signs of leakage. In time, the synthetic rubber pipe will tend to deteriorate, and will eventually leak. Apart from the obvious fire risk, the evaporating fuel will affect fuel economy. If the pipe is to be renewed, always use the correct replacement type to ensure a good leak-proof fit. Never use natural tubing because this will tend to break up when in contact with petrol and will obstruct the carburettor jets.

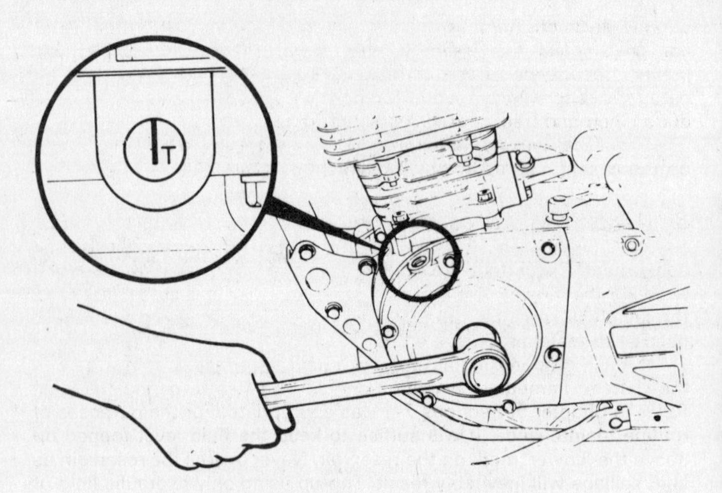

Alternator flywheel mark denoting TDC

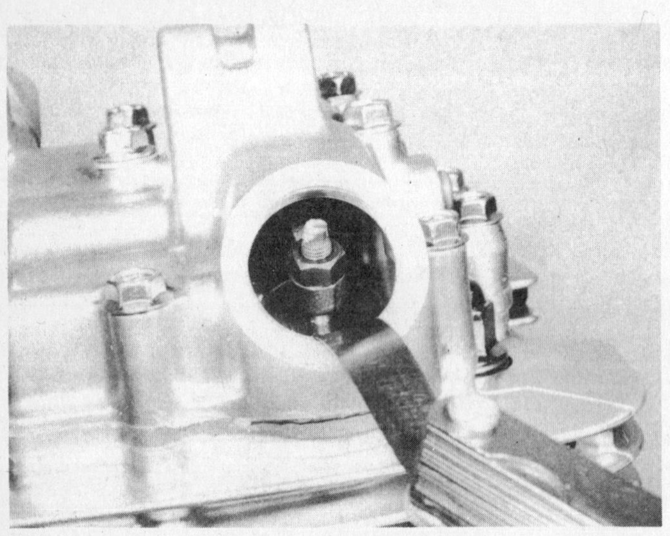
Valve clearances are checked with feeler gauges ...

... and adjusted using spanner and screwdriver

6 Check the clutch adjustment

While the clutch can be adjusted at two points, at the operating mechanism and at the cable itself, it will suffice for the purposes of Routine Maintenance to regard the operating mechanism as set and to make all normal adjustments using the cable length adjusters. The clutch is adjusted correctly when there is 4 mm (0.16 in) of free play in the cable, the free play being measured between the butt end of the clutch handlebar lever and its handlebar clamp.

Reset the cable free play, if necessary, using the adjusters provided. Use first the lower cable adjuster, reserving the handlebar adjuster for minor alterations.

In the event that adjustment is no longer possible with the cable adjusters, the crankcase right-hand cover must be removed as described in Section 9 of Chapter 1 and the operating mechanism reset as described in Section 37 of the same Chapter. Screw in fully the cable adjusters to achieve the maximum free play in the cable before the mechanism setting is altered.

Complete clutch maintenance by applying a few drops of oil to the lever pivots, to the adjuster threads, and to all exposed lengths of inner cable.

7 Check the battery

It is essential that the battery is maintained in excellent condition to prolong its life. In addition to the check of the electrolyte level, the condition of the terminals should be examined. The exposed terminals employed on the battery fitted to these machines are prone to corroding, producing a variety of faults in the electrical system if allowed to go unchecked. Clean away all traces of dirt and corrosion, scraping the terminals and connections with a knife and using emery cloth to finish off. Remake the connections while the joint is still clean, and then smear the assembly with petroleum jelly (not grease) to prevent recurrence of the corrosion. Finish off by checking that the battery is securely clamped in its mountings and that the vent tube is quite clean and free from kinks or blockages.

The electrolyte level must be maintained between the level marks on the casing. Top up, if necessary, using only distilled water.

8 Adjust the brakes and check brake pad or shoe wear

The hydraulic front brake of the GS125 ES model requires no adjustment. Check that the fluid level in the master cylinder reservoir is above the 'Lower' level mark. Remember that in a hydraulic brake, the level will sink only gradually, as the pads wear and more fluid is needed to maintain pressure.

A rapid drop in the fluid level is indicative of a leak somewhere in the system; immediate attention should be given to curing the leak. Refer to Chapter 5, Sections 7-11 as appropriate. For the purposes of routine maintenance, it will suffice to keep the fluid level topped up above the 'Lower' mark on the reservoir. Never overfill the reservoir, as fluid spillage will inevitably result. Top-up using only hydraulic fluid of the correct specifications.

The cable-operated front brakes fitted to the other models will require regular adjustment to compensate for shoe wear and for variations in the cable itself. To check that the adjustment is correct, apply firmly the front brake and measure the distance between the handlebar lever ball end and the throttle twistgrip. The distance should be 20-30 mm (0.8-1.2 in) when the handlebar lever is firmly applied. If adjustment is necessary, it should be made at the adjusting nut on the lower end of the GS125 cable, and at the adjuster set in the brake backplate of the DR125 S models. Once the front brake has been correctly adjusted, spin the front wheel and apply the front brake hard to settle the cable and brake components. Check that the adjustment has not been altered, re-setting it if necessary, and ensure that the adjuster locknuts (where fitted) are securely tightened and that all the rubber cable protecting sleeves are correctly replaced. Check that the front wheel is free to rotate easily.

The rear drum brake fitted to all models is adjusted by means of a single nut at the rear end of the brake operating rod. Turn the nut clockwise to reduce free play, if necessary, to measure 20 - 30 mm (0.8 - 1.2 in) at the brake pedal tip. Check that the rear wheel rotates freely and that the stop lamp is functioning properly. Remember that the stop lamp switch height must be adjusted every time the rear brake adjustment is altered. To adjust the switch height, turn its plastic sleeve nut as required until the stop lamp bulb lights when the brake pedal free play has been taken up and the rear brake shoes are just beginning to engage the brake drum.

Complete brake maintenance by oiling all lever pivot points, all exposed lengths of cable, cable nipples and the rear brake linkage with a few drops of oil from a can. Remember not to allow excessive oil onto the operating linkage, in case any surplus should find its way into the brake drum or onto the tyre.

Regular checks must be made to ensure that the friction material of the brake pads or shoes is not worn down to a dangerous level, and to ensure that worn items are renewed in good time to maintain peak brake efficiency.

All drum brakes have external wear indicators which function as follows. A line is scribed across the outside end of the brake camshaft and aligns with a line cast in the brake backplate, as shown in the accompanying illustration. If, when the brake is correctly adjusted and applied hard, the camshaft line is seen to be outside the arc of the backplate line, the friction material is worn to beyond permissible limits and the shoes must be renewed. This task requires the removal of the front or rear wheel, as applicable, and is described in Sections 4, 12 and 15 of Chapter 5.

The disc brake of the GS125 ES model is also fitted with an external wear indicator, which takes the form of a groove inscribed around the periphery of each pad's friction material and painted red. The grooves are visible looking down on the caliper from in front of the fork leg; if the pads are worn to the point where the groove is touching the disc, the pads must be renewed as a pair. Similarly, if the pads are so fouled with dirt that no marks can be seen, they must be removed for cleaning.

Removal of the pads for renewal or for inspection and cleaning is a relatively simple operation which can be carried out without disturbing any of the brake hose connections. Slacken and remove the two fork lower leg/caliper mounting bracket bolts and withdraw the caliper from the machine. Carefully invert the caliper without twisting the brake hose. Push the caliper mounting bracket towards the main caliper body as far as possible and withdraw both brake pads and the pad spring.

If the pads are to be renewed, the old items may now be discarded. If, however, the pads have not yet worn down to the red inscribed wear limit marks, they should now be cleaned thoroughly, using a fine wire brush that is completely free of oil or grease. Remove all traces of road dirt and corrosion from both the pads and from the caliper assembly. Be especially careful to use a sharply-pointed instrument to clean out the groove in each of the pad friction surfaces and to pick out any particles of foreign matter embedded in the friction material. Examine the surface of each pad. Any areas of glazing may be eased down with emery cloth, but any contamination of the friction material by oil or grease can only be cured by renewal of the pads irrespective of the amount of friction material remaining. **Note:** the asbestos dust from brake pads and shoes is harmful to health if inhaled. Wear a dust mask and attend to the pads in a well ventilated area.

Obtain a small amount of silicone- or PBC-based brake caliper grease and apply a thin smear of this grease to the caliper mounting bracket at its brake pad retaining pins. Check that the caliper mounting bracket slides easily on the two spindles and that the rubber grommets around these axle pins are in good condition. If new pads are to be fitted, the caliper piston must now be pushed back as far as possible into the caliper bore to provide the clearance necessary to accommodate the unworn pads. It should be possible to do this with hand pressure alone. If any undue stiffness is encountered the caliper assembly should be dismantled for examination as described in Section 11 of Chapter 5. While pushing the piston back, maintain a careful watch on the fluid level in the handlebar reservoir. If the reservoir has been overfilled, the surplus fluid will prevent the piston returning fully and must be removed by soaking it up with a clean cloth. Take great care to prevent fluid spillage.

Apply a thin smear of caliper grease to the outer edge and rear surface of the moving pad. Take care to apply caliper grease to the metal backing of the pad only and not to allow any grease to contaminate the friction material. Carefully fit the pad anti-rattle spring, ensuring that it is correctly located, and insert the moving pad into its aperture in the caliper mounting bracket. Check that the pad is free to slide in the mounting bracket. Carefully depress the pad spring and insert the fixed pad, ensuring that it is correctly located on the two pad retaining pins of the caliper mounting bracket.

Replace the caliper assembly on the machine, taking care not to twist the brake hose, and ensure that the brake pads and the caliper mounting bracket are aligned correctly on the brake disc and on the fork lower leg. Replace the caliper mounting bracket/fork lower leg mounting bolts, tightening them to a torque setting of 1.5 - 2.5 kgf m (11 - 18 lbf ft). Apply the front brake lever gently and repeatedly to bring the pad firmly into contact with the disc and to restore full brake pressure, being careful to watch the fluid level in the handlebar reservoir. Do not let the level drop below the 'Lower' level mark on the reservoir or there is a risk of air entering the system. New hydraulic brake fluid of the correct type must be added, if necessary, to prevent the level dropping below this point. If new brake pads have been fitted the level should be restored, by topping-up if necessary, to the 'Upper' level mark.

Before taking the machine out on the road, be careful to check for fluid leaks from the system, and that the front brake is working correctly. Remember also that new pads, and to a lesser extent, cleaned pads will require a bedding-in period before they will function at peak efficiency. Where new pads are fitted use the brake gently but firmly for the first 50-100 miles to enable the pads to bed in fully.

9 Check and adjust the steering head bearings

Place the machine on the centre stand so that the front wheel is clear of the ground. If necessary, place blocks below the crankcase to prevent the motorcycle from tipping forwards.

Grasp the front fork legs near the wheel spindle and push and pull firmly in a fore and aft direction. If play is evident between the upper and lower steering yokes and the head lug casting, the steering head bearings are in need of adjustment. Imprecise handling or a tendency for the front forks to judder may be caused by this fault.

Bearing adjustment is correct when the adjuster ring is tightened, until resistance to movement is felt and then loosened $\frac{1}{8}$ to $\frac{1}{4}$ of a turn. The adjuster ring should be rotated by means of a C-spanner after

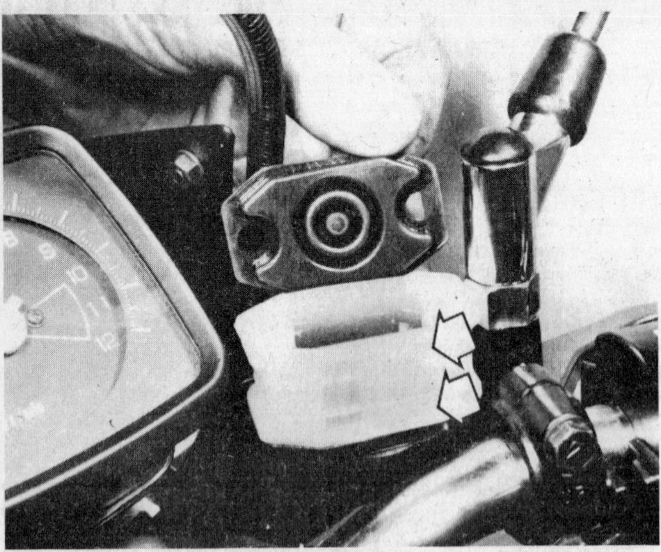

Hydraulic fluid level must be maintained between 'Upper' and 'Lower' level marks (arrowed) on reservoir body

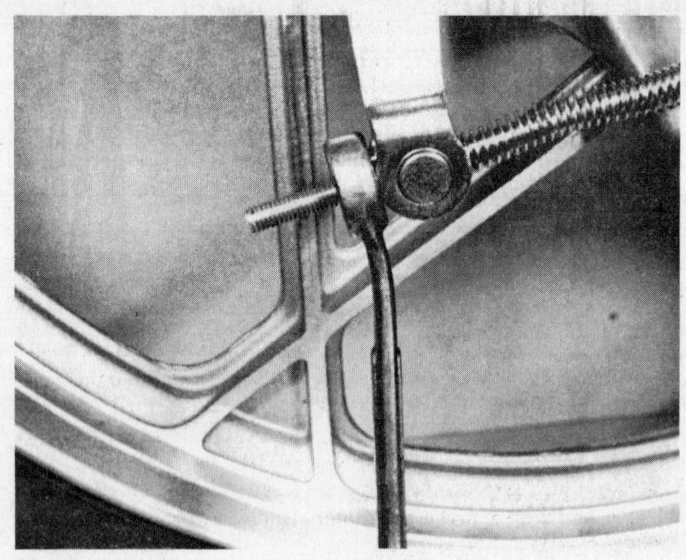

Rear brake is adjusted as shown – all models

GS125 ES model only – remove caliper mounting bolts to release caliper

Slide caliper mounting bracket towards caliper body to release fixed pad ...

... then displace moving pad – note correct position of pad spring

Pad surface and centre groove must be cleaned – note wear limit groove (arrowed)

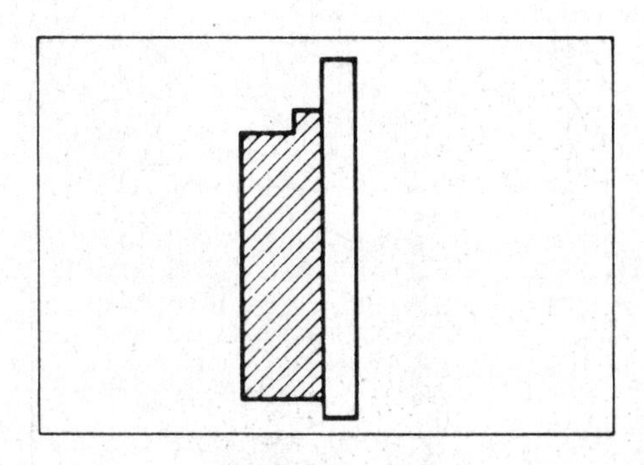

Stepped area of brake pad material indicates usable range

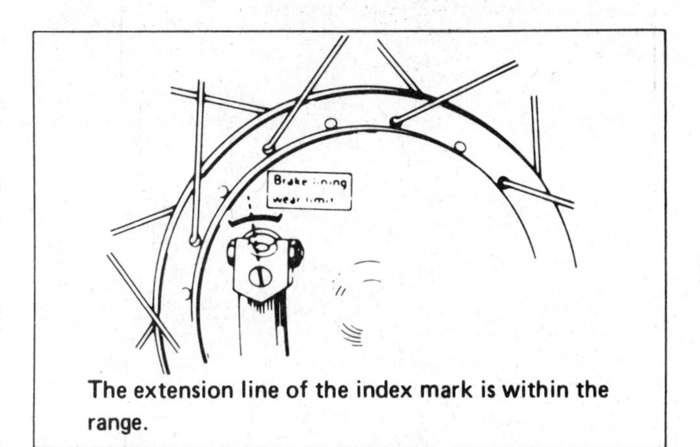

The extension line of the index mark is within the range.

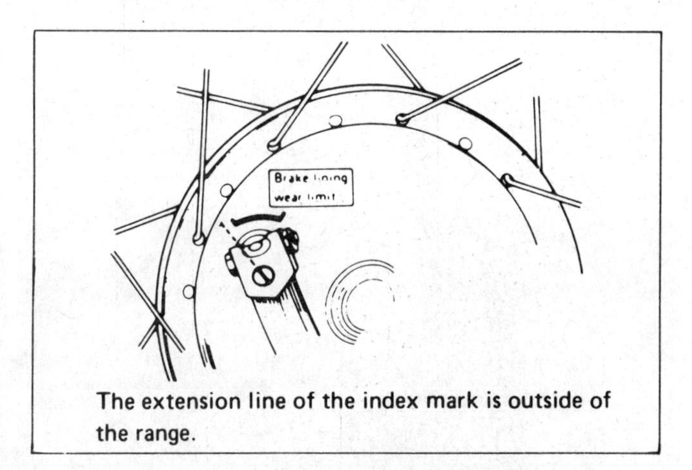

The extension line of the index mark is outside of the range.

Drum brake wear indicator plate range

slackening the steering stem top bolt.

Take great care not to overtighten the nut. It is possible to place a pressure of several tons on the head bearings by overtightening even though the handlebars may seem to turn quite freely. Overtight bearings will cause the machine to roll at low speeds and give imprecise steering. Adjustment is correct if there is no play in the bearings and the handlebars swing to full lock either side when the machine is on the centre stand with the front wheel clear of the ground. Only a light tap on each end should cause the handlebars to swing. Secure the adjuster ring by tightening the steering stem top bolt to a torque setting of 3.5 - 5.5 kgf m (25.5 - 40 lbf ft), then check that the setting has not altered.

10 Check the condition of the suspension and wheels

At the same time as the steering head bearings are checked, take the opportunity closely to examine the front and rear suspension. Ensure that the front forks work smoothly and progressively by pumping them up and down whilst the front brake is held on. Any faults revealed by this check should be investigated further, as any deterioration in the stability of the machine can have serious consequences. Check carefully for signs of leaks around the front fork oil seals. If any damage is found, it must be repaired immediately as described in the relevant Sections of Chapter 4. Examine the rear suspension in the same way and check for wear in the swinging-arm pivot by pushing and pulling horizontally at its rear end. There should be no discernible play at the pivot.

Following the instructions given in Sections 2 and 3 (as applicable) of Chapter 5, check the wheels thoroughly, looking for axial play at the rims which would denote wheel bearing wear, damage or distortion of the rim itself, incorrect spoke tension (if applicable) and any other sign of damage or wear. Similarly make a thorough check of the tyres, as described under the daily checks.

11 Grease rear suspension linkage bearing surfaces – DR125 S only

Since the 'Full-Floater' suspension system incorporates so many important bearing surfaces, each of which is largely exposed to the dirt and debris thrown up by the rear wheel, it is essential to remove the complete assembly at regular intervals so that all bearing components can be cleaned and re-greased thoroughly. Refer to Sections 8 and 9 of Chapter 4 for a description of the necessary work. It is worth thinking about the fitting of a grease nipple to every bearing so that lubrication can be carried out more easily. If the bearings are not lubricated regularly, they will wear rapidly producing a marked deterioration in the machine's handling and stability. Do not forget that the linkage must be in good, unworn, condition if the machine is to pass an MOT test.

12 General checks and lubrication

Proceed methodically round the machine, checking all nuts, bolts, and screws, tightening securely any that may have worked loose. Be careful not to overtighten any component, and make use of the torque wrench settings given in the Specifications Section of the various Chapters of this Manual to ensure that this does not happen. Pay particular attention, if working on a DR125 S model, to the security of the rear sprocket retaining bolts, and on all models, to the security of the engine mounting bolts, exhaust system fasteners, cylinder head and barrel retaining nuts, and to the security of any fastener which is known from past experience to slacken off regularly.

Using a spout-type oil can, proceed again around the machine, applying a few drops of oil to all control cables and operating linkages, and to all stand and footrest pivots (where applicable). Ensure that all cable end nipples and any exposed adjuster threads are also lightly lubricated. Similarly, use a water dispersant lubricant such as WD40 to protect and lubricate all exposed electrical components such as the horn and stop lamp switches. Use the long plastic nozzle supplied with the aerosol can to lubricate the internal contacts of the handlebar switches, ignition switch and the fuel tank lock. The above checks are an essential part of the routine maintenance procedure, and play an important role in preserving the smooth and safe operation of the handlebar controls and in offsetting the effects of wear and corrosion.

At regular intervals the centre stand and prop stand (whichever is fitted to the machine in question), the rear brake pedal, and the throttle twistgrip, must be dismantled so that all traces of corrosion and dirt can be removed, and the various components greased. This operation must be carried out to prevent excessive wear and to ensure that the various components can be operated smoothly and easily, in the interests of safety. The opportunity should be taken to examine closely each component, renewing any that show signs of excessive wear or of any damage.

The twistgrip is removed by unscrewing the screws which fasten both halves of the handlebar right-hand switch assembly. The throttle cable upper end nipple can be detached from the twistgrip with a suitable pair of pliers and the twistgrip slid off in the handlebar end. Carefully clean and examine the handlebar end, the internal surface of the twistgrip, and the two halves of the switch cluster. Remove any rough burrs with a fine file, and apply a coating of grease to all the bearing surfaces. Slide the twistgrip back over the handlebar end, insert the throttle cable end nipple into the twistgrip flange, and reassemble the switch cluster. Check that the twistgrip rotates easily and that the throttle snaps shut as soon as it is released. Tighten the switch retaining screws securely, but do not overtighten them.

Although the regular daily checks will ensure that the control cables are lubricated and maintained in good order, it is recommended that a positive check is made on each cable at this mileage/time interval to ensure that any faults will not develop unnoticed to the point where smooth and safe control operation is impaired. If any doubt exists about the condition of any of the cables, the component in question should be removed from the machine for close examination. Check the outer cables for signs of damage, then examine the exposed portions of the inner cables. Any signs of kinking or fraying will indicate that renewal is required. To obtain maximum life and reliability from the cables they should be thoroughly lubricated. To do the job properly and quickly use one of the hydraulic cable oilers available from most motorcycle shops. Free one end of the cable and assemble the cable oiler as described by the manufacturer's instructions. Operate the oiler until oil emerges from the lower end, indicating that the cable is lubricated throughout its length. This process will expel any dirt or moisture and will prevent its subsequent ingress.

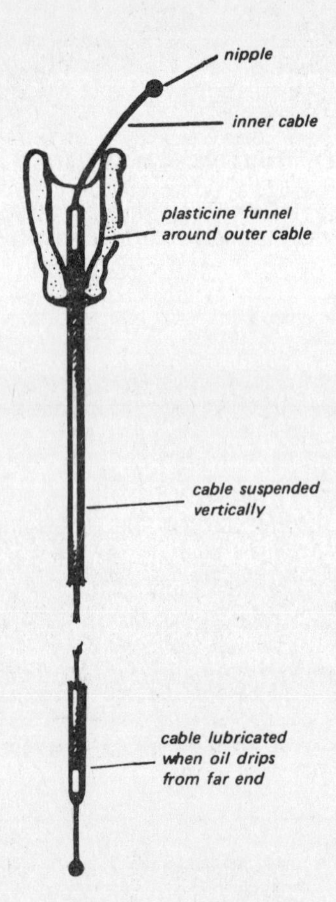

Oiling a control cable

If a cable oiler is not available, an alternative is to remove the cable from the machine. Hang the cable upright and make up a small funnel arrangement using plasticene or by taping a plastic bag around the upper end. Fill the funnel with oil and leave it overnight to drain through. Note that where nylon-lined cables are fitted, they should be used dry or lubricated with a silicone-based lubricant suitable for this application. On no account use ordinary engine oil because this will cause the liner to swell, pinching the cable.

Check all pivots and control levers, cleaning and lubricating them to prevent wear or corrosion. Where necessary, dismantle and clean any moving part which may have become stiff in operation.

When refitting the cables onto the machine, ensure that they are routed in easy curves and that full use is made of any guide or clamps that have been provided to secure the cable out of harm's way. Adjustment of the individual cables is described under previous Routine Maintenance tasks.

Be very careful to ensure that all controls are properly adjusted and are functioning correctly before taking the machine out on the road.

Annually, or every 6000 miles (10 000 km)

This service interval is in the nature of a major overhaul; first complete all the tasks listed under the previous mileage/time headings, then carry out the following:

1 Clean oil pump filter gauze

The oil pump pick-up filter gauze should be removed for cleaning at every other oil filter element/engine oil change. When the oil has drained fully, during the course of the oil change described under the six-monthly/3000 mile service heading, remove the three retaining bolts and withdraw the triangular cast plate from the crankcase underside. Unscrew the two retaining screws and displace the filter gauze.

Wash the filter gauze and the triangular plate, using petrol and a soft bristle brush to remove any stubborn traces of contamination. Be sure to observe the necessary fire precautions when carrying out this cleaning procedure.

Carefully inspect the filter gauze for any signs of damage. If the gauze is split or holed, it must be renewed as it no longer forms an effective barrier between the sump and oil pump.

Fit the filter gauze back into the crankcase housing, taking care to tighten fully its retaining screws. Check that the sealing O-ring is in good condition, renewing it if necessary, wipe away any surplus oil, then refit the triangular plate, rotating it so that the cast arrow mark faces to the front. Refit and tighten securely the three retaining bolts, then refit the drain plug and the oil filter paper element as previously described, and refill the crankcase with oil. Check carefully for signs of oil leakage.

Tighten securely filter gauze retaining screws – renew sealing O-ring if damaged

2 Renew the spark plug

It is recommended that the spark plug is renewed at this interval, regardless of its apparent condition. This will assist in maintaining the ignition system at peak efficiency and will minimise the risk of spark plug failure through extended use. It will also prevent the unnecessary waste of fuel. Refer to Section 10 of Chapter 3; always fit the recommended make and type of spark plug, and check that it is correctly gapped before fitting.

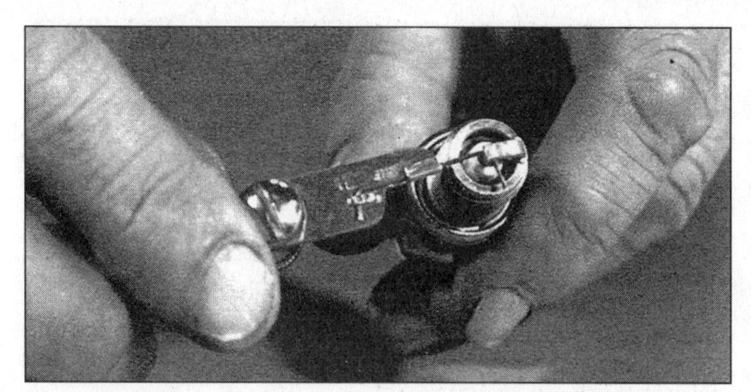

Electrode gap check - use a wire type gauge for best results

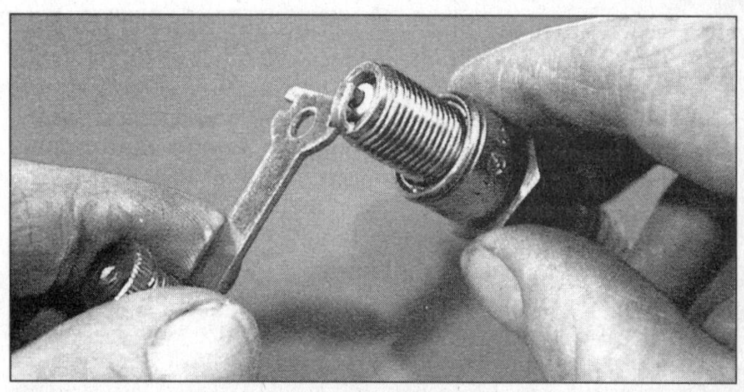

Electrode gap adjustment - bend the side electrode using the correct tool

Normal condition - A brown, tan or grey firing end indicates that the engine is in good condition and that the plug type is correct

Ash deposits - Light brown deposits encrusted on the electrodes and insulator, leading to misfire and hesitation. Caused by excessive amounts of oil in the combustion chamber or poor quality fuel/oil

Carbon fouling - Dry, black sooty deposits leading to misfire and weak spark. Caused by an over-rich fuel/air mixture, faulty choke operation or blocked air filter

Oil fouling - Wet oily deposits leading to misfire and weak spark. Caused by oil leakage past piston rings or valve guides (4-stroke engine), or excess lubricant (2-stroke engine)

Overheating - A blistered white insulator and glazed electrodes. Caused by ignition system fault, incorrect fuel, or cooling system fault

Worn plug - Worn electrodes will cause poor starting in damp or cold weather and will also waste fuel

3 Clean the fuel filter

Switch the fuel tap to the 'Off' position and remove the filter bowl from the bottom of the fuel tap by unscrewing it with a close-fitting ring spanner, then pick out the sealing O-ring. Remove all traces of dirt from the filter bowl and check the condition of the sealing O-ring, renewing it if it is worn or damaged. Refit the O-ring and the filter bowl, tightening the bowl by just enough to nip the O-ring tight.

Note that if excessive traces of dirt or water appear in the filter bowl, the fuel tank must be drained, removed from the machine and flushed out, as described in Section 2 of Chapter 2, to prevent blockages in the fuel system. If such dirt or water gets through to the carburettor, it should be removed from the machine and cleaned out, as described in Sections 5 and 6 of Chapter 2. If the draining of the fuel tank is necessary, the opportunity should be taken to dismantle the tap for cleaning, as described in Section 3 of the same Chapter, this being necessary to gain access to the top filter stack.

4 Change the fork oil

This is an important task which must be carried out to ensure the continuing stability and safety of the machine on the road. Fork oil gradually degenerates as it loses viscosity and contaminated by water and dirt, which produces a very gradual loss of damping. This can occur over a long period of time, thus being completely unnoticed by the rider until the machine is in a dangerous condition. Regular changes of the fork oil will eliminate this possibility. Refer to Chapter 4, Sections 2,3 and 4 for details of removal and refitting. Note that as all GS125 machines are not fitted with front fork drain plugs, each fork leg must be removed from the machine and inverted to drain the oil. This does, however, offer an excellent opportunity to examine the forks for wear or damage and fits in well with the other maintenance operations that are necessary at this interval.

On DR125 S machines only, drain plugs are provided. To use these, place a suitable container at the side of the front wheel, some distance from the drain plug to be removed and lay a sheet of cardboard or newspaper against the wheel to prevent oil getting on to the brake or tyre. Remove the drain plug and pump gently on the forks to eject the oil by applying the front brake and leaning on the handlebars. When all the oil has been pumped out, repeat the process on the other leg. Leave the machine for a while to allow as much oil as possible to drain to the bottom of the fork legs, then repeat the pumping action to expel the remainder. Refit the drain plugs, tightening them carefully, then remove the fork top plugs and the fork springs as described in Section 3 of Chapter 4. Take care to support the machine so that it cannot topple over when the springs are removed. Have the full amount of fork oil ready for each leg but remember that as there will be some oil left in the fork leg, a certain amount will be surplus to requirements and that it is more important to have the oil level correct.

Using the dipstick described in Section 3 of Chapter 4, add oil until it is the specified distance (see Chapter 4 Specifications) from the top of the stanchion when the stanchion is fully compressed and the fork springs are removed. Gently pump the forks up and down to distribute the oil around each fork leg, then recheck the oil level, adding oil if necessary. When the level is correct, raise the front of the machine to extend fully the forks, then refit the springs and top plugs as described.

5 Grease the (drum) brake camshaft

If the brake(s) have not been stripped and rebuilt before, during the course of Routine Maintenance, this should now be done so that the brake camshaft can be cleaned and greased, thus preserving the smoothness of operation and efficiency of the brakes. Remove the wheel as described in Section 4 or 15 (as applicable) of Chapter 5, and overhaul the brake assembly as described in Section 12 of the same Chapter.

6 Grease the wheel bearings, speedometer drive and instrument drive cables

While the wheels are removed from the machine during the course of the previous maintenance operation, take the opportunity to pack some grease into the speedometer drive, as described in Section 14 of Chapter 4, and to check that the wheel bearings are in good condition and well lubricated, as described in Section 5 of Chapter 5.

The speedometer and tachometer cables are secured at both ends by large knurled rings. Using a suitable pair of pliers, unscrew each knurled ring to free the cable from its instrument or drive mechanism. It should then be possible to withdraw the cable carefully from its mountings and remove it from the machine. Remove the inner cable by pulling it out from the bottom of the outer. Carefully examine the inner cable for signs of fraying, kinking, or for any shiny areas which will indicate tight spots, and the outer cable for signs of cracking, kinking or any other damage. Renew the complete cable if necessary. To lubricate the cable, smear a small quantity of grease onto the lower length only of the inner. Do not allow any grease on the top six inches of the cable as the grease will work its way rapidly up the length of the cable as it rotates and get into the instrument itself. This will rapidly ruin the instrument which will then have to be renewed. Insert the inner cable in the outer and refit the cable.

Additional routine maintenance

Certain aspects of routine maintenance make it impossible to place operations under specific mileage or calendar headings, or may necessitate modification of these headings. A good example of the latter is the effect of a dusty environment on certain maintenance operations, notably cleaning the air filter element. Similarly a machine ridden over rough or dirty roads will require more frequent attention to the cycle parts, ie suspension components and wheels, and to the chain. A machine ridden at constant high speeds will need attention to the brakes and engine/transmission components far more than a machine ridden at excessively slow speeds. The problem of how to achieve the correct balance between too little maintenance, which will result in premature and expensive damage to the machine, and too much, is a delicate one which is unfortunately only resolved by personal experience. This experience is best gained by strict adherence to the specified mileage/time headings until the owner feels qualified to alter them to suit his own machine.

Some tasks are recommended for safety reasons due to the fact that materials can deteriorate through old age alone, irrespective of usage or mileage; four such tasks are given below.

1 Grease the steering head and swinging arm pivot bearings

If the steering head bearings have not been dismantled for any other reason, they should be stripped for examination and greasing every two years or 12 000 miles (20 000 km). This task would fit in conveniently with the fork oil change, especially on all GS125 models, and is described in Chapter 4, Sections, 2, 5 and 6.

Similarly, on GS125 models only, the swinging arm should be removed for the pivot bearings to be examined and greased, if this has not been done previously.

2 Renew the brake fluid – GS125 ES only

If the brake fluid is not completely changed during the course of routine maintenance, it should be changed at least every two years. Brake fluid is hygroscopic, which means that it absorbs moisture from the air. Although the system is sealed, the fluid will gradually deteriorate and must be renewed before contamination lowers its boiling point to an unsafe level.

Before starting work, obtain a full can of new DOT 3 or 4 or SAE J1703 hydraulic fluid and read Chapter 5, Section 8. Prepare the clear plastic tube and glass in the same way as for bleeding the hydraulic system, open the bleed nipple by unscrewing it 1/4 – 1/2 a turn with a spanner and apply the front brake lever gently and repeatedly. This will pump out the old fluid. Keep the master cylinder reservoir topped up at all times, otherwise air may enter the system and greatly lengthen the operation. The old brake fluid is invariably much darker in colour than the new, making it much easier to see when the old fluid has been pumped out and the new fluid has completely replaced it.

When the new fluid appears in the clear plastic tubing with no traces of old fluid contaminating it, close the bleed nipple, remove the plastic tubing and refit the rubber dust cap on the nipple. Top the master cylinder reservoir up to the 'Upper' level mark. Carefully dry the diaphragm with a clean lint-free cloth, fold it into its compressed state, and refit the diaphragm and the reservoir cover, tightening securely the two retaining screws. Wash off any surplus fluid with fresh water and check for any fluid leaks which may subsequently appear. Remember to check that full brake pressure is restored and that the front brake is working properly before taking the machine out on the road.

3 Renew the fuel feed pipe

Because the fuel feed pipe from the fuel tank to the carburettor is constructed of thin-walled synthetic rubber which will be affected by heat and the elements over a period of time, Suzuki recommend that, to anticipate any risk of fuel leakage, the pipe is renewed every four years.

4 Renew the hydraulic brake hose – GS125 ES only

Suzuki recommend that in the interest of safety, the hydraulic hose of the front disc brake should be renewed every four years. This is because constant contact of the hose with road salts, moisture, etc, will cause the hose material to harden and eventually split, which in turn will greatly increase the risk of fluid leakage. Renew the hose regardless of its apparent condition.

5 Cleaning the machine

Keeping the motorcycle clean should be considered as an important part of the routine maintenance, to be carried out whenever the need arises. A machine cleaned regularly will not only succumb less speedily to the inevitable corrosion of external surfaces, and hence maintain its market value, but will be far more approachable when the time comes for maintenance or service work. Furthermore, loose or failing components are more readily spotted when not partially obscured by a mantle of road grime and oil.

Surface dirt should be removed using a sponge and warm, soapy water; the latter being applied copiously to remove the particles of grit which might otherwise cause damage to the paintwork and polished surfaces.

Oil and grease is removed most easily by the application of a cleaning solvent such as 'Gunk' or 'Jizer'. The solvent should be applied when the parts are still dry and worked in with a stiff brush. Large quantities of water should be used when rinsing off, taking care that water does not enter the carburettors, air cleaners or electrics.

If desired a polish such as Solvol Autosol can be applied to the aluminium alloy parts to restore the original lustre. This does not apply in instances, much favoured by Japanese manufacturers, where the components are lacquered. Application of a wax polish to the cycle parts and a good chrome cleaner to the chrome parts will also give a good finish. Always wipe the machine down if used in the wet, and make sure the chain is well oiled. There is less chance of water getting into control cables if they are regularly lubricated, which will prevent stiffness of action.

Conversion factors

Length (distance)						
Inches (in)	X	25.4	= Millimetres (mm)	X	0.0394	= Inches (in)
Feet (ft)	X	0.305	= Metres (m)	X	3.281	= Feet (ft)
Miles	X	1.609	= Kilometres (km)	X	0.621	= Miles
Volume (capacity)						
Cubic inches (cu in; in³)	X	16.387	= Cubic centimetres (cc; cm³)	X	0.061	= Cubic inches (cu in; in³)
Imperial pints (Imp pt)	X	0.568	= Litres (l)	X	1.76	= Imperial pints (Imp pt)
Imperial quarts (Imp qt)	X	1.137	= Litres (l)	X	0.88	= Imperial quarts (Imp qt)
Imperial quarts (Imp qt)	X	1.201	= US quarts (US qt)	X	0.833	= Imperial quarts (Imp qt)
US quarts (US qt)	X	0.946	= Litres (l)	X	1.057	= US quarts (US qt)
Imperial gallons (Imp gal)	X	4.546	= Litres (l)	X	0.22	= Imperial gallons (Imp gal)
Imperial gallons (Imp gal)	X	1.201	= US gallons (US gal)	X	0.833	= Imperial gallons (Imp gal)
US gallons (US gal)	X	3.785	= Litres (l)	X	0.264	= US gallons (US gal)
Mass (weight)						
Ounces (oz)	X	28.35	= Grams (g)	X	0.035	= Ounces (oz)
Pounds (lb)	X	0.454	= Kilograms (kg)	X	2.205	= Pounds (lb)
Force						
Ounces-force (ozf; oz)	X	0.278	= Newtons (N)	X	3.6	= Ounces-force (ozf; oz)
Pounds-force (lbf; lb)	X	4.448	= Newtons (N)	X	0.225	= Pounds-force (lbf; lb)
Newtons (N)	X	0.1	= Kilograms-force (kgf; kg)	X	9.81	= Newtons (N)
Pressure						
Pounds-force per square inch (psi; lbf/in²; lb/in²)	X	0.070	= Kilograms-force per square centimetre (kgf/cm²; kg/cm²)	X	14.223	= Pounds-force per square inch (psi; lbf/in²; lb/in²)
Pounds-force per square inch (psi; lbf/in²; lb/in²)	X	0.068	= Atmospheres (atm)	X	14.696	= Pounds-force per square inch (psi; lbf/in²; lb/in²)
Pounds-force per square inch (psi; lbf/in²; lb/in²)	X	0.069	= Bars	X	14.5	= Pounds-force per square inch (psi; lbf/in²; lb/in²)
Pounds-force per square inch (psi; lbf/in²; lb/in²)	X	6.895	= Kilopascals (kPa)	X	0.145	= Pounds-force per square inch (psi; lbf/in²; lb/in²)
Kilopascals (kPa)	X	0.01	= Kilograms-force per square centimetre (kgf/cm²; kg/cm²)	X	98.1	= Kilopascals (kPa)
Millibar (mbar)	X	100	= Pascals (Pa)	X	0.01	= Millibar (mbar)
Millibar (mbar)	X	0.0145	= Pounds-force per square inch (psi; lbf/in²; lb/in²)	X	68.947	= Millibar (mbar)
Millibar (mbar)	X	0.75	= Millimetres of mercury (mmHg)	X	1.333	= Millibar (mbar)
Millibar (mbar)	X	0.401	= Inches of water (inH₂O)	X	2.491	= Millibar (mbar)
Millimetres of mercury (mmHg)	X	0.535	= Inches of water (inH₂O)	X	1.868	= Millimetres of mercury (mmHg)
Inches of water (inH₂O)	X	0.036	= Pounds-force per square inch (psi; lbf/in²; lb/in²)	X	27.68	= Inches of water (inH₂O)
Torque (moment of force)						
Pounds-force inches (lbf in; lb in)	X	1.152	= Kilograms-force centimetre (kgf cm; kg cm)	X	0.868	= Pounds-force inches (lbf in; lb in)
Pounds-force inches (lbf in; lb in)	X	0.113	= Newton metres (Nm)	X	8.85	= Pounds-force inches (lbf in; lb in)
Pounds-force inches (lbf in; lb in)	X	0.083	= Pounds-force feet (lbf ft; lb ft)	X	12	= Pounds-force inches (lbf in; lb in)
Pounds-force feet (lbf ft; lb ft)	X	0.138	= Kilograms-force metres (kgf m; kg m)	X	7.233	= Pounds-force feet (lbf ft; lb ft)
Pounds-force feet (lbf ft; lb ft)	X	1.356	= Newton metres (Nm)	X	0.738	= Pounds-force feet (lbf ft; lb ft)
Newton metres (Nm)	X	0.102	= Kilograms-force metres (kgf m; kg m)	X	9.804	= Newton metres (Nm)
Power						
Horsepower (hp)	X	745.7	= Watts (W)	X	0.0013	= Horsepower (hp)
Velocity (speed)						
Miles per hour (miles/hr; mph)	X	1.609	= Kilometres per hour (km/hr; kph)	X	0.621	= Miles per hour (miles/hr; mph)
Fuel consumption*						
Miles per gallon, Imperial (mpg)	X	0.354	= Kilometres per litre (km/l)	X	2.825	= Miles per gallon, Imperial (mpg)
Miles per gallon, US (mpg)	X	0.425	= Kilometres per litre (km/l)	X	2.352	= Miles per gallon, US (mpg)

Temperature

Degrees Fahrenheit = (°C x 1.8) + 32

Degrees Celsius (Degrees Centigrade; °C) = (°F - 32) x 0.56

**It is common practice to convert from miles per gallon (mpg) to litres/100 kilometres (l/100km), where mpg (Imperial) x l/100 km = 282 and mpg (US) x l/100 km = 235*

Chapter 1 Engine, clutch and gearbox

Refer to Chapter 7 for information relating to DR125 SF, SH, SJ 'Raider', GN125 and GZ125 'Marauder' models

Contents

Specifications

Engine

Type	Air-cooled, single-cylinder, OHC
Bore	57.0 mm (2.24 in)
Stroke	48.8 mm (1.92 in)
Capacity	124 cc (7.56 cu in)
Compression ratio	9.5:1

Compression pressure

GS125, GS125 ES	142 – 199 psi (10.0 – 14.0 kg/cm^2)
Service limit	114 psi (8.0 kg/cm^2)
DR125 S	171 – 228 psi (12.0 – 16.0 kg/cm^2)
Service limit	142 psi (10.0 kg/cm^2)

Valve clearances – engine cold

Inlet and exhaust	0.08 – 0.13 mm (0.003 – 0.005 in)

Camshaft and rocker gear

Camshaft lobe height:	
Inlet	33.830 – 33.870 mm (1.332 – 1.334 in)
Service limit	33.530 mm (1.320 in)
Exhaust	32.990 – 33.030 mm (1.299 – 1.300 in)
Service limit	32.690 mm (1.287 in)
Valve timing:	
Inlet opens	35° BTDC
Inlet closes	57° ABDC
Exhaust opens	56° BBDC
Exhaust closes	24° ATDC
Camshaft journal OD	21.959 – 21.980 mm (0.864 – 0.865 in)
Camshaft journal oil clearance	0.032 – 0.066 mm (0.001 – 0.003 in)
Service limit	0.150 mm (0.006 in)
Camshaft maximum runout	0.100 mm (0.004 in)
Rocker ID	12.000 – 12.018 mm (0.4724 – 0.4732 in)

Rocker shaft OD:	
Z models	11.966 – 11.984 mm (0.4711 – 0.4718 in)
All later models	11.977 – 11.995 mm (0.4715 – 0.4722 in)

Camshaft drive chain

Number of links	98
Maximum length of 20 links	129.9 mm (5.114 in)

Cylinder head

Gasket face maximum warpage:	
Cylinder head/cylinder barrel	0.05 mm (0.002 in)
Cylinder head/cylinder head cover	0.05 mm (0.002 in)
Cylinder head cover/cylinder head	0.05 mm (0.002 in)
Camshaft bearing surface ID	22.012 – 22.025 mm (0.866 – 0.867 in)

Valves, guides and springs

Valve head diameter:		
Inlet	30.000 mm (1.181 in)	
Exhaust	26.000 mm (1.024 in)	
Valve lift:		
Inlet	7.500 mm (0.295 in)	
Exhaust	6.500 mm (0.256 in)	
Inlet valve stem OD:		
Z models	5.460 – 5.475 mm (0.2150 – 0.2156 in)	
All later models	5.475 – 5.490 mm (0.2156 – 0.2161 in)	
Exhaust valve stem OD:		
Z models	5.445 – 5.460 mm (0.2144 – 0.2150 in)	
All later models	5.455 – 5.470 mm (0.2148 – 0.2154 in)	
Valve guide ID – Inlet and exhaust	5.500 – 5.512 mm (0.2165 – 0.2170 in)	
Inlet valve guide/valve stem clearance:		
Z models	0.025 – 0.052 mm (0.0010 – 0.0020 in)	
All later models	0.010 – 0.037 mm (0.0004 – 0.0015 in)	
Service limit – all models	0.35 mm (0.0138 in)	
Exhaust valve guide/valve stem clearance:		
Z models	0.040 – 0.067 mm (0.0016 – 0.0026 in)	
All later models	0.030 – 0.057 mm (0.0012 – 0.0022 in)	
Service limit – all models	0.35 mm (0.0138 in)	
Valve stem maximum runout	0.05 mm (0.002 in)	
Valve head maximum runout	0.03 mm (0.001 in)	
Valve head minimum thickness	0.5 mm (0.020 in)	
Valve seat width	0.9 – 1.1 mm (0.035 – 0.043 in)	
Valve stem tip/collet groove length*:		
Early type	4.22 mm (0.166 in)	
Service limit	3.8 mm (0.15 in)	
Late type	3.14 mm (0.124 in)	
Service limit	2.6 mm (0.10 in)	
Inlet valve spring minimum free length:		
Inner	35.1 mm (1.382 in)	
Outer	39.8 mm (1.567 in)	
Inlet valve spring pressure:		
Inner	7.1 – 8.3 kg at 32.5 mm (15.653 – 18.298 lbs at 1.28 in)	
Outer	17.0 – 20.3 kg at 36.0 mm (37.478 – 44.753 lbs at 1.42 in)	
Exhaust valve spring minimum free length:	**Early**	**Late**
Inner	35.1 mm (1.38 in)	36.0 mm (1.42 in)
Outer	39.8 mm (1.57 in)	39.2 mm (1.54 in)
Exhaust valve spring pressure*:		
Inner – early type	7.1 – 8.3 kg at 32.5 mm (15.653 – 18.298 lbs at 1.28 in)	
Inner – late type	7.8 – 9.2 kg at 32.5 mm (17.196 – 20.282 lbs at 1.28 in)	
Outer – early type	17.0 – 20.3 kg at 36.0 mm (37.478 – 44.753 lbs at 1.42 in)	
Outer – late type	18.9 – 22.3 kg at 36.0 mm (41.667 – 49.162 lbs at 1.42 in)	
Valve springs – paint colour coding:	**Early**	**Late**
Inlet	Yellow	Yellow
Exhaust	Yellow	None

** Note: Late type valves and exhaust valve springs fitted as standard to all GS125 ESF, ESK and ESM models*

Cylinder barrel

Standard bore size	57.000 – 57.015 mm (2.244 – 2.245 in)
Service limit:	
Z models	57.085 mm (2.247 in)
All later models	57.110 mm (2.248 in)
Piston/cylinder clearance:	
Z models	0.045 – 0.055 mm (0.0018 – 0.0022 in)
All later models	0.020 – 0.030 mm (0.0008 – 0.0012 in)
Service limit – all models	0.120 mm (0.0047 in)
Gasket face maximum warpage	0.05 mm (0.002 in)

Piston and piston rings

Piston standard OD:	
Z models	56.950 – 56.965 mm (2.242 – 2.243 in)
All later models	56.975 – 56.990 mm (2.243 – 2.244 in)
Service limit – all models	56.880 mm (2.239 in)
Gudgeon pin bore	14.002 – 14.008 mm (0.551 – 0.552 in)
Service limit	14.030 mm (0.553 in)
Gudgeon pin OD	13.994 – 14.002 mm (0.550 – 0.551 in)
Service limit	13.980 mm (0.550 in)
Top compression ring:	
End gap – free	Approx 7.0 mm (0.276 in)
Service limit	5.6 mm (0.221 in)
End gap – installed	0.10 – 0.25 mm (0.004 – 0.010 in)
Service limit	0.70 mm (0.028 in)
Thickness	1.175 – 1.190 mm (0.046 – 0.047 in)
Piston groove width	1.210 – 1.230 mm (0.048 – 0.049 in)
Piston ring/ring groove maximum clearance	0.180 mm (0.007 in)
Second compression ring:	
End gap – free	Approx 7.5 mm (0.295 in)
Service limit	6.0 mm (0.236 in)
End gap – installed	0.10 – 0.25 mm (0.004 – 0.010 in)
Service limit	0.70 mm (0.028 in)
Thickness	1.170 – 1.190 mm (0.046 – 0.047 in)
Piston groove width	1.210 – 1.230 mm (0.048 – 0.049 in)
Piston ring/ring groove maximum clearance	0.150 mm (0.006 in)
Oil scraper ring:	
Piston groove width	2.51 – 2.53 mm (0.098 – 0.099 in)

Crankshaft

Connecting rod small-end ID	14.004 – 14.012 mm (0.551 – 0.552 in)
Service limit	14.040 mm (0.553 in)
Maximum connecting rod deflection (at small-end)	3.0 mm (0.118 in)
Connecting rod big-end side clearance	0.10 – 0.45 mm (0.004 – 0.018 in)
Service limit	1.0 mm (0.039 in)
Connecting rod big-end width	15.95 – 16.00 mm (0.628 – 0.629 in)
Maximum runout	0.05 mm (0.002 in)
Width across flywheels	52.9 – 53.1 mm (2.083 – 2.091 in)

Primary drive

Type	Helical gear
Reduction ratio	3.471:1 (59/17T)

Clutch

Type	Wet, multi-plate
Clutch friction plates:	
Number	5
Thickness	2.9 – 3.1 mm (0.114 – 0.122 in)
Service limit	2.6 mm (0.102 in)
Tongue width	11.8 – 12.0 mm (0.465 – 0.472 in)
Service limit	11.0 mm (0.433 in)
Clutch plain plates:	
Number	4
Thickness	1.55 – 1.65 mm (0.061 – 0.065 in)
Maximum warpage	0.10 mm (0.004 in)
Clutch springs:	
Number	5
Minimum free length	29.5 mm (1.161 in)

Gearbox

	GS125, GS125 ES	DR125 S
Type	5-speed, constant mesh	6-speed, constant mesh
Gear ratios (number of teeth):		
1st	3.000:1 (33/11T)	3.000:1 (33/11T)
2nd	1.857:1 (26/14T)	1.857:1 (26/14T)
3rd	1.368:1 (26/19T)	1.368:1 (26/19T)
4th	1.095:1 (23/21T)	1.095:1 (23/21T)
5th	0.913:1 (21/23T)	0.913:1 (21/23T)
6th	N/App	0.800:1 (20/25T)
Selector fork claw end/pinion groove clearance	0.10 – 0.30 mm (0.004 – 0.012 in)	
Service limit	0.50 mm (0.020 in)	
Gear pinion selector fork groove width:		
Output shaft pinions	5.0 – 5.1 mm (0.197 – 0.201 in)	
Input shaft pinion	5.5 – 5.6 mm (0.217 – 0.221 in)	
Selector fork claw end thickness:		
Output shaft forks	4.8 – 4.9 mm (0.189 – 0.193 in)	
Input shaft fork	5.3 – 5.4 mm (0.209 – 0.213 in)	

Input shaft length (1st – 2nd gear):	
GS125, GS125 ES	87.8 – 88.0 mm (3.457 – 3.465 in)
DR125 S	88.0 – 88.1 mm (3.465 – 3.469 in)

Final drive

Type	Chain and sprocket
Chain size	428 ($\frac{1}{2}$ x $\frac{5}{16}$)
Number of links:	
GS125, GS125 ES	116
DR125 S	122
20 link length:	
Standard	254.0 mm (10.000 in)
Service limit	259.0 mm (10.197 in)
Reduction ratio:	
GS125, GS125 ES	3.071:1 (43/14T)
DR125 S	3.357:1 (47/14T)

Torque settings

Component	kgf m	lbf ft
Cylinder head cover retaining bolts	0.9 – 1.0	6.5 – 7
Cylinder head retaining nuts:		
8 mm	1.5 – 2.0	11 – 14.5
6 mm	0.7 – 1.1	5 – 8
Cylinder barrel retaining nuts	0.7 – 1.1	5 – 8
Camshaft sprocket mounting bolts	1.0 – 1.3	7 – 9
Alternator rotor retaining nut	3.0 – 4.0	22 – 29
Primary drive gear retaining nut	4.0 – 6.0	29 – 43
Clutch centre retaining nut	3.0 – 5.0	22 – 36
Gearbox sprocket retaining nut	8.0 – 10.0	58 – 72
Starter clutch Allen bolts	1.5 – 2.0	11 – 14.5
Oil drain plug	1.8 – 2.0	13 – 14.5
Engine mounting bolts:		
80 mm long	3.7 – 4.5	27 – 33
All others	2.8 – 3.4	20 – 25
Exhaust pipe/cylinder head retaining nuts or bolts	0.9 – 1.2	6.5 – 8.5
Exhaust system mounting bolts	0.9 – 1.2	6.5 – 8.5

1 General description

The Suzuki GS125 and DR125 models are fitted with an air-cooled, single-cylinder, four-stroke engine that is built in unit with the transmission components, all major castings being of light aluminium alloy.

The engine is straightforward in design; a pressed-up crankshaft assembly rotating on two ball journal main bearings supports a connecting rod which has a needle roller bearing at its big-end and bears directly on the gudgeon pin at its small-end. The piston is fitted with two plain compression rings and a single three-piece oil control ring, and moves inside a cylinder barrel formed by a cast-iron liner surrounded by a finned light alloy sleeve. Two valves are employed, these being opened via rocker arms by an overhead camshaft. The camshaft runs in bearing surfaces formed by the cylinder head and cylinder head cover castings and is driven by a single row roller-type chain from the crankshaft left-hand end. Ignition and generator components are mounted on the crankshaft left-hand end, while the oil pump is driven by a gear train from the crankshaft right-hand end.

The engine's output is transmitted via the primary drive gear to a wet, multi-plate clutch. The gearbox is of the constant mesh type and has five speeds on the GS125 and GS125 ES models, and six speeds on the DR125 S model.

Starting is effected by a kickstart on the GS125 and DR125 S models and by an electric starter motor on the GS125 ES model.

2 Operations with the engine/gearbox unit in the frame

The following components can be removed for repair or renewal with the engine/gearbox unit in the frame, although if several operations need to be undertaken simultaneously it is usually worthwhile taking the unit out of the frame to gain better access and more comfortable working conditions.

a) Cylinder head cover
b) Camshaft and rocker gear
c) Cylinder head and valves
d) Cylinder barrel
e) Cam chain and tensioner components
f) Piston and piston rings
g) Alternator and ignition components
h) Starter motor and drive components – GS125 ES only
i) Gearbox sprocket and neutral indicator/gear position indicator switch
j) Oil pump, oil filter element and oil filter screen
k) Clutch assembly and operating mechanism
l) Primary drive gear, kickstart drive gear and idler gear (where fitted)
m) Gear selector mechanism – external components only

3 Operations with the engine/gearbox unit removed from the frame

It will be necessary to remove the engine/gearbox unit from the frame to gain access to the following components:

a) Crankshaft and main bearings
b) Gearbox components
c) Gear selector drum and forks
d) Kickstart shaft and return spring – GS125 and DR125 S only

Engine/gearbox unit removal is necessary to permit the crankcase halves to be separated. it is a relatively simple task which can be carried out by one person alone and should take no more than one hour.

4 Removing the engine/gearbox unit from the frame

1 When working on a DR125 S model, a stand must be available which holds the machine securely in the upright position. To allow adequate access the stand must bear on the frame horizontal bracing tube which passes between the footrest mounting points; such a stand can be fabricated or purchased from a supplier advertising in a

motorcycle press. GS125 models are fitted with a centre stand.

2 Place the machine on its stand on a suitable working surface noting that work is easier if the machine is raised to a convenient height on an hydraulic ramp or a stout table or a platform constructed from wooden planks supported on blocks. Secure the machine with blocks or ropes to prevent it from falling.

3 Place a suitable container underneath the crankcase, remove the oil filler and drain plugs, and allow the engine oil to drain completely, draining being quicker and more efficient if the engine is warmed up to normal operating temperature so that the oil is thin. Release the three screws or nuts retaining the oil filter chamber cap to the crankcase right-hand cover, then withdraw the cap, the spring, the filter element and the two O-rings, catching any oil.

4 Carefully pull away the sidepanels which are retained by moulded prongs engaging with rubber grommets set in the frame or in the base of the fuel tank. The seat is secured to the frame by two bolts, but on DR125 S models only, two additional bolts secure a strap which passes across the seat. Remove the bolts and withdraw the strap (where applicable) and the seat.

5 Note that it is a good practice to prevent the loss of nuts, bolts and washers by refitting all fasteners to their original positions once the component has been removed.

6 Switch the fuel tap to the 'Off' position, disengage the spring clip securing the fuel pipe upper end, and pull the pipe off the tap spigot. Unscrew the two bolts securing the fuel tank on GS125 models, or the single bolt on DR125 S models. Lift the tank up at the rear and pull it backwards to release it from the front mounting rubbers. Place the tank to one side so that it cannot be scratched or dented and take suitable precautions to prevent the risk of fire.

7 Disconnect the battery at its negative (–) lead by unscrewing the terminal retaining bolt. If the machine is to be out of service for some time, disconnect also the battery positve (+) lead and remove the battery from the machine. As described in Chapter 6 give the battery regular refresher charges. Store the battery in a safe place to prevent damage.

8 Remove the pinch bolt securing the clutch operating lever to its shaft and lever carefully the operating lever off the shaft splines. On GS125 models only, slide back the rubber adjuster cover, slacken the adjuster locknut and unscrew the clutch cable adjuster. Disengage the cable from the retaining lug set in the crankcase top surface and release the clamp securing the cable to the engine front mounting top bolt. Allow the cable to hang down in front of the engine.

9 Pull the spark plug cap off the spark plug and hang the HT lead over the frame top tube, then unscrew the tachometer cable lower end retaining ring, pull the cable out of its housing and allow the cable to hang down the frame front downtube. On GS125 ES models only, disconnect the black heavy-gauge starter motor cable from its terminal on the starter motor, this being secured by a nut and lock washers. Disengage the spring clip and pull the breather hose off its crankcase stub, then secure the breather hose clear of the engine.

10 On GS125 models, remove the complete exhaust system, which is retained by two Allen bolts at the exhaust pipe/cylinder head joint and by a single bolt at the pillion footrest mounting. On DR125 S models, remove only the exhaust pipe, which is secured by two nuts at the cylinder head/exhaust pipe joint and by a single clamp bolt at the exhaust pipe/silencer joint; remove the two nuts and the bolt, then carefully tap forwards the exhaust pipe to release it.

11 Slacken the clamping screws securing the air filter hose to the rear of the carburettor body and the inlet stub to the front of the carburettor body, then carefully manoeuvre the carburettor clear of the hose and stub. Drain any surplus fuel from the float chamber by unscrewing the float chamber drain screw, then secure the carburettor assembly to the frame top tube so that it is out of the way.

12 Remove the kickstart lever (where fitted) by unscrewing its retaining pinch bolt and pulling the lever off its shaft splines. On DR125 S models only, remove the three retaining bolts and withdraw the crankcase bashplate.

13 Remove the gearchange lever on DR125 S models by unscrewing the pinch bolt and pulling the lever off the shaft splines. On all GS125 models remove the circlip and washer securing the gearchange linkage to its frame pivot point, unscrew the pinch bolt securing the linkage front arm to the gearchange shaft, and withdraw the complete linkage. Remove the three bolts securing the gearbox sprocket cover and withdraw the cover.

14 Knock back the raised tab of the sprocket retaining nut lock washer, apply hard the rear brake and unscrew the sprocket retaining nut. Pull the sprocket off the output shaft splines, disengage it from the chain and allow the chain to hang over the swinging arm pivot. Remove the crankcase retaining bolt which also secures the engine earth lead, disconnect the earth lead, and refit temporarily the bolt. Starting from the crankcase top, trace the engine electrical leads up the frame tubes, releasing any clamps which secure the leads to the frame, and disconnect the leads at the multi-pin block connectors or individual snap connectors.

15 Remove the four bolts securing the engine front mounting plate and withdraw the mounting plate. Remove the engine lower rear mounting bolt and unscrew the swinging arm pivot bolt retaining nut and washer. Using a hammer and a suitable drift, tap out the swinging arm pivot bolt until it is clear of the crankcase lug, then withdraw the drift until it also is clear of the crankcase lug. Both the pivot bolt and the drift must remain in place to support fully both pivots of the swinging arm.

16 Remove the retaining nuts and the right-hand plate of the engine top mounting/cylinder head steady assembly, supporting the engine/gearbox unit to prevent it from falling. Support the unit with one hand, withdraw the left-hand plate and engine mounting bolts and lower the engine/gearbox unit to the ground, manoeuvring it to clear the footrests and brake pedal.

4.3 Oil drain plug is screwed into triangular plate on crankcase underside

4.7 Disconnect the battery to prevent short circuits

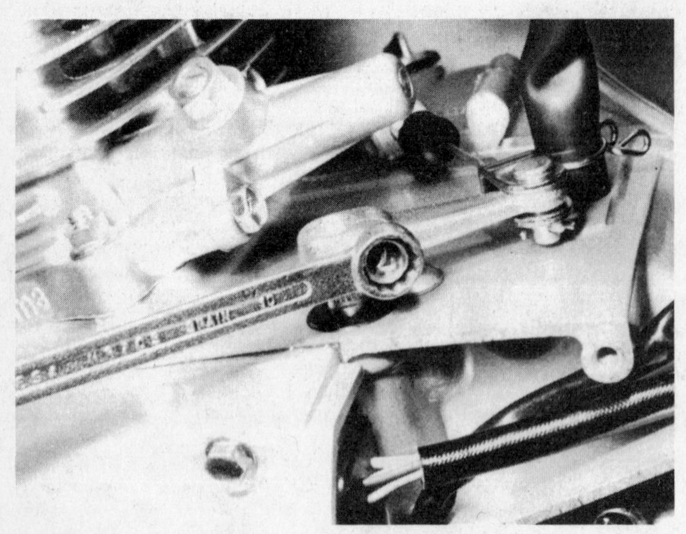

4.8 Clutch cable is disconnected by removing operating lever from release shaft

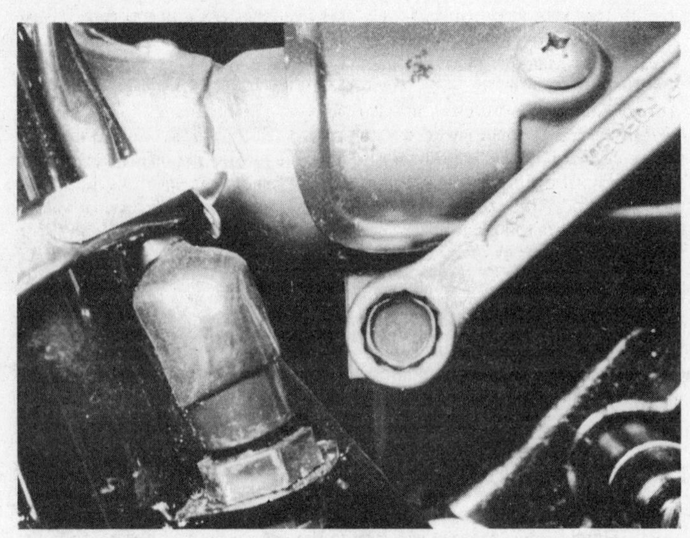

4.10 DR125 S model – slacken clamp bolt to release exhaust pipe only

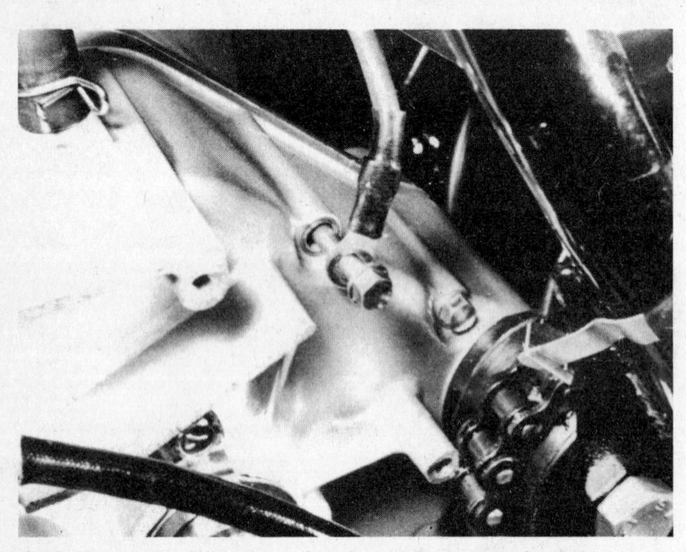

4.14a Engine earth lead is retained by crankcase securing bolt

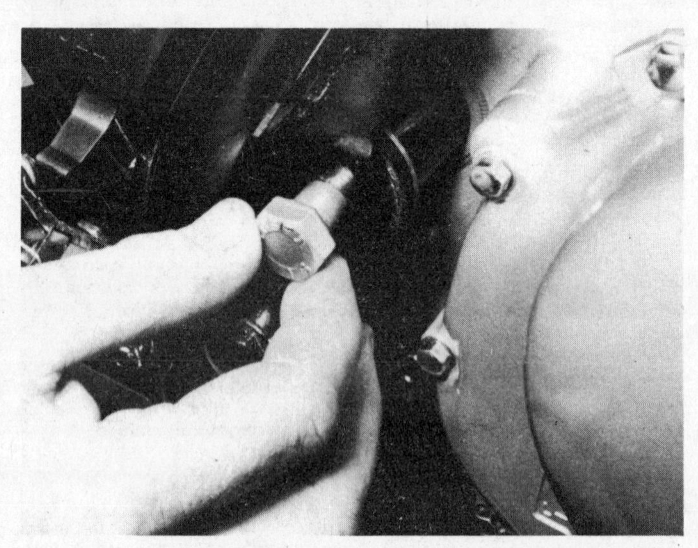

4.15 Insert drift or bolt temporarily to maintain swinging arm alignment

5 Dismantling the engine/gearbox unit: preliminaries

1 Before any dismantling work is undertaken, the external surfaces of the unit should be thoroughly cleaned and degreased. This will prevent the contamination of the engine internals, and will also make working a lot easier and cleaner. A high flash point solvent, such as paraffin (kerosene) can be used, or better still, a proprietary engine degreaser such as Gunk. Use old paintbrushes and toothbrushes to work the solvent into the various recesses of the engine castings. Take care to exclude solvent or water from the electrical components and inlet and exhaust ports. The use of petrol (gasoline) as a cleaning medium should be avoided, because the vapour is explosive and can be toxic if used in a confined space.

2 When clean and dry, arrange the unit on the workbench, leaving a suitable clear area for working. Gather a selection of small containers and plastic bags so that parts can be grouped together in an easily identifiable manner. Some paper and a pen should be on hand to permit notes to be made and labels attached where necessary. A supply of clean rag is also required.

3 Before commencing work, read through the appropriate section so that some idea of the necessary procedure can be gained. When removing the various engine components it should be noted that great force is seldom required, unless specified. In many cases, a component's reluctance to be removed is indicative of an incorrect approach or removal method. If in any doubt, re-check with the text.

6 Dismantling the engine/gearbox unit: removing the cylinder head cover, camshaft and the cylinder head

1 The above components can be removed whether the engine/gearbox unit is in the frame or not, but in the former case the sidepanels, the seat, the fuel tank, the carburettor and the exhaust system or exhaust pipe must be removed, and the tachometer cable and spark plug cap must be disconnected. The engine top mounting/cylinder head steady assembly also must be dismantled. All operations are described in Section 4 of this Chapter.

2 Remove the two valve adjuster inspection caps, the spark plug, and the hexagon-headed and circular inspection caps screwed into the crankcase left-hand cover. Applying a spanner to the alternator rotor retaining nut via the inspection aperture in the side of the crankcase left-hand cover, rotate the engine until it is at TDC on the compression stroke. This is found by checking that both rocker arms are free when the index mark (a straight line with the letter 'T' or 'O') stamped on the

rim of the alternator rotor and visible via the inspection aperture in the top of the crankcase left-hand cover is aligned exactly with the arrow cast on the crankcase cover (GS125 models) or is exactly in the middle of the aperture (DR125 S model). This relieves the pressure on the rocker gear and prevents the risk of warping the cover as it is removed.
3 Remove the single screw which secures the tachometer driven gear housing to the cylinder head cover, then carefully lever the housing out of its recess in the cover, taking great care not to crack or damage it. Withdraw the tachometer driven gear.
4 Following the sequence shown in the accompanying illustration, progressively slacken and then remove the ten cylinder head cover retaining bolts. Do not disturb the two rocker shaft retaining bolts which are identified by the conical recesses in their heads. Tap lightly the cover with a soft-faced mallet to release the seal and lift away the cover. Remove the rubber camshaft end cap.
5 Remove its two retaining bolts and withdraw the cam chain tensioner assembly from the rear of the cylinder barrel.
6 Knock back the raised tab of the camshaft sprocket mounting bolt lock washer, then remove the first sprocket mounting bolt. Rotate the crankshaft until the second mounting bolt is exposed, then remove this in the same way. Be careful not to drop either a bolt or the lock washer into the cam chain tunnel. Tap the sprocket to the left, off the camshaft locating shoulder and pin, disengage the sprocket from the cam chain, and withdraw together the camshaft and sprocket, being careful not to allow any component to fall into the cam chain tunnel. Remove the camshaft locating half ring from its groove in the cylinder head/camshaft left-hand bearing surface.
7 Prevent the cam chain from dropping into the crankcase by wiring it to a convenient point on the outside of the engine/gearbox unit or by passing a rod through it.
8 If the cylinder head is to be removed with the engine/gearbox unit in the frame, remove the cam chain tensioner blade mounting bolt from the cylinder head left-hand side, and pull the tensioner blade out of the cam chain tunnel.
9 Working in a diagonal sequence progressively slacken, then remove the cylinder head retaining nuts. Note that there are six of these; do not omit the two 6 mm nuts on the cylinder head left-hand side. Tap lightly the cylinder head with a soft-faced mallet to break the seal, then lift the cylinder head off its mounting studs, preventing the cam chain from dropping into the crankcase as the head is removed.
10 Remove the cylinder head gasket and the two locating dowels, then lift out the cam chain guide blade. Unless the cam chain is to be removed as described in Section 8 of this Chapter, secure it to prevent it from dropping into the crankcase.

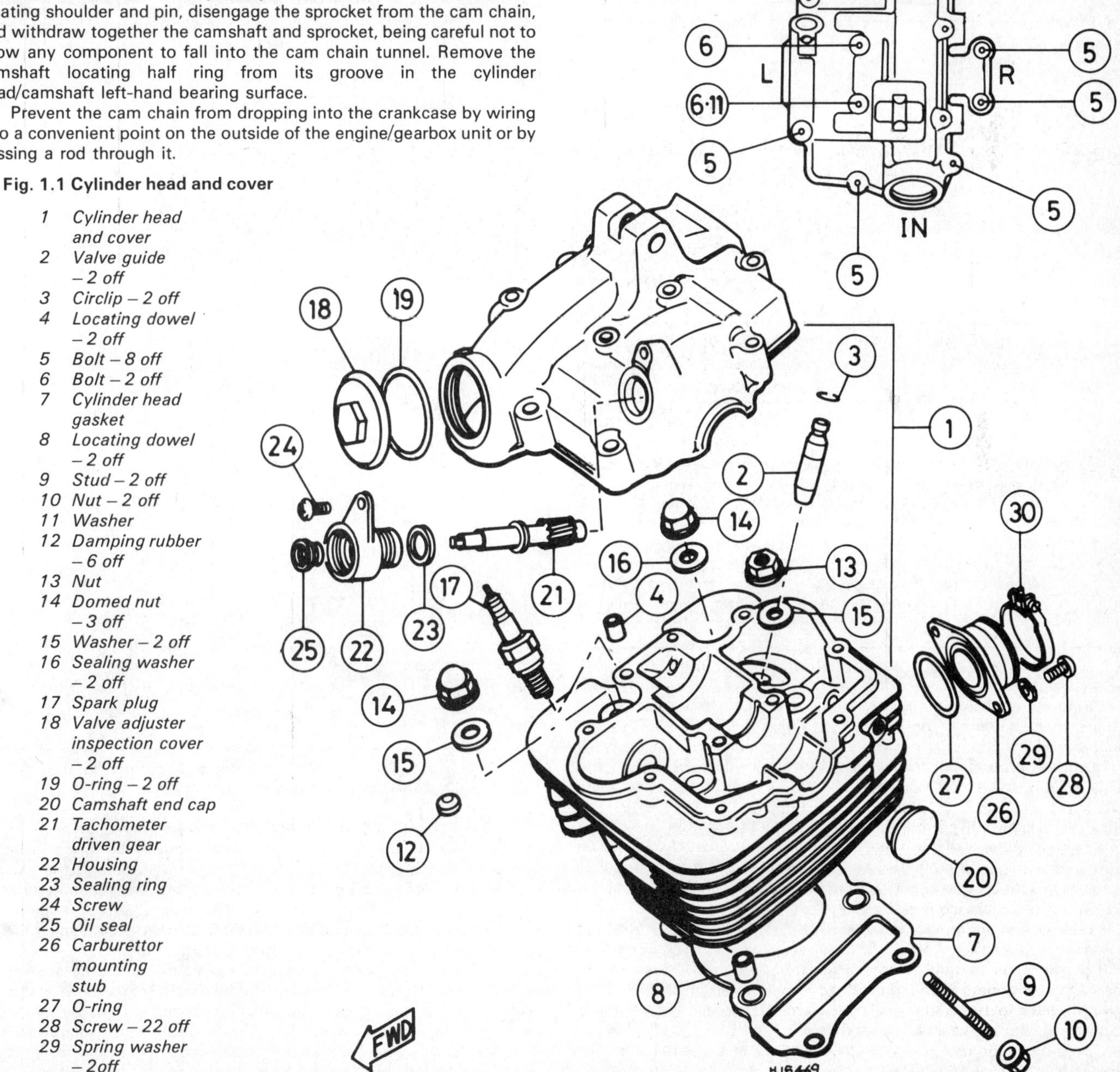

Fig. 1.1 Cylinder head and cover

1 Cylinder head and cover
2 Valve guide – 2 off
3 Circlip – 2 off
4 Locating dowel – 2 off
5 Bolt – 8 off
6 Bolt – 2 off
7 Cylinder head gasket
8 Locating dowel – 2 off
9 Stud – 2 off
10 Nut – 2 off
11 Washer
12 Damping rubber – 6 off
13 Nut
14 Domed nut – 3 off
15 Washer – 2 off
16 Sealing washer – 2 off
17 Spark plug
18 Valve adjuster inspection cover – 2 off
19 O-ring – 2 off
20 Camshaft end cap
21 Tachometer driven gear
22 Housing
23 Sealing ring
24 Screw
25 Oil seal
26 Carburettor mounting stub
27 O-ring
28 Screw – 22 off
29 Spring washer – 2off
30 Clamp

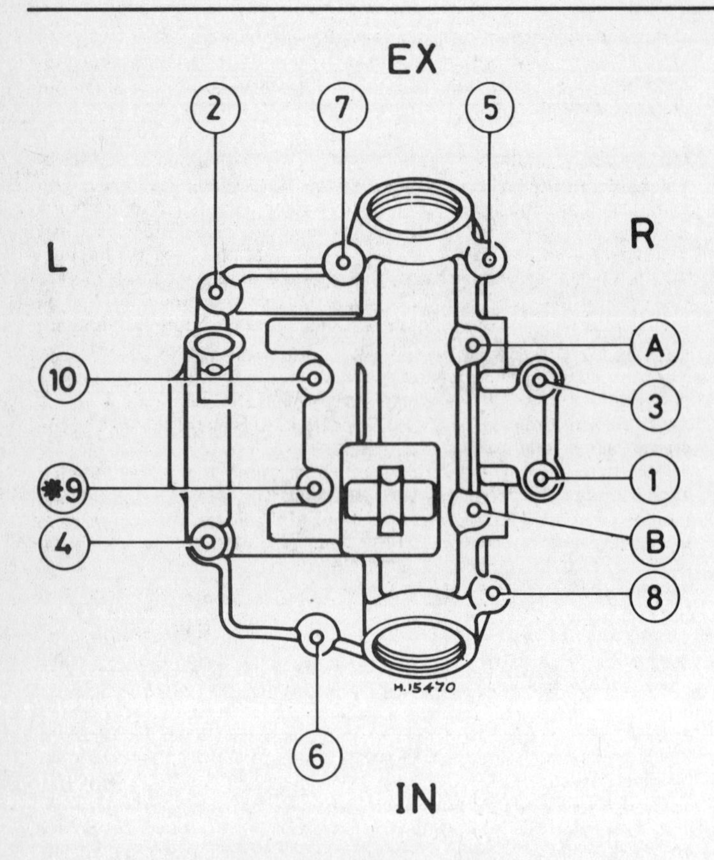

Fig. 1.2 Cylinder head cover retaining bolt slackening sequence

A and B – do not disturb
** – sealing washer under this bolt*

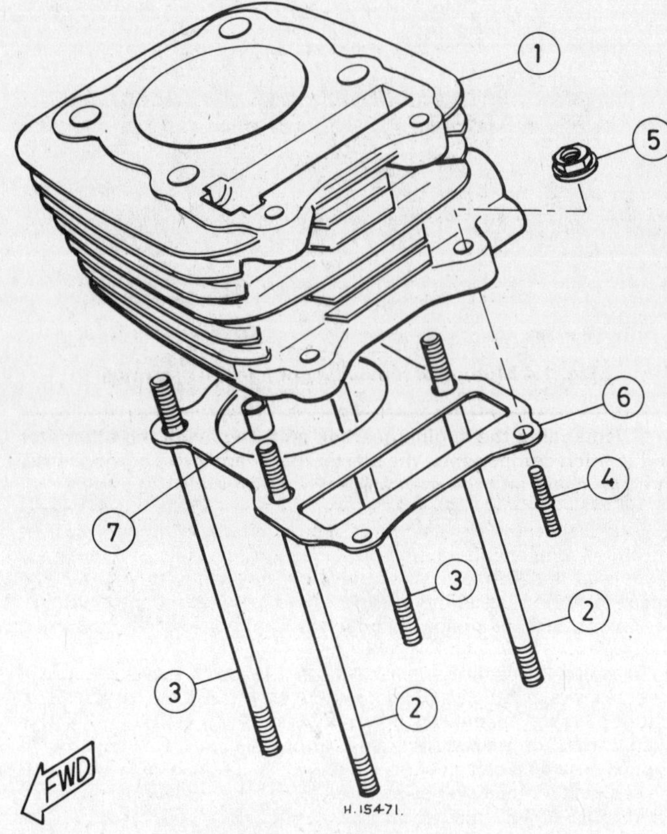

Fig. 1.3 Cylinder barrel

1 *Cylinder barrel*
2 *Stud – 2 off*
3 *Stud – 2 off*
4 *Stud – 2 off*
5 *Nut – 2 off*
6 *Cylinder base gasket*
7 *Locating dowel – 2 off*

7 Dismantling the engine/gearbox unit: removing the cylinder barrel and piston

1 These components can be withdrawn with the engine/gearbox unit in or out of the frame, but in both cases the work described in the previous Section must be carried out first.

2 Rotate the crankshaft until the piston is at TDC, being careful to keep taut the cam chain so that it does not jam between the crankshaft sprocket and crankcase wall. Remove the two cylinder barrel retaining nuts, tap lightly the barrel with a soft-faced mallet to break the seal, and lift the barrel until the bottom of the piston skirt is revealed.

3 To prevent dirt or debris from falling into the crankcase, pack tightly the cam chain tunnel aperture and the crankcase mouth with clean rag. Allow the cam chain to drop on to the rag and remove fully the barrel. Remove also the cylinder base gasket and the two cylinder locating dowels.

4 Use a sharp-pointed instrument to prise out one of the gudgeon pin circlips, press out the gudgeon pin and remove the piston. If the gudgeon pin is a tight fit, soak a rag in boiling water, wring it out, and wrap it around the piston. The heat will expand the piston sufficiently to release its grip on the gudgeon pin. If the pin is still tight, it may be tapped out using a hammer and a suitable drift, but care must be taken to support firmly the connecting rod while this is being done.

5 The piston rings are removed by holding the piston in both hands and prising gently the ring ends apart with the thumbnails until the rings can be lifted out of their grooves and on to the piston lands, one side at a time. The rings can then be slipped off the piston and put to one side for cleaning and examination. The bottom compression ring is supported by a very thin expander which must be prised carefully from the groove with a sharp-pointed instrument. If the rings are stuck in their grooves by excessive carbon deposits, use three strips of thin metal sheet to remove them, as shown in the accompanying illustration.

7.4 Remove circlip using suitable instrument

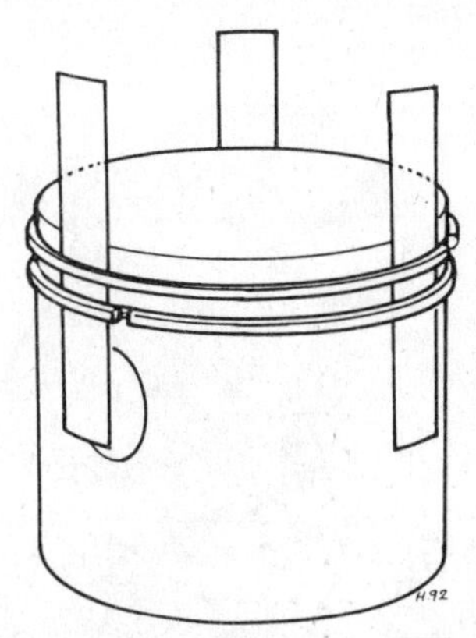

Fig. 1.4 Method of removing gummed piston rings

8 Dismantling the engine/gearbox unit: removing the alternator and ignition components, the starter motor and drive components, the cam chain and the neutral/gear position indicator switch

1 The crankcase left-hand cover can be removed to gain access to the above components whether the engine/gearbox unit is in the frame or not. In the former case it will be necessary to remove the sidepanels, the seat and the fuel tank to allow the electrical leads to be disconnected, to remove the gearbox sprocket cover, and to drain the oil, before work can begin.
2 Remove the crankcase left-hand cover retaining screws; there are seven screws on GS125 ES models and six on GS125 and DR125 S models. Simplify identification on reassembly by pressing each screw into a cardboard template as it is removed, so storing each screw in its original position. Tap lightly the cover with a soft-faced mallet to break the seal and lift the cover away. Note that there is one locating dowel fitted to GS125 ES models only.
3 If the cover proves difficult to remove, do not lever it from position, but remove instead the circular inspection cap and pull the cover away by inserting a finger into the inspection aperture and tapping lightly around the joint area with a soft-faced mallet.
4 Remove the neutral indicator switch (GS125, DR125 S models) or the gear position indicator switch (GS125 ES model) by removing the two retaining screws and lifting away the switch. Withdraw the sealing O-ring, the switch contact, and the contact spring from the end of the selector drum

GS125 and DR125 S models

5 To prevent the crankshaft from rotating while the rotor retaining nut is removed, one of the following methods can be adopted; select top gear and apply hard the back brake to lock the crankshaft via the transmission, apply a holding tool such as a strap wrench to the rotor itself, or place a close-fitting metal bar through the connecting rod small-end eye and support the bar on two wooden blocks placed across the crankcase mouth. Having selected the appropriate method, remove the rotor retaining nut and the lock washer and plain washer beneath it.
6 Rotor removal is to be carried out using only the Suzuki service tool 09930-30161 and the slide hammer 09930-30102 or using a pattern centre bolt flywheel puller which will suit most small-capacity Japanese machines and is available from any good motorcycle dealer. Do not risk damaging the rotor by attempting to remove it by any other method.
7 If the Suzuki tools are available, screw fully the adaptor into the rotor centre, noting that a left-hand thread is employed, and screw the slide hammer shaft into the tool body. One or two sharp taps with the slide hammer will easily release the rotor. If the slide hammer is not available it can be replaced by a 16 mm (thread size) metric bolt and the tool used in the same way as the pattern centre bolt puller, this being described below.
8 Remembering that a left-hand thread is employed, screw fully the tool body into the rotor centre, then, firmly tighten down against the crankshaft end the tool centre bolt. Tap smartly on the bolt head with a hammer to jar free the rotor. If unsuccessful, tighten further the centre bolt and repeat the process.
9 Withdraw from the crankshaft end taper the rotor locating Woodruff key and store this with the rotor. Remove the three stator plate retaining screws and withdraw the stator plate.
10 The cam chain can be withdrawn once the alternator rotor and stator and the cylinder head have been removed. Drop the chain into the crankcase, disengage it from the crankshaft sprocket and withdraw it.

GS125 ES model

11 Apply an open-ended spanner to the flats of the alternator rotor centre boss to prevent the rotor from turning while its retaining nut is unscrewed.
12 To prevent damage to the rotor, it must be removed using only the Suzuki service tool 09930-30180/1. This is normally used in conjunction with a slide hammer (Suzuki part no 09930-30102), but if the slide hammer is not available it can be replaced by a 16 mm (thread size) metric bolt and the tool used as a conventional centre bolt flywheel puller. The procedure of rotor removal is as described in paragraphs 8 and 9 above, but the starter clutch driven gear must be removed with the rotor and kept in position to prevent the loss of the starter clutch rollers.
13 Withdraw from the crankshaft end taper the rotor locating Woodruff key and store this with the rotor. The alternator stator and ignition pulser coil are secured by screws to the inside of the crankcase left-hand cover; remove the screws to release either of these components.
14 Withdraw the cam chain as described in paragraph 10 of this Section.
15 Remove the two starter motor retaining bolts and withdraw the starter motor from the front of the crankcases. Withdraw the thick spacer, the starter idler gear shaft and the idler gear itself.

9 Dismantling the engine/gearbox unit: removing the crankcase right-hand cover

1 The crankcase right-hand cover can be removed with the engine/gearbox unit in the frame or removed from it, but in the former case the engine oil must be drained and the oil filter element removed, and the kickstart lever (GS125 and DR125 S models only) removed, the work necessary being described in Section 4.
2 Progressively slacken, then remove, the ten bolts which retain the cover, noting the sealing washer fitted under the head of the bolt above and to the right of the filler plug orifice. Store the bolts in a cardboard template as shown in the accompanying photograph. Do not forget to remove the oil filter chamber cap, if this has not been done.
3 Tap around the joint with a soft-faced mallet to break the seal, then lift away the cover, ensuring that the thrust washer fitted over the end of the kickstart shaft (GS125 and DR125 S only) remains in place. Remove the cover gasket and two locating dowels.

10 Dismantling the engine/gearbox unit: removing the clutch assembly

1 When the crankcase right-hand cover has been removed as described in the previous Section, the clutch assembly can be withdrawn whether the engine/gearbox unit is in the frame or not.
2 Working in a diagonal sequence progressively slacken, then remove, the clutch spring retaining bolts. Withdraw the clutch pressure plate and release mechanism, the clutch springs, and the clutch friction and plain plates. Knock back the raised portion of the clutch centre retaining nut lock washer.
3 To prevent the clutch centre from rotating, fabricate a holding tool from two metal strips as shown in the accompanying photograph. An alternative method is to select top gear and to lock the transmission by applying hard the rear brake by fabricating a tool to hold securely the gearbox sprocket. Remove the clutch centre retaining nut and its lock washer.
4 Lift away the clutch centre, the first thrust washer, the clutch outer drum and the second thrust washer. Note that on 'Z' (1982) models only, the clutch outer drum rotates on a separate bush which must be removed before the second thrust washer is displaced.
5 Pull the clutch pushrod out of the input shaft centre. Remove the single screw which is threaded into the crankcase top surface to retain the clutch release shaft oil seal, then dig out the seal using a sharp-pointed instrument and taking great care not to damage the crankcase casting. Lift out the release shaft and its locating washer.

8.12 Hold rotor with spanner while retaining nut is unscrewed – GS125 ES only

8.13a GS125 ES only – special tool shown is essential for rotor removal

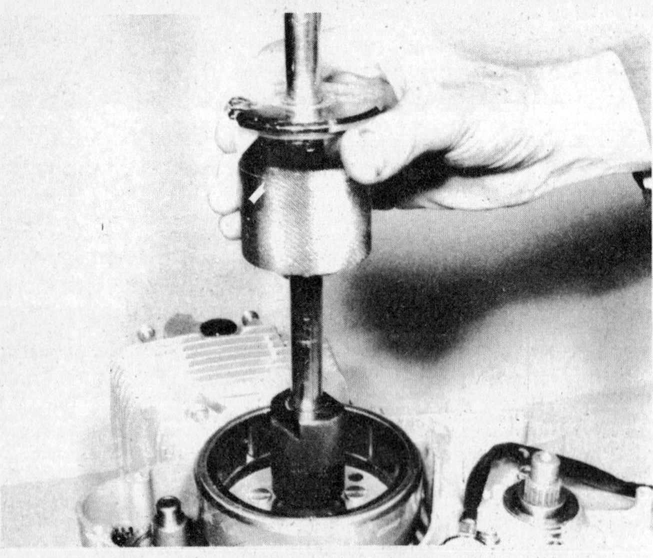

8.13b Slide hammer is useful but not essential – similar tool can be fabricated or alternative method used (see text)

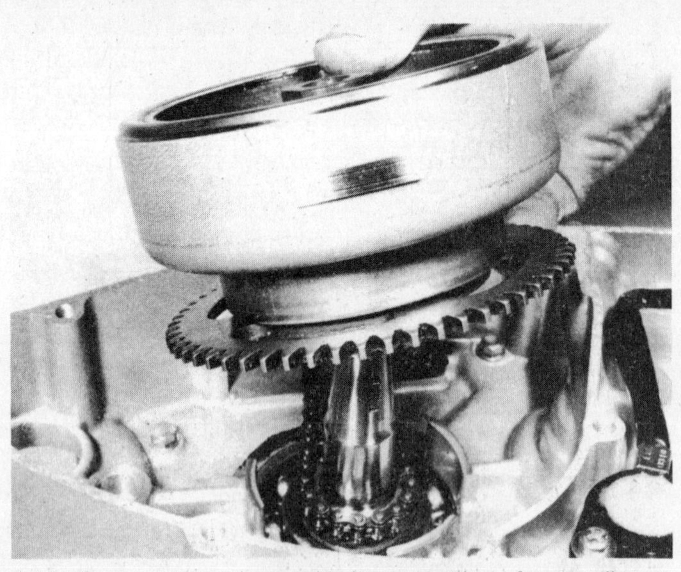

8.13c Keep starter clutch assembled to prevent loss of any small components

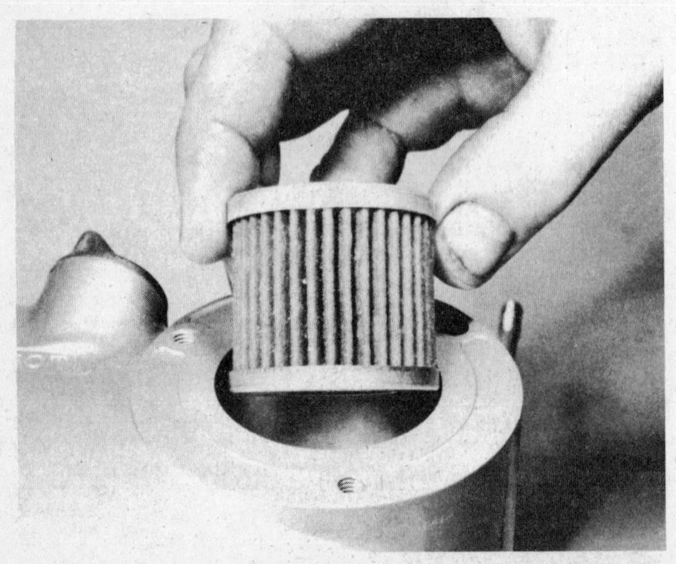

9.1 Oil filter element must be removed before crankcase cover is withdrawn

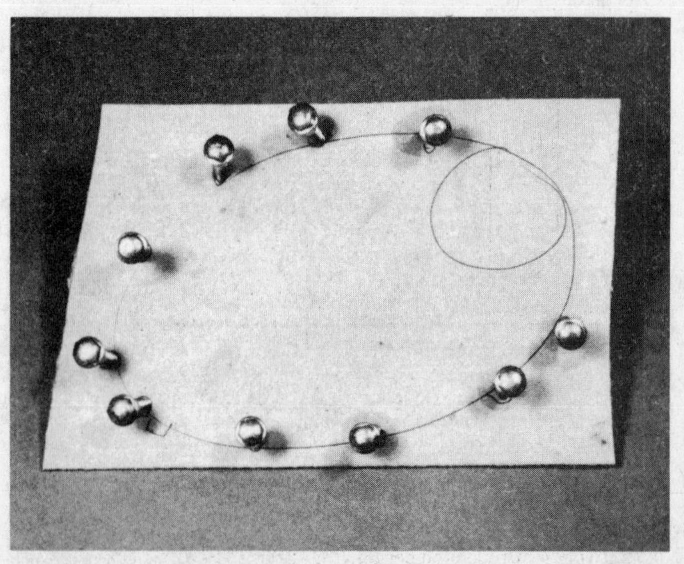

9.2 Fabricate cardboard template as shown to retain crankcase cover screws in correct positions

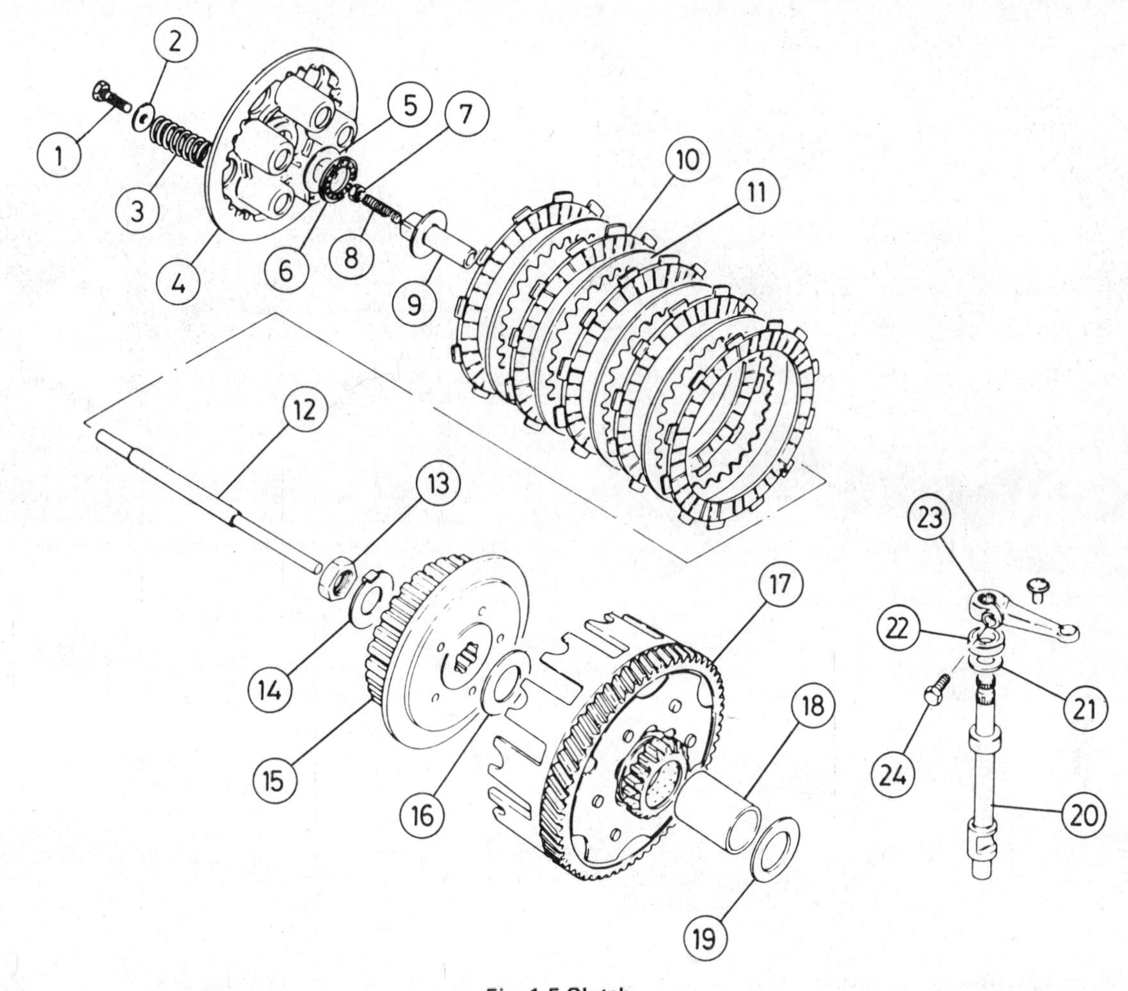

Fig. 1.5 Clutch

1	*Bolt – 5 off*	7	*Locknut*	13	*Nut*	19	*Thrust washer*
2	*Washer – 5 off*	8	*Adjusting screw*	14	*Lock washer*	20	*Release shaft*
3	*Spring – 5 off*	9	*Pushrod end*	15	*Clutch centre*	21	*Washer*
4	*Pressure plate*	10	*Friction plate – 5 off*	16	*Thrust washer*	22	*Oil seal*
5	*Thrust washer*	11	*Plain plate – 4 off*	17	*Outer drum*	23	*Operating lever*
6	*Thrust bearing*	12	*Pushrod*	18	*Bush – Z models only*	24	*Bolt*

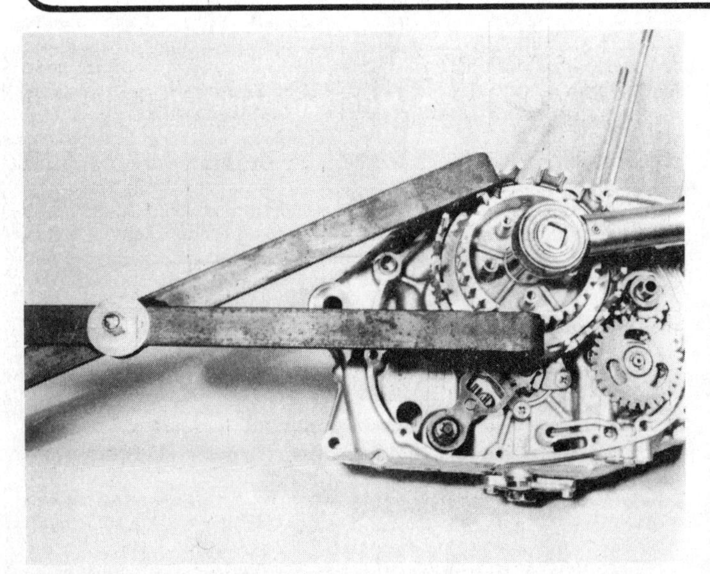

10.2 Tool locking clutch centre while retaining nut is slackened or tightened

10.5a Remove retaining screw ...

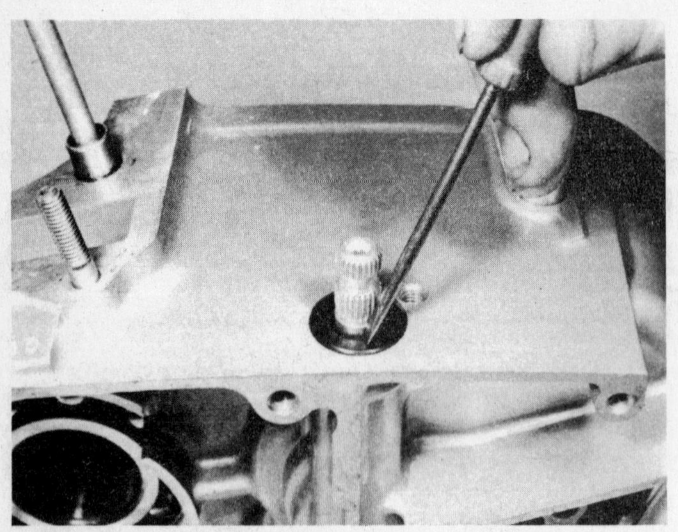

10.5b ... and prise out oil seal to remove clutch release shaft

11 Dismantling the engine/gearbox unit: removing the oil pump, the primary drive gear and the kickstart drive gears

1 The above components can be removed after the crankcase right-hand cover has been withdrawn, whether the engine/gearbox unit is in the frame or not. While the oil pump and primary drive gear can be removed with the clutch in situ, clutch removal will be necessary before the kickstart drive gears (GS125 and DR125 S models only) can be withdrawn. Refer to Sections 9 and 10 of this Chapter.
2 To remove the oil pump, rotate the crankshaft until the raised portion of the primary drive gear retaining nut lock washer is adjacent to the oil pump driven gear, then displace the driven gear retaining circlip and withdraw the driven gear from the oil pump shaft. Use an impact driver to release the three oil pump retaining screws and withdraw the pump assembly. Take care not to lose the small pin which passes through the oil pump shaft.
3 To remove the primary drive gear, if the clutch assembly is in place lock the crankshaft by selecting top gear and applying hard the rear brake or by fabricating a tool to hold securely the gearbox sprocket. If the clutch is removed, lock the crankshaft by applying a holding tool such as a strap wrench to the alternator rotor or by passing a close-fitting metal bar through the connecting rod small-end eye and supporting the bar on two wooden blocks placed across the crankcase mouth. An alternative method is to refit temporarily the clutch outer drum and to sprag the primary drive gears by wedging between them a wooden or soft alloy bar or a tightly-wadded piece of rag.
4 **Note**: the primary drive gear retaining nut employs a left-hand thread. Knock back the raised portion of the lock washer and unscrew **clockwise** the retaining nut. Withdraw the lock washer, the oil pump drive gear and the primary drive gear, noting which way round the gears are fitted. Prise the Woodruff key from its slot in the crankshaft.
5 Withdraw the thrust washer and the kickstart pinion from the kickstart shaft (GS125 and DR125 S models only), then displace the retaining circlip and remove from the output shaft right-hand end the first thrust-washer, the kickstart idler pinion and the second thrust washer.

12 Dismantling the engine/gearbox unit: removing the gear selector external components

1 Those components of the gear selector mechanism which can be removed without separating the crankcases are the gearchange shaft and return spring and the selector ratchet assembly; the crankcase right-hand cover and clutch assembly must be removed to gain access to them. The work necessary can be carried out with the engine/gearbox unit in the frame or removed from it, and is described in Sections 9 and 10 of this Chapter. In addition the neutral detent plunger assembly can be removed, a task which requires no additional work other than to prevent the loss of oil.
2 Remove the gearchange lever or linkage front arm from the gearchange shaft left-hand end and pull the shaft out to the right. Use an impact driver to release the screws retaining the ratchet assembly locating plate and the pawl guide plate, then withdraw the two plates.
3 Being careful to restrain the two pawls so that they cannot fly out, lift the selector ratchet assembly out of the selector drum and secure the two pawls with an elastic band so that they cannot be lost.
4 The neutral detent plunger assembly is situated underneath the crankcase, next to the triangular crankcase filter cover. Unscrew the plunger cap and withdraw the spring and plunger.

13 Dismantling the engine/gearbox unit: separating the crankcase halves

1 The crankcase halves can be separated only after the engine/gearbox unit has been removed from the frame as described in Section 4 of this Chapter and after all the components have been removed as described in Sections 5-12.
2 On GS125 ES models only, remove from the output shaft right-hand end the circlip and thrust washer.
3 Unscrew the three bolts which retain the triangular crankcase filter cover, then withdraw the cover and its sealing O-ring. Remove also the filter gauze which is retained by two screws. Place the engine/gearbox unit on the bench with its left-hand side uppermost and support it on two wooden blocks.
4 Working in a diagonal sequence from the outside inwards, progressively slacken and remove the crankcase securing screws; there are twelve screws on GS125 ES models, and eleven on GS125 and DR125 S models. Store the screws in a cardboard template as previously described. Invert the engine/gearbox unit and support it on the wooden blocks so that its right-hand side is now uppermost.
5 To prevent damage to the gearchange selector drum and selector stop arm during crankcase separation, the crankcase right-hand half must be withdrawn to leave intact the crankshaft and gearbox components in the crankcase left-hand half.
6 If it is available, the Suzuki service tool Part Number 09910-80115 should be used to separate the crankcase halves. It is fitted to the crankcase right-hand half as shown in the accompanying photograph and pressure is applied to the crankshaft and input shaft by tightening down the T-handle. Ensure that the two rectangular bars are exactly square to the crankcase at all times and check that the crankcase right-hand half lifts evenly, using a soft-faced mallet to tap around the joint face and on the shaft ends to assist the tool.
7 If the tool is not available, an alternative method can be employed, using only a soft-faced mallet. Never use a metal hammer to strike directly the castings or other components, the risk of damage is too great.
8 Grasping firmly the crankcase right-hand half, tap lightly and repeatedly all around the joint area and on the protruding right-hand ends of the crankshaft, the input and output shafts and the selector drum. The regular, light, tapping will jar apart the components without any excessive force being necessary. When the crankcase right-hand half is loose enough, lift it away noting the presence of the two large locating dowels; unless these are firmly fixed, remove them and store them safely. Check also that the thrust washer and bush fitted to the output shaft right-hand end are in place and have not stuck to the crankcase half.
9 If the crankcases are reluctant to separate, never be tempted to tap ever harder in an effort to release them; check again that all fasteners etc. have been removed. It may be that corrosion has formed around the locating dowels, requiring the application of penetrating fluid to release the two castings. If the crankcase halves separate but then stick, ensure that they are square to one another and not tying. If necessary, tap the two halves back together and start again. Never use levers between the mating surfaces to aid separation.

14 Dismantling the engine/gearbox unit: removing the crankshaft and gearbox components

1 On GS125 and DR125 S models only, remove the kickstart shaft from the crankcase right-hand half by displacing the circlip from the shaft left-hand end, withdrawing the spring guide, and releasing the spring. Unhook the spring outer end from the crankcase and allow the spring to return slowly to its relaxed position, then remove the spring inner end from the shaft and push the shaft out to the right.

2 Unhook the selector stop arm spring from the post set in the crankcase wall, remove the stop arm pivot bolt, and withdraw the stop arm assembly.
3 Withdraw the two selector fork shafts, then remove the two selector forks that engage on the output shaft gear pinions. As each fork is withdrawn, use a spirit-based felt marker to mark its upper, or right-hand, surface so that it is clearly identifiable and cannot be confused or refitted wrongly. Remove the selector drum complete with the camplate and washer on its left-hand end, then withdraw the remaining selector fork. Refit the selector forks on their correct shafts.
4 To prevent the loss of any components, tighten an elastic band over the output shaft right-hand end. Remove as a single unit the two gear clusters, using a soft-faced mallet to tap on the output shaft left-hand end.
5 Support the crankcase left-hand half on two wooden blocks so that its left-hand side is uppermost. The two blocks must be placed as close as possible around the crankshaft to support securely the casting, and must be of sufficient height to permit the crankshaft to be removed from underneath. Refit temporarily the rotor retaining nut, screwing the nut down until it is flush with the crankshaft end. Using only a soft-faced mallet, tap the crankshaft downwards out of its main bearing, withdrawing the nut as soon as the crankshaft is released so that the crankshaft can be pulled through the main bearing. Do not allow the crankshaft to drop clear and note that excessive force will not be required, a few firm taps should suffice.

11.2 Rotate crankshaft so that raised portion of locking tab is in position shown

11.3 Note method used to lock crankshaft while retaining nut is slackened or tightened – primary drive gear retaining nut is a left-hand thread

12.3 Prevent pawls (arrowed) from flying out as ratchet assembly is removed

13.3 Do not forget to remove gauze filter before crankcases are separated

13.6 Manufacturer's crankcase separating tool is assembled as shown – useful but not essential

15 Dismantling the engine/gearbox unit: removing oil seals and bearings

1 Before removing any oil seal or bearing, check that it is not secured by a retaining plate. If this is the case, use an impact driver or spanner, as appropriate, to release the securing screws or bolts and lift away the retaining plate. Pull the gearbox sprocket spacer out of the output shaft oil seal.
2 Oil seals are easily damaged when disturbed and, thus, should be renewed as a matter of course during overhaul. Prise them out of position using the flat of a screwdriver and taking care not to damage the alloy seal housings.
3 The crankshaft and gearbox bearings are a press fit in their respective crankcase locations. To remove a bearing the crankcase casting must be heated so that it expands and releases its grip on the bearing which can then be drifted or pulled out.
4 To prevent casting distortion, it must be heated evenly to a temperature of about 100°C by placing it in an oven; if an oven is not available, place the casting in a suitable container and carefully pour boiling water over it until it is submerged.
5 Taking care to prevent personal injury when handling heated components, lay the casting on a clean surface and tap out the bearing using a hammer and a suitable drift. If the bearing is to be re-used apply the drift only to the bearing outer race, where this is accessible, to avoid damaging the bearing. In some cases it will be necessary to apply pressure to the bearing inner race; in such cases inspect closely the bearing for signs of damage before using it again. When drifting a bearing from its housing it must be kept square to the housing to prevent its tying in the housing with the resulting risk of damage. Where possible, use a tubular drift such as a socket spanner which bears only on the bearing outer race; if this is not possible, tap evenly around the outer race to achieve the same result.
6 In some cases, bearings are pressed into blind holes in the castings. These bearings must be removed by heating the casting and tapping it face downwards on to a clean wooden surface to dislodge the bearing under its own weight. If this is not successful the casting should be taken to a motorcycle service engineer who has the correct internally expanding bearing puller.

16 Examination and renovation: general

1 Before examining the parts of the dismantled engine unit for wear it is essential that they should be cleaned thoroughly. Use a petrol/paraffin mix or a high flash-point solvent to remove all traces of old oil and sludge which may have accumulated within the engine. Where petrol is included in the cleaning agent normal fire precautions should be taken and cleaning should be carried out in a well ventilated place.
2 Examine the crankcase castings for cracks or other signs of damage. If a crack is discovered it will require a specialist repair.
3 Examine carefully each part to determine the extent of wear, checking with the tolerance figures listed in the Specifications section of this Chapter or in the main text. If there is any doubt about the condition of a particular component, play safe and renew.
4 Use a clean lint free rag for cleaning and drying the various components. This will obviate the risk of small particles obstructing the internal oilways, and causing the lubrication system to fail.
5 Various instruments for measuring wear are required including a vernier gauge or external micrometer and a set of standard feeler gauges. The machine's manufacturer recommends the use of Plastigage for measuring radial clearance between the camshaft and its bearing surfaces. Plastigage consists of a fine strand of plastic material manufactured to an accurate diameter. A short length of Plastigage is placed between the two surfaces, the clearance of which is to be measured. The surfaces are assembled in their normal working positions and the securing bolts fastened to the correct torque loading; the surfaces are then separated. The amount of compression to which the gauge material is subjected and the resultant spreading indicates the clearance. This is measured directly, across the width of the Plastigage, using a pre-marked indicator supplied with the Plastigage kit. If Plastigage is not available both an internal and external micrometer will be required to check wear limits. Additionally, although not absolutely necessary, a dial gauge and mounting bracket is invaluable for accurate measurement of end float, and play between components of very low diameter bores – where a micrometer cannot reach.
6 After some experience has been gained the state of wear of many components can be determined visually or by feel and thus a decision on their suitability for continued service can be made without resorting to direct measurement.

17 Examination and renovation: engine cases and covers

1 Small cracks or holes in aluminium castings may be repaired with an epoxy resin adhesive, such as Araldite, as a temporary expedient. Permanent repairs can only be effected by argon-arc welding, and a specialist in this process is in a position to advise on the viability of proposed repair.
2 Damaged threads can be economically reclaimed by using a diamond section wire insert, of the Helicoil type, which is easily fitted after drilling and re-tapping the affected thread. Most motorcycle dealers and small engineering firms offer a service of this kind.
3 Sheared studs or screws can usually be removed with screw extractors, which consist of tapered, left-hand thread screws, of very hard steel. These are inserted by screwing anti-clockwise, into a pre-drilled hole in the stud, and usually succeed in dislodging the most stubborn stud or screw. If a problem arises which seems to be beyond your scope, it is worth consulting a professional engineering firm before condemning an otherwise sound casing. Many of these firms advertise regularly in the motorcycle papers.

18 Examination and renovation: bearings and oil seals

1 The crankshaft and gearbox bearings can be examined while they are still in place in the crankcase castings. Wash them thoroughly to remove all traces of oil, then feel for free play by attempting to move the inner race up and down, then from side-to-side. Examine the bearing balls or rollers and the bearing tracks for pitting or other signs of wear, then spin hard the bearing. Any roughness caused by defects in the bearing balls or rollers or in the bearing tracks will be felt and heard immediately.
2 If any signs of free play or wear are discovered, or if the bearing is not free and smooth in rotation but runs roughly and slows down jerkily, the bearing must be renewed. Bearing removal is described in Section 15 of this Chapter, and refitting in Section 32.
3 To prevent oil leaks occurring in the future, all oil seals and O-rings should be renewed whenever they are disturbed during the course of an overhaul, regardless of their apparent condition. Do not forget the O-ring set in the gearbox sprocket spacer.

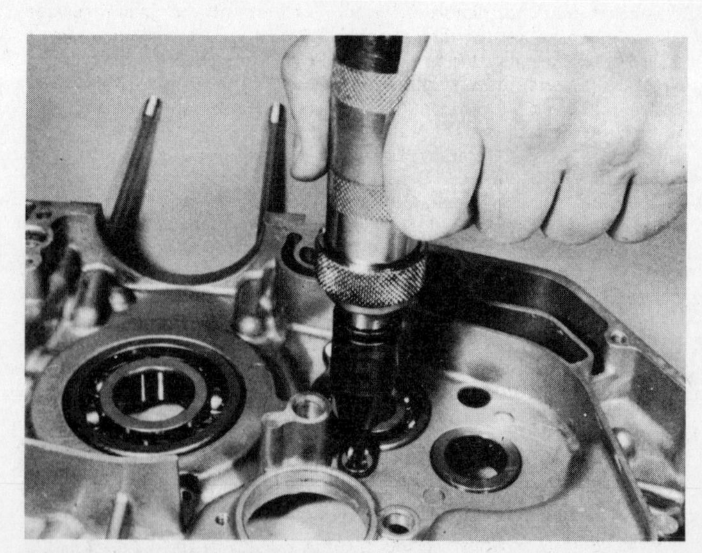

15.1 Remove retaining plates (where fitted) so that ...

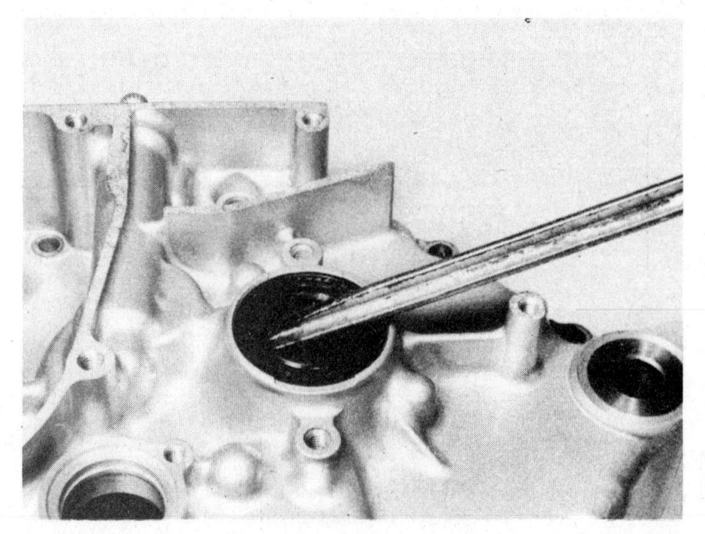

15.2 ... oil seals can be levered out of their housings ...

15.3 ... and bearings can be drifted out after heating crankcase casting

19 Examination and renovation: camshaft and rocker gear

1 The camshaft and rocker gear must be very closely inspected, particularly if engine oil changing at the correct interval has been neglected or if the oil level has been allowed to drop too far, even if the engine has been dismantled for some other reason.
2 Examine the camshaft visually for signs of wear. The camshaft lobes should have a smooth surface and be entirely free from scuff marks or indentations. Wear will probably be most evident on the ramps of each cam and where the cam contour changes sharply. It is unlikely that severe wear will be encountered during the normal service life of the machine unless the lubrication system has failed, causing the case hardened surface to wear through. If necessary, check with the Specification given at the beginning of this Chapter and measure the cam lobe height in each case. If either of the cam lobes is below the service limit, the camshaft must be renewed.
3 The camshaft runs directly on bearings formed by the material of the cylinder head and cylinder head cover; accordingly these two castings must be checked also when assessing bearing wear. Examine the bearing surfaces; signs of anything but light scoring or scuffing will mean that the component concerned should be renewed.
4 When assessing camshaft bearing wear, first check that the cylinder head and cylinder head cover are mating correctly and are not distorted. Lay each casting on a flat surface such as a sheet of plate glass and check for distortion at points all around the gasket surface using feeler gauges. If distortion of 0.05 mm (0.002 in) or more is found on either gasket face, the cylinder head and cylinder head cover must be renewed as a single unit, althouth it may be possible in less severe cases to flatten the gasket face by rubbing it with a rotary motion on a sheet of 400 grade emery paper laid on the flat surface. Be careful not to remove too much material or the camshaft oil clearance will be reduced excessively.
5 When the gasket faces are known to be flat, and are completely clean use Plastigage as described in Section 16 of this Chapter to measure the camshaft radial, or oil, clearance which must not exceed 0.15 mm (0.006 in) with the cylinder head cover retaining bolts tightened to a torque setting of 0.9 – 1.0 kgf m (6.5 – 7 lbf ft). If the clearance is excessive, measure the overall diameter of the camshaft bearing journals; if these are worn to less than 21.959 mm (0.864 in) the camshaft must be renewed, and if the camshaft is within tolerances, the cylinder head and cylinder head cover must be renewed as a single unit.
6 Note that if excessive camshaft bearing wear is found, there are several light engineering companies advertising in the national motorcycle press who will reclaim such damage by fabricating and fitting separate plain or roller bearings, thus saving the cost of a new cylinder head/cylinder head cover assembly.
7 If signs of oil starvation are evident such as dry, badly scored, bearing surfaces, deposits of burnt oil, or excessive heating (or blueing) on any component, the cause must be found and rectified before the engine is reassembled. Examine and clean thoroughly all oilways, using compressed air to clean out any oilways that cannot be reached by other means.
8 Dismantle separately the rocker arm assemblies to prevent the accidental interchange of components. The various parts of each assembly must be kept separate for the same reason. Wear in the rocker gear is revealed by a light tapping noise which should not be confused with the similar noise produced by excessive valve clearances.
9 Remove the rocker shaft retaining bolt, noting the sealing washer under its head, to release each rocker shaft, which can be pulled out of the cylinder head with a pair of pliers. On later models only, note carefully the position of the wave washer fitted on the rocker shaft to reduce the rocker arm endfloat. Remove and discard the sealing O-ring fitted around each rocker shaft; these must be renewed whenever the shaft is disturbed.
10 Check the fit of the rocker shafts in the rocker arm and cylinder head cover; if excessive play is apparent measure the outside diameter of each rocker shaft and the matching bore of each rocker arm, comparing the measurements obtained with those given in the Specifications Section of this Chapter and renewing any component that is worn.
11 Check the tip of each rocker arm at the point where the arm makes contact with the cam. If signs of cracking, scuffing or breakthrough in the case hardened surface are evident, fit a new replacement.
12 Check the thread of the tappet adjusting screw, the thread of the rocker arm into which it fits and the thread of the locknut. The hardened end of the tappet adjuster must also be in good condition.
13 On reassembly, renew the O-ring fitted to each rocker shaft and the sealing washer fitted under the head of each shaft retaining bolt. Smear grease over each O-ring and check that all components are being refitted in their original positions and are well lubricated. Insert each rocker shaft into the cylinder head cover until it just protrudes on the inside, then refit the wave washer (later models only) and the rocker arm. Push the shaft fully into place, rotating it as necessary to align the threaded hole drilled to accept the shaft retaining bolt. Apply a few drops of thread locking compound to the retaining bolt threads, then refit and tighten securely the retaining bolt, not forgetting its sealing washer.

20 Examination and renovation: cylinder head, valves, valve seats and guides

NOTE: Later models of both the GS and DR125 may have been fitted with a revised cylinder head assembly. The modification concerns the relocation of the valve guides so that they sit lower in the head, with corresponding changes to the position of the valve stem collet groove and to the collet halves themselves. There are no details of engine numbers from which this modification applies, so make sure that the old valve components are taken as patterns when ordering new parts.

19.2 Measuring camshaft lobe height

19.5 Plastigage is used as described in text to measure camshaft oil clearance

19.9 Remove rocker shaft retaining bolts – identified by recessed heads

19.13 Rotate shaft to align threaded hole with retaining bolt – O-rings should be renewed whenever disturbed

If it is necessary to fit a new head, only the new (modified) type is now supplied. It follows that it will be necessary to purchase new valves and collet halves as well as the head itself, if the cylinder head is of the older pattern. The exhaust valve spring was also changed, the latest pattern being longer and having a higher spring rate than the original. The revised spring only is supplied as a replacement part, and it is quite in order to use this as a direct replacement for the early pattern. Old type springs must **never** be fitted to the new cylinder head assembly, however. Note that the spring must be installed with the tighter coils nearest the cylinder head. Refer to the Specifications at the beginning of this Chapter when checking the valves and springs.

1 It is best to remove all carbon deposits from the combustion chamber before the valves are withdrawn for inspection, but do not forget to pay equal attention to cleaning the intake and exhaust ports once the valves have been removed. Use a blunt-ended scraper to avoid damaging the soft alloy, and finish off with a metal polish to achieve a smooth surface.

2 Clean the cylinder head (to cylinder barrel) gasket surface and check it for distortion with a straight-edge and a set of feeler gauges. The straight-edge should be laid diagonally across the mating surface from corner to corner and any clearance between its lower edge and the mating surface checked with the gauges. Repeat the procedure with the straight-edge laid between the opposite corners and then at several positions in between the four corners. If the largest clearance found exceeds 0.05 mm (0.002 in), Suzuki recommend that the cylinder head be replaced with a new item. It is worth noting however, that if the amount of distortion found is only slightly greater than the limit given, it could well be worth seeking the advice of a competent motorcycle engineer who can advise on whether skimming the mating surface flat is possible without the subsequent risk of the pistons coming into contact with the valve heads once the engine is reassembled and started.

3 If the cylinder head is seen to be only slightly warped, the mating surface may be lapped on a surface plate or a sheet of plate glass. Place a sheet of 400 or 600 grit abrasive paper on the surface plate. Lay the cylinder head, gasket-face down, on the paper and gently rub it with an oscillating motion to remove any high spots. Lift the head at frequent intervals to inspect the progress of the operation, and take care to remove the minimum amount of material necessary to restore a flat sealing surface. It should be remembered that most cases of cylinder head warpage can be traced to unequal tensioning of the cylinder head nuts and bolts by tightening them in incorrect sequence or by using incorrect or unmeasured torque settings.

4 Finally, make sure that the external cooling fins of the cylinder head are not clogged with oil or road dirt which will prevent the free flow of air and cause the engine to overheat, and check the condition

of the spark plug thread; if damaged this can be repaired by the use of a Helicoil thread insert.

5 Obtain two marked containers so that the appropriate valve components can be kept separate once removed. A valve spring compression tool must be used to compress each set of valve springs in turn, thereby allowing the split collets to be freed from the valve stem and the spring retaining plate, the springs, the spring seat and the valve guide oil seal to be removed from the cylinder head along with the valve itself. As each valve is removed, check that it will pass through the guide bore without resistance. Use fine abrasive paper to polish away any burrs or raised edges from the collet groove until the valve can be removed easily. Keep separate the inlet and exhaust valve spring components; while there is no possibility of confusing the valves, since their heads are marked and of different sizes, the springs, collets, and spring retainers are identical and may be refitted wrongly if care is not taken.

6 Using a micrometer, measure the diameter of the valve stem at various points along its length. If any one of the measurements obtained is less than the service limit given in the Specifications Section of this Chapter, the valve should be renewed. Check the amount of valve stem runout by supporting the valve stem on two V-blocks and slowly rotating it whilst measuring the amount of runout with a dial gauge. Use a similar method to measure the amount of radial runout of the valve head and compare both sets of measurements obtained with the service limits given in the Specifications. In either case, if these measurements are beyond limits, then the valve must be renewed.

7 Examine the valve head and seating face. If the face is burnt, pitted, or damaged, the valve must be renewed; do not attempt to restore a damaged valve seating face by excessive grinding-in. Measure the thickness of the valve head as shown in the accompanying illustration, if the head is worn to 0.5 mm (0.02 in) or less, the valve must be renewed.

8 Check that the valve stem tip is not indented or otherwise damaged; slight wear can be repaired by grinding down the stem, provided that this does not reduce the collet groove/stem tip length to less than the service limit. If the stem is ground, ensure that this distance is measured carefully, and on refitting the valve, check that the stem tip protrudes above the collets, as shown in the accompanying illustration.

9 Check the valve stem/valve guide clearance using a dial gauge and a valve that is either new or known to be within the specified wear limits. Insert the valve into the guide and push it down until it is in the maximum lift position. Clamp the dial gauge to the cylinder head so that its pointer rests against the valve head. Take two measurements, one at 90° to the other, and renew the valve guide if the amount of wear indicated is greater than the service limit given in the Specifications.

10 To renew a valve guide, place the cylinder head in an oven to heat it evenly to about 100°C (212°F). The old guide can now be tapped out from the cylinder side. The correct drift should be shouldered with the smaller diameter the same size as the valve stem and the larger diameter slightly smaller than the OD of the valve guide. If a suitable drift is not available a plain brass drift may be utilized with great care. Before removing old guides scrape away any carbon deposits which have accumulated on the guide where it projects into the port. If in doubt, seek the advice of a Suzuki specialist. Note that each valve guide is fitted with a metal locating ring which must be discarded along with the guide.

11 Clean out the valve guide holes in the cylinder head with an 11.20 mm (0.44 in) reamer, clean away any swarf and lubricate the surface of each hole. Failure to keep the hole lubricated during insertion of the guide may result in damage to the guide or head. Ensure the cylinder head is well supported on a clean flat worksurface during insertion of the guides and is heated to the same temperature as for removal. Fit a new locating ring to each new guide and drive each guide into its hole until the locating ring abuts against its recess in the cylinder head. New valve guides, once installed, must be finished to the correct dimensions using a 5.5 mm (0.216 in) reamer. Be careful to remove all traces of swarf after reaming, and oil the valve guide bore before inserting the valve. Note that if a new valve guide is fitted, the valve seat must be recut to centre the seat with the guide axis, this task being described in paragraphs 14 – 18 of this Section.

12 If the valve face and seat are in good condition, with only light pitting, they may be restored by grinding-in. Commence by smearing a trace of fine valve grinding compound (carborundum paste) on the valve seat and apply a suction tool to the head of the valve. Oil the valve stem and insert the valve in the guide so that the two surfaces to be ground in make contact with one another. With a semi-rotary motion, grind in the valve head to the seat, using a backward and forward action. Lift the valve occasionally so that the grinding compound is distributed evenly. Repeat the application until an unbroken ring of light grey matt finish is obtained on both valve and seal. This denotes the grinding operation is now complete. Use only fine grinding compound, never coarse, and be careful to remove only the barest minimum of material necessary to achieve a good seating. Carefully wash off the valve and seat so that all traces of grinding compound are removed.

13 Measure the width of the valve seating ring, which should be 0.9 – 1.1 mm (0.035 – 0.043 in). If the ring is wider or narrower than this the seat must be recut. Also check that the ring is exactly in the centre of the valve face, as shown in the accompanying illustration; if too high or too low, the seat position must be altered, again by recutting, this task being described in the following paragraphs.

14 If the valve/valve seat contact area cannot be restored correctly by grinding-in, or if new valve guides have been fitted, the valve seats must be recut. This is a task requiring some skill and expensive equipment and should be entrusted, therefore, to a competent Suzuki dealer. If it is decided to attempt this task at home, obtain the Suzuki valve seat cutting set, Part Number 09916-21110, and proceed as follows.

15 Insert the solid pilot into the valve guide with a rotary motion until the pilot shoulder is about 10 mm (0.4 in) from the guide end, fit the 45° cutter attachment to the tool handle and place the cutter over the pilot. Using firm hand pressure, rotate the cutter through one or two full turns to clean the seat, then withdraw the cutter and examine the seat. If the seat is continuous and unmarked by pitting, proceed to the next step, but if pitting is still evident refit the cutter and repeat the procedure until all pitting has been removed. Be very careful to remove only the bare minimum of material necessary.

16 When all pitting has been removed, and a continuous seating ring obtained, remove the cutter and pilot. Coat the valve seat evenly and lightly with Engineers Blue (or fine valve grinding compound), oil the valve stem and attach a suction tool to the valve head. Insert the valve stem into the guide and press the valve on to its seat. Rotate the valve once (or more often if valve grinding compound is being used) and withdraw it. The marks left by the Engineers Blue or valve grinding compound should form a continuous unbroken ring that is of the correct width on both valve face and seat, and in the centre of the valve face, as described in paragraph 13 above.

17 If the contact area is too narrow, it may be widened by using the 45° cutter. If it is too wide, it may be narrowed by using either the 15° or 75° cutter; note that the latter two cutters will raise or lower the contact area on the valve face, and must be used as necessary. Remove only the minimum of material to prevent the valve from seating too deeply in the head. If this is allowed to happen, performance will be reduced, valve gear wear will be increased, and difficulties will be encountered when adjusting the valve clearances.

18 Once the correct valve seat width and position is achieved, lightly skim the seat surface with the 45° cutter in order to remove any burrs caused by the cutting procedure. The finished seat surface should be matt in appearance and have a smooth finish, thus providing the ideal surface for correct bedding in of the valve once the engine is started. It is not necessary to grind the valve in on completion of the recutting procedure. Be careful to wash off all traces of swarf and foreign matter once the seat recutting operation is finished.

19 Examine the condition of the valve collets and the groove on the valve stem in which they seat. If there is any sign of damage, new parts should be fitted. Check that the spring retaining plate is not cracked. If the collets work loose or the plate splits whilst the engine is running, a valve could drop into the cylinder and cause extensive damage.

20 Check the free length of each of the valve springs. The springs have reached their serviceable limit when they have compressed to the limit readings given in the Specifications Section of this Chapter.

21 Reassemble the valve and valve springs by reversing the dismantling procedure. Fit new oil seals to each valve guide and oil both the valve stem and the valve guide with a molybdenum disulphide based lubricant prior to reassembly. Take special care to ensure the valve guide oil seal is not damaged when the valve is inserted. As a final check after assembly, give the end of each valve stem a light tap with a hammer, to make sure the split collets have located correctly. Note that each spring must be fitted with its close coil end nearest the cylinder head.

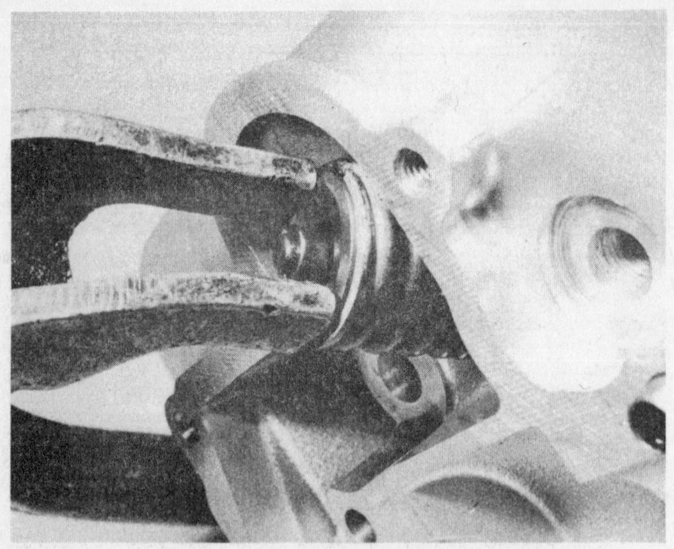

20.5 Apply valve spring compressor to release split collets

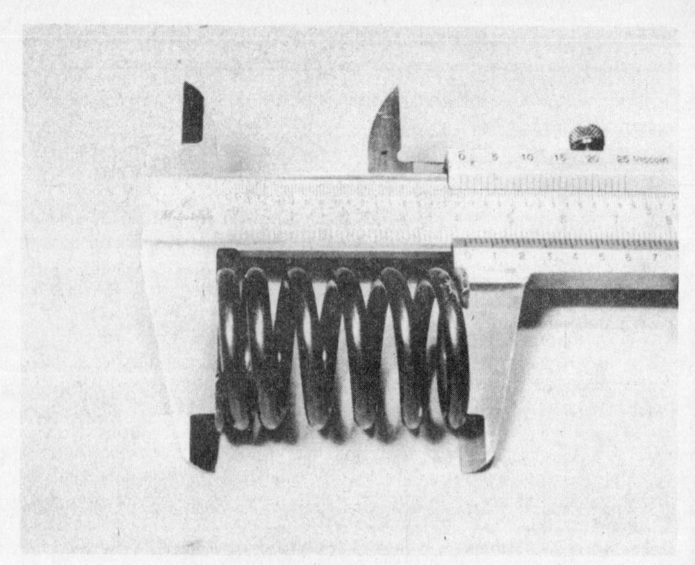

20.20 Measuring valve spring free length

20.21a Always renew valve guide oil seals

20.21b Refit valve spring seat as shown ...

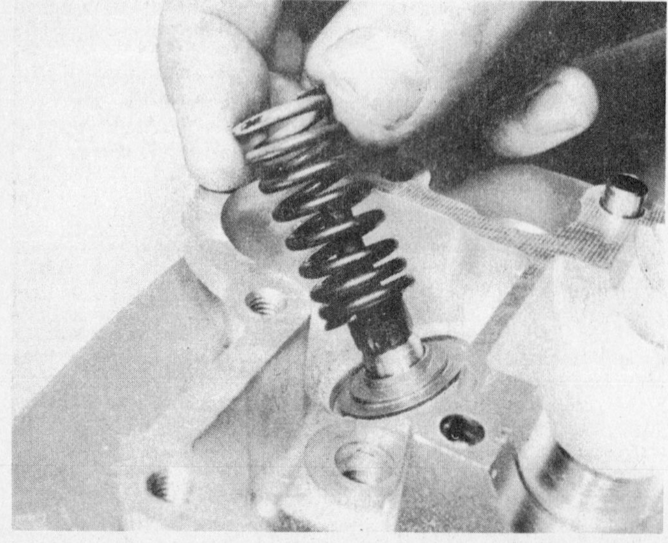

20.21c ... followed by valve inner spring ...

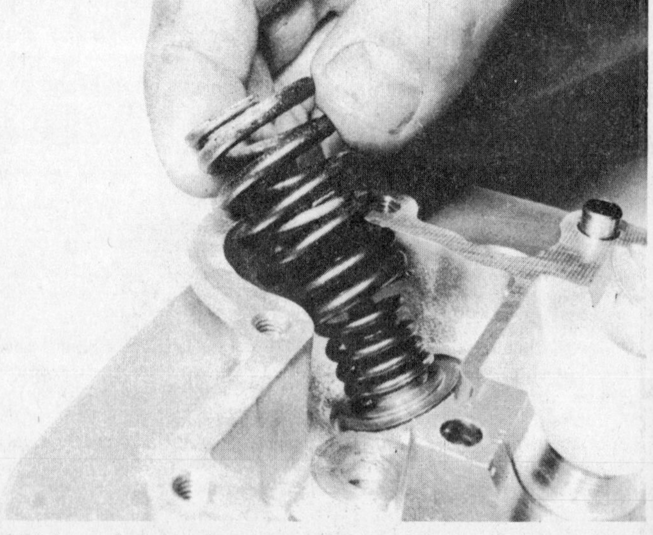

20.21d ... and outer spring – note close-spaced coils next to cylinder head

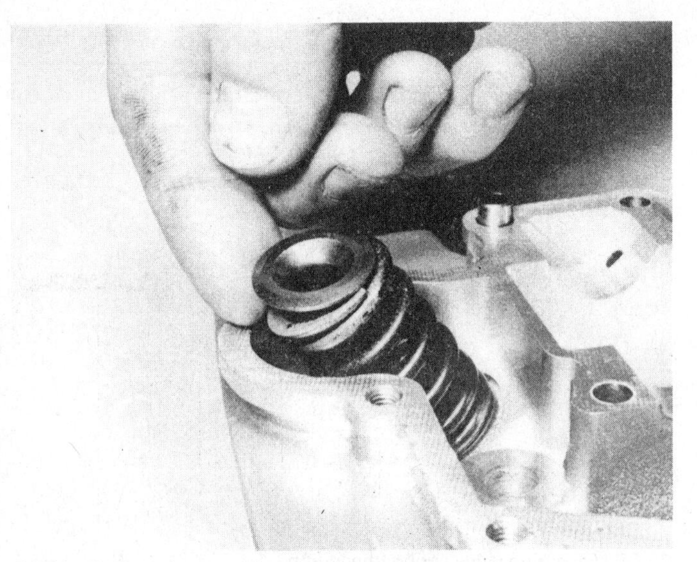

20.21e Refit spring retaining collar

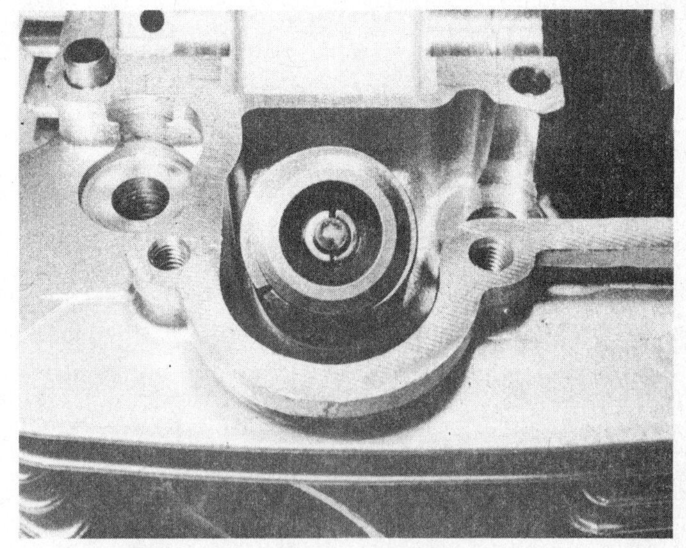

20.21f Check that split collets are correctly located

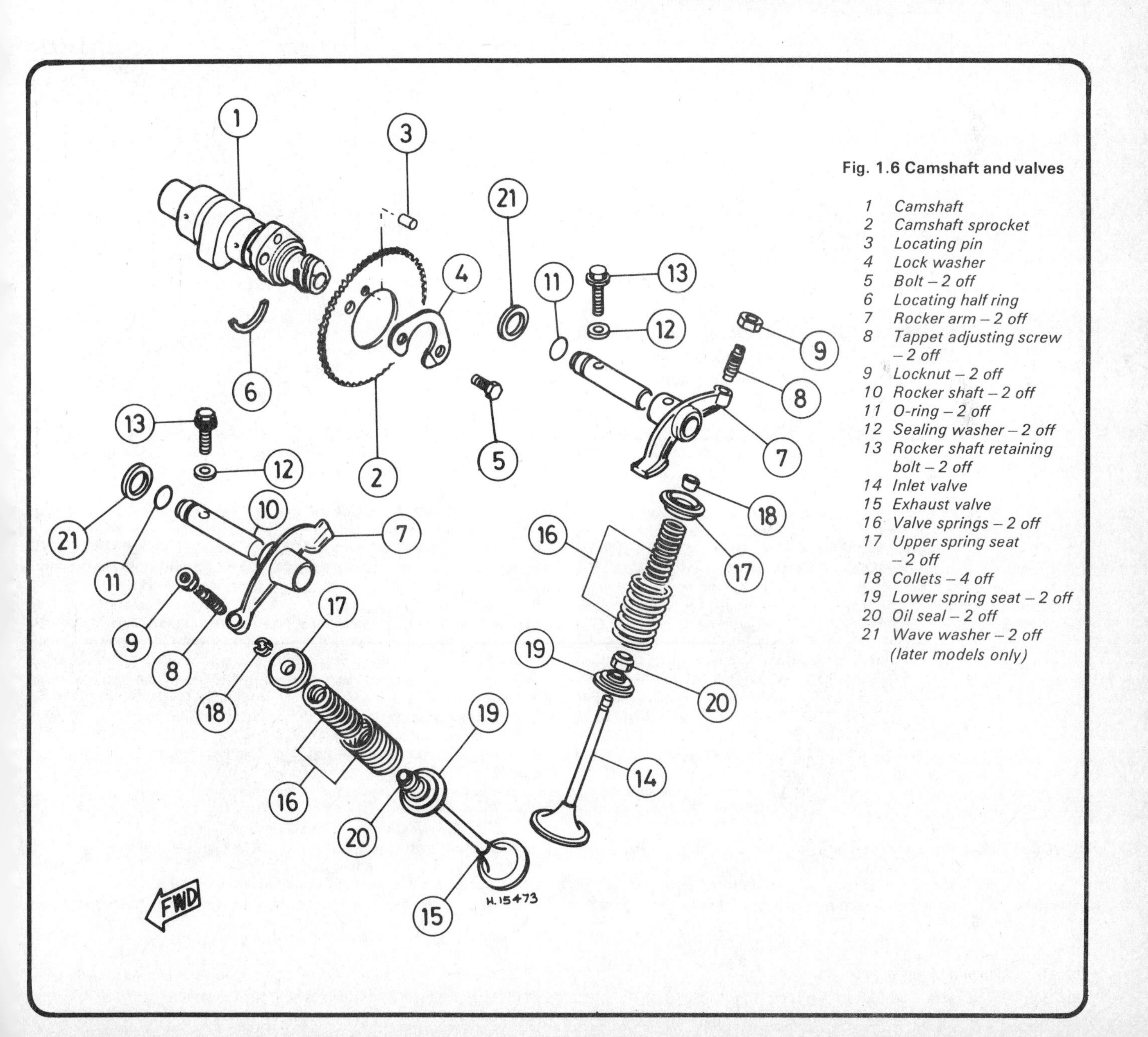

Fig. 1.6 Camshaft and valves

1. *Camshaft*
2. *Camshaft sprocket*
3. *Locating pin*
4. *Lock washer*
5. *Bolt – 2 off*
6. *Locating half ring*
7. *Rocker arm – 2 off*
8. *Tappet adjusting screw – 2 off*
9. *Locknut – 2 off*
10. *Rocker shaft – 2 off*
11. *O-ring – 2 off*
12. *Sealing washer – 2 off*
13. *Rocker shaft retaining bolt – 2 off*
14. *Inlet valve*
15. *Exhaust valve*
16. *Valve springs – 2 off*
17. *Upper spring seat – 2 off*
18. *Collets – 4 off*
19. *Lower spring seat – 2 off*
20. *Oil seal – 2 off*
21. *Wave washer – 2 off (later models only)*

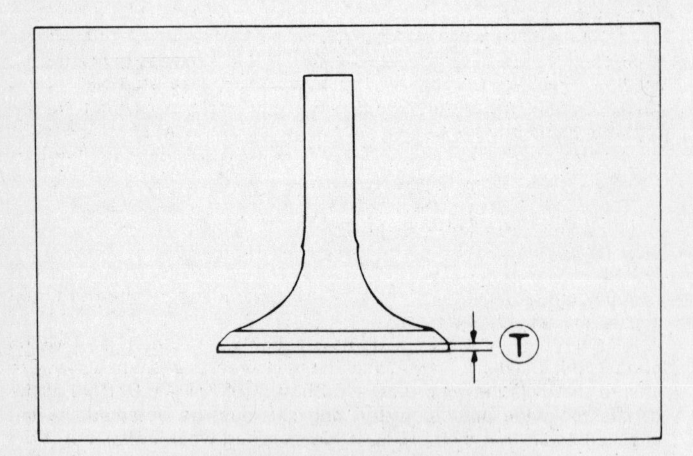

Fig. 1.7 Valve head thickness measurement

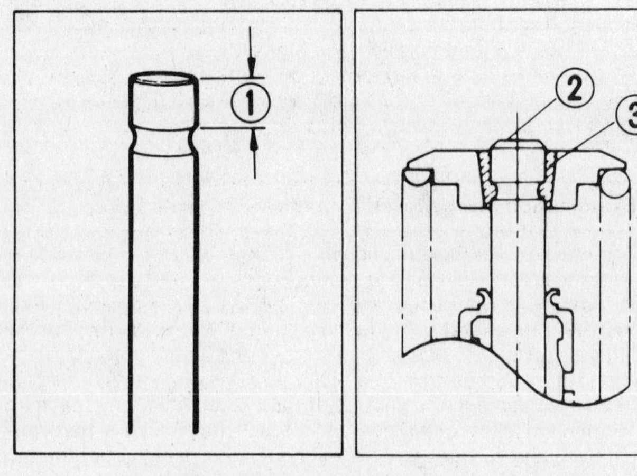

Fig. 1.8 Correct fitting of a refaced valve stem

1 *Not less than 3.8 mm (0.14 in) – early models, 2.6 mm (0.10 in) – late models*
2 *Stem tip*
3 *Collets*

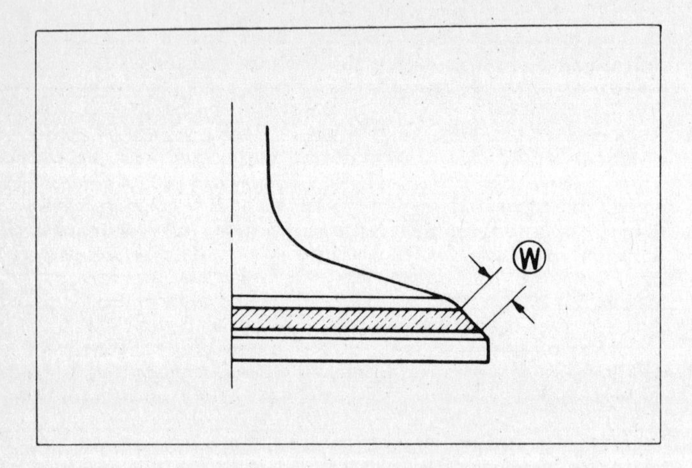

Fig. 1.9 Valve seating ring width measurement

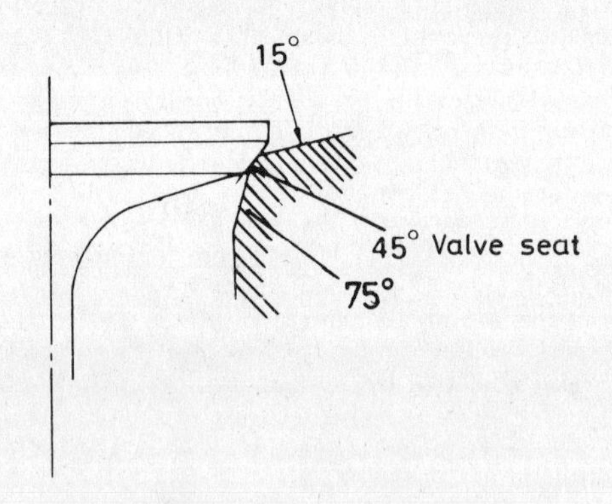

Fig. 1.10 Valve seat recutting angles

21 Examination and renovation: cylinder barrel

1 The usual indication of a badly worn cylinder bore and piston is excessive smoking from the exhaust, high crankcase compression which causes oil leaks, and piston slap, a metallic rattle that occurs when there is little or no load on the engine. If the top of the cylinder bore is examined carefully, it will be found that there is a ridge at the front and back, the depth of which will indicate the amount of wear which has taken place. This ridge marks the limit of travel of the top piston ring.
2 Since there is a difference in cylinder wear in different directions, side to side and back to front measurements should be made. Take measurements at three different points down the length of the cylinder bore, starting at a point just below the wear ridge and following this with measurements half way down the bore and at a point just above the lower edge of the bore. If any of these measurements exceed the service limit given in the Specifications Section of this Chapter, the cylinder must be rebored and fitted with an oversize piston. If the barrel has been rebored previously, do not forget to compensate for this by adding the total amount of the rebore to the cylinder standard bore size and service limit.
3 Oversize pistons and piston rings are available in two sizes of 0.5 mm (0.020 in) and 1.0 mm (0.040 in)
4 If measuring equipment is not available, a reasonable indication of bore wear can be obtained using a set of feeler gauges and a piston that is either new or is known from accurate measurements (See Section 22) to be within the specified wear limits. Insert the piston into the bore from the top, ensuring that the arrow mark cast on the piston crown is facing to the front of the barrel, and push it down until the base of the piston skirt is 20 – 25 mm ($\frac{3}{4}$ – 1 in) below the top surface of the bore. If it is possible to insert a feeler gauge of 0.12 mm (0.0047 in) thickness or greater between the forward side of the piston skirt (ie the thrust face) and the cylinder wall, the cylinder barrel should be considered worn and taken to a Suzuki dealer or similar expert for accurate measurement.
5 Check that the surface of the cylinder bore is free from score marks or other damage that may have resulted from an earlier engine seizure or a displaced gudgeon pin. A rebore will be necessary to remove any deep scores, irrespective of the amount of bore wear that has taken place, otherwise a compression leak will occur.
6 When fitting new piston rings to be run in a part-worn cylinder bore, the bore surface must be prepared first by glaze-busting. This is a process which involves the use of a cylinder-bore honing tool, usually in conjunction with an electric drill, to break down the surface glaze which forms on any cylinder bore in normal use. The prepared bore surface will then have a very lightly roughened finish which will assist the new piston rings to bed in rapidly and fully. Furthermore, the lip at the top of the bore, which will have been formed by the bore wearing, must be removed. If this is not done the new top ring will come into contact with the edge and shatter. Most motorcycle dealers have glaze-busting equipment and will be able to carry out the necessary work for a small charge.
7 Clean the cylinder barrel/gasket cylinder head surface and check it for distortion with a straight-edge and a set of feeler gauges. The straight-edge should be laid diagonally across the surface from corner to corner and any clearance between its lower edge and the surface checked with gauges. Repeat the procedure with the straight-edge laid between the opposite corners and then at several positions in between the four corners. If the largest clearance found exceeds 0.05 mm (0.002 in), Suzuki recommend that the cylinder block be replaced with a new item. It is worth noting however, that if the amount of distortion found is only slightly greater than the limit given, it could well be worth seeking the advice of a competent motorcycle engineer who can

advise on whether skimming the surface flat is possible without the subsequent risk of the piston coming into contact with the valve heads once the engine is reassembled and started.
8 Finally, make sure the external cooling fins of the cylinder block are not clogged with oil or road dirt which will prevent the free flow of air and cause the engine to overheat.

22 Examination and renovation: piston and piston rings

1 Attention to the piston and piston rings can be overlooked if a rebore is necessary, since new components will be fitted.
2 If a rebore is not necessary, examine the piston carefully. Reject pistons that are scored or badly discoloured as the result of exhaust gases by-passing the rings.
3 Remove all carbon from the piston crowns, using a blunt scraper, which will not damage the surface of the piston. A scraper made from soft aluminium alloy or hardwood is ideal; never use a hard metal scraper with sharp edges as this will almost certainly gouge the alloy surface of the piston. Clean away carbon deposits from the valve cutaways and finish off with metal polish so that a smooth, shining surface is achieved. Carbon will not adhere so readily to a polished surface.
4 Small high spots on the back and front areas of the piston can be carefully eased back with a fine swiss file. Dipping the file in methylated spirits or rubbing its teeth with chalk will prevent the file clogging and eventually scoring the piston. Only very small quantities of material should be removed, and never enough to interfere with the correct tolerances. Never use emery paper or cloth to clean the piston skirt; the fine particles of emery are inclined to embed themselves in the soft aluminium and consequently accelerate the rate of wear between bore and piston.
5 The piston outside diameter is measured at right angles to the gudgeon pin axis, at a point 15 mm (0.6 in) above the base of the piston skirt on all 'Z' models and at a point 12 mm (0.5 in) above the base of the skirt on all later models. If the piston is worn to 56.88 mm (2.239 in) or less, it must be renewed. The piston/cylinder clearance can be calculated by subtracting the piston diameter from the maximum bore measurement; if the clearance figure derived is 0.12 mm (0.0047 in) or greater, the barrel must be rebored and an oversize piston fitted.
6 Check that the gudgeon pin is a tight press fit in the piston; if any slackness is apparent, measure the gudgeon pin outside diameter and the inside diameter of each piston boss, renewing any component that is worn beyond the service limits set in the Specifications Section of this Chapter.
7 Discard the gudgeon pin circlips; these should never be re-used once they have been disturbed. Obtain new ones to be fitted on reassembly.
8 The piston ring/piston groove clearance must be checked, to ensure that the piston compression rings do not have excessive sidefloat. Either measure the clearance with the rings fitted to the piston, using feeler gauges, or measure the rings and the piston grooves and subtract the ring thickness figure from the groove width figure to calculate the clearance. If this clearance is greater than the service limit given in the Specifications Section of this Chapter, determine by careful measurement and comparison with the figures given which component is worn; if either the piston rings or the piston are worn to beyond the set service limits, they must be renewed.
9 Finally, check that the gudgeon pin circlip grooves are undamaged and that the piston rings and piston grooves are free from carbon; a short length of old piston ring ground to a chisel profile and fitted with a handle is ideal for scraping carbon deposits from piston ring grooves.
10 Piston ring wear is checked by measuring the end gap. First check the elasticity of each compression ring by measuring its free end gap, which must not be less than the service limit given in the Specifications Section of this Chapter, then insert each ring in turn into the cylinder bore having previously established that the bore is within the specified wear limits (see Section 21); the test results will not be accurate if the test is carried out in an excessively worn cylinder bore. Use the piston crown to position each ring squarely in the bore at about 10 mm ($\frac{1}{2}$ in) from the bottom. Using feeler gauges, measure the ring installed end gap which must not exceed 0.7 mm (0.028 in); if the end gap is excessive the piston rings must be renewed.
11 It is considered by many good practice to renew the rings as a matter of course, regardless of their apparent condition; this course is recommended, particularly if the machine has covered a high mileage. When ordering new pistons or rings, note that if the cylinder has been rebored, the oversize will be indicated by markings stamped on the piston crown, while oversize compression rings are stamped with a number on their upper edge adjacent to the end gap. The number 50 denotes an oversize ring of 0.5 mm whereas the number 100 denotes an oversize ring of 1.0 mm. Oversize oil ring spacers are identified by the following colour codes.

Red denotes 0.5 mm oversize
Yellow denotes 1.0 mm oversize

The oil ring side rail oversizes are determined by measurement across the outside diameter.
12 Do not assume when fitting new rings that their end gaps will be correct; the installed end gap must be measured as described above to ensure that it is between 0.10 – 0.25 mm (0.004 – 0.010 in); if the gap is too wide another piston ring set must be obtained (having checked again that the bore is within specified wear limits), but if the gap is too narrow it must be widened by the careful use of a fine file.

23 Examination and renovation: crankshaft assembly

1 The most likely areas of crankshaft failure are as follows. Main bearing failure, accompanied by a low rumbling noise as the engine is running, or big-end bearing failure, accompanied by a pronounced click or knock from the crankcases. In both the above cases the noise will increase gradually and will be accompanied by increasingly severe levels of vibration which will be felt through the frame and footrests. Excessive wear in the small-end bearing is revealed by an annoying light metallic rattle. Crankshaft wear is unlikely to take place until a very high mileage has been covered, or unless routine maintenance, in the form of regular oil and filter changes has been neglected.
2 Wash off the crankshaft to remove all traces of oil, then make a close visual inspection of the assembly, using a magnifying glass if necessary. If any obvious signs of damage are encountered, for example bent or distorted mainshafts or connecting rod, cracks or nicks in the connecting rod, damaged mainshaft threads, worn keyways or damaged cam chain sprocket teeth, take the assembly to a good Suzuki dealer for his advice; some damage may be reclaimed whereas in other cases the only solution will be the renewal of the crankshaft assembly.
3 Check for wear in the big-end bearing by arranging the connecting rod in the TDC position and by pushing and pulling the connecting rod. No discernible movement will be evident in an unworn bearing, but care must be taken not to confuse end float, which is normal, and bearing wear. If a dial gauge is readily available, a further test may be carried out by setting the gauge pointer so that it abuts against the upper edge of the periphery of the small-end eye. Measurement may then be taken of the amount of side-to-side deflection of the connecting rod. If this measurement exceeds the service limit of 3.0 mm (0.12 in) then the big-end bearing must be renewed. Check also the big-end side clearance; while a certain amount is intentional, it should not exceed 1.0 mm (0.039 in).
4 Any measurement of crankshaft deflection can only be made with the crankshaft assembly removed from the crankcase and set up on V-blocks which themselves have been positioned on a completely flat surface. The amount of deflection should be measured with a dial gauge at a point just inboard of the threaded portion of the left-hand mainshaft. If the amount of deflection shown by the gauge needle exceeds the service limit of 0.05 mm (0.002 in), then the assembly must be replaced with a serviceable item.
5 Check that the gudgeon pin is a tight press fit in the connecting rod small-end eye. If any play is detected measure the outside diameter of the gudgeon pin and the internal diameter of the small-end eye. If either is worn beyond the set limit, renewal of that component will be required.
6 If any fault is found or suspected in any of the components comprising the crankshaft assembly, it is recommended that the complete crankshaft assembly is taken to a Suzuki Service Agent, who will be able to confirm the worst, and supply a new or service-exchange assembly. The task of dismantling and reconditioning the big-end assembly is a specialist task, and it is considered to be beyond the scope and facilities of the average owner.

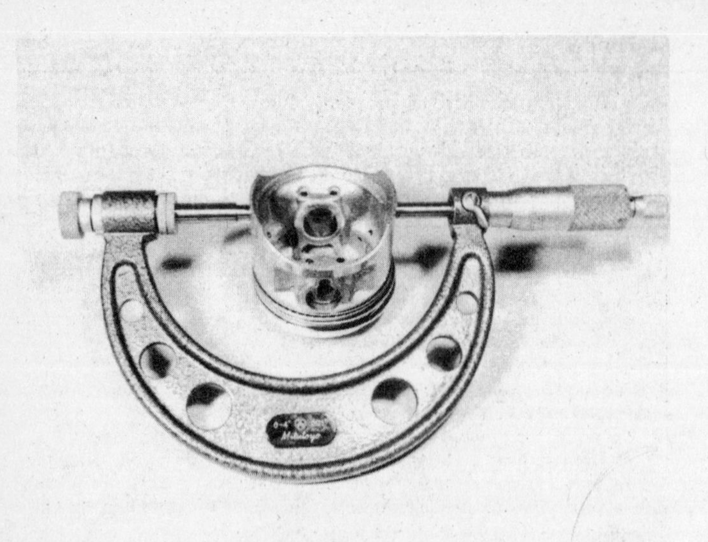

22.5 Measuring piston outside diameter

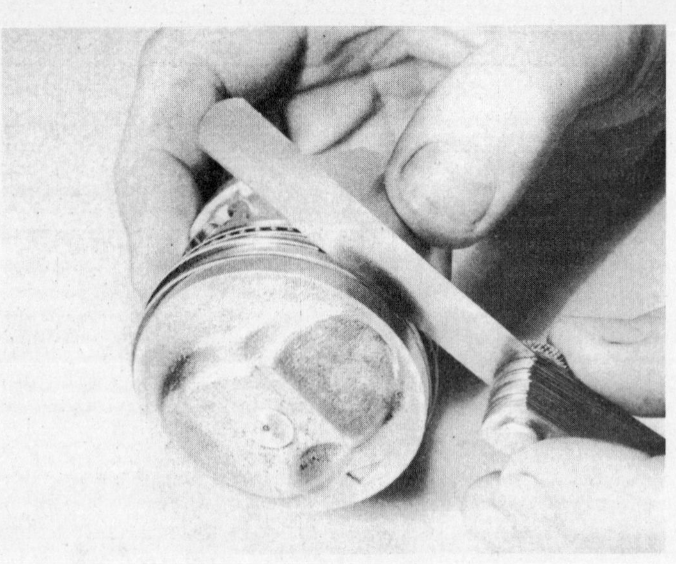

22.8a Measuring piston ring/ring groove clearance using feeler gauges

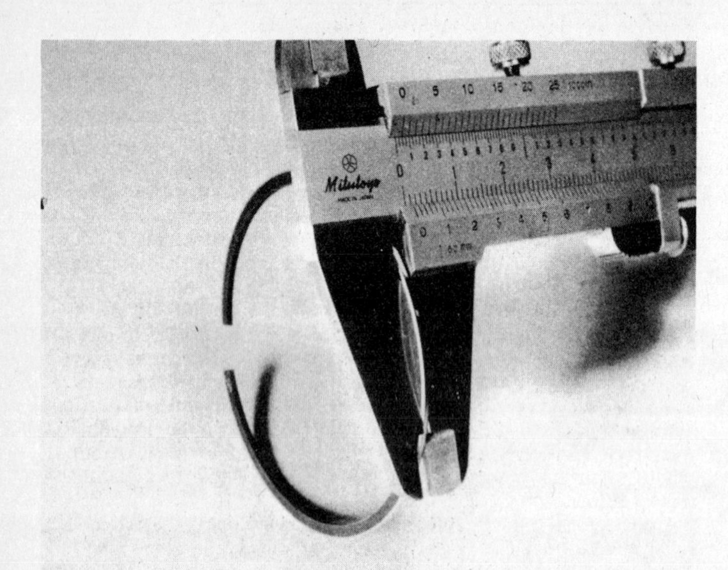

22.8b Measuring piston ring thickness

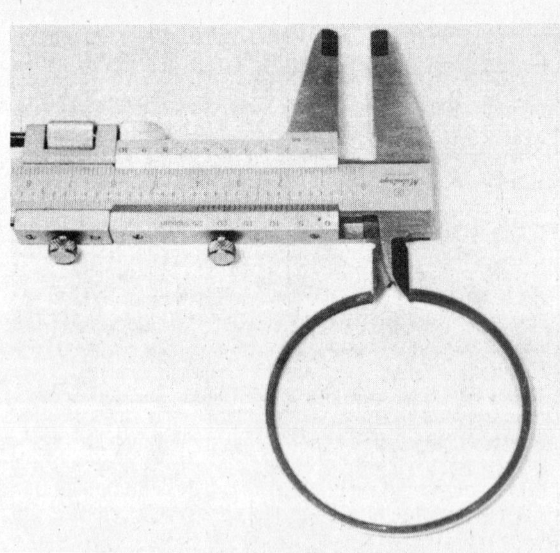

22.10a Measuring piston ring free end gap

22.10b Measuring piston ring installed end gap

23.3 Using feeler gauges to measure big-end side clearance

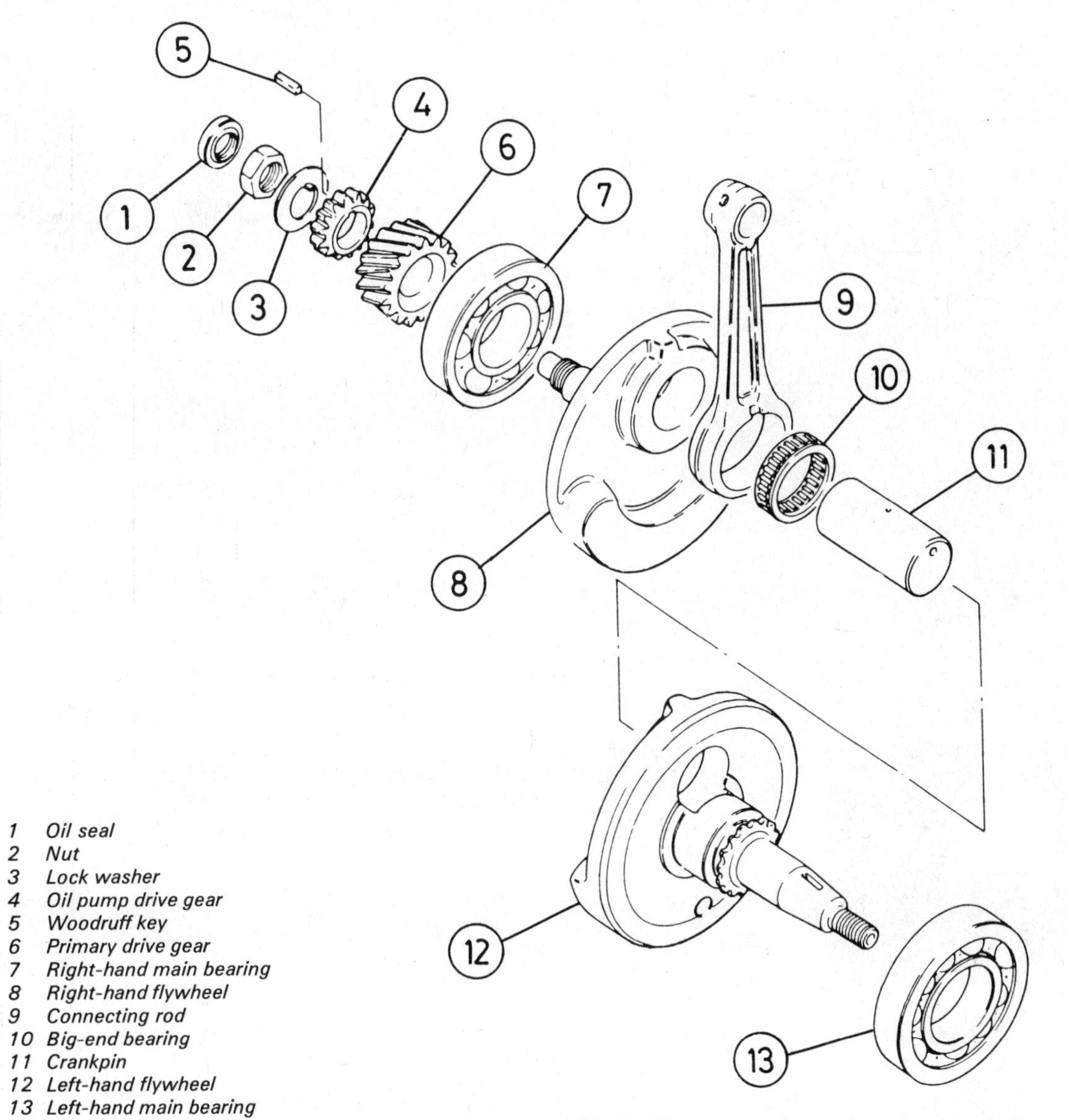

1 Oil seal
2 Nut
3 Lock washer
4 Oil pump drive gear
5 Woodruff key
6 Primary drive gear
7 Right-hand main bearing
8 Right-hand flywheel
9 Connecting rod
10 Big-end bearing
11 Crankpin
12 Left-hand flywheel
13 Left-hand main bearing

Fig. 1.11 Crankshaft

24 Examination and renovation: cam chain and tensioner components

1 Check the amount of wear in the cam chain by washing it thoroughly to remove all traces of oil, then laying it on a flat surface and pulling it tight. Mark any pin along the length of the chain, count off 21 pins and measure the distance between the two; if this distance exceeds 129.9 mm (5.114 in), the chain must be renewed.

2 Examine closely the chain, looking for signs of wear or damage such as cracked, broken, or missing rollers, fractured sideplates, and links that are either stiff or unduly sloppy in action; any such damage will mean that the chain must be renewed. Note also that if the chain was rattling before the engine was dismantled and could not be quietened by tensioner adjustment, or if the tensioner was found to be near the limit of its adjustment range, the chain should be renewed regardless of its apparent condition. It is considered by many a worthwhile precaution to renew the chain as a matter of course, especially if a high mileage has been covered.

3 Renew the camshaft sprocket if its teeth are hooked, chipped or otherwise worn, and examine closely the sprocket retaining bolts and lock washer. The bolt threads must be clean and in good condition, and the lock washer must be renewed if its locking tabs are weakened by excessive use.

4 The chain tensioner, like the chain, normally lives a trouble-free life. The mechanism should not, however, be neglected. Check the tensioner blade and the guide blade for wear and for separation of the rubber coating from the backing piece. No specifications are laid down for acceptable blade wear, but it is suggested that the components are renewed if wear has reduced the thickness of rubber to less than 50%. If the rubber has begun to separate from the blade, the component should be renewed as a matter of course.

5 Dismantle the tensioner assembly by slackening the locknut and removing the adjuster screw, preventing the tensioner pushrod from flying out under spring pressure. Renew as a matter of course the sealing O-ring fitted to the adjusting screw. Check that the threads of the adjuster screw and locknut are in good condition, renewing the screw, the locknut or the tensioner body if the threads are distorted through overtightening. Remove all traces of corrosion from all components.

6 Check that the tensioner pushrod is straight, then use a fine file and emery paper to polish away any burrs or raised edges. Check that the pushrod slides easily in the tensioner body. The tensioner spring can be checked only by comparison with a new component; renew it if there is the slightest doubt about its efficiency.

7 On reassembly, refit the spring to the pushrod and insert the pushrod into the tensioner body until the spring is fully compressed, then rotate as necessary the pushrod until the flat on its shank is facing the threaded hole for the adjuster screw. Check that the sealing O-ring has been renewed and is correctly fitted, then refit the adjuster screw and tighten it by just enough to retain securely the pushrod; do not overtighten the adjuster screw. Refit the plain washer and locknut.

24.4 Cam chain tensioner and guide blades must be unworn, as shown

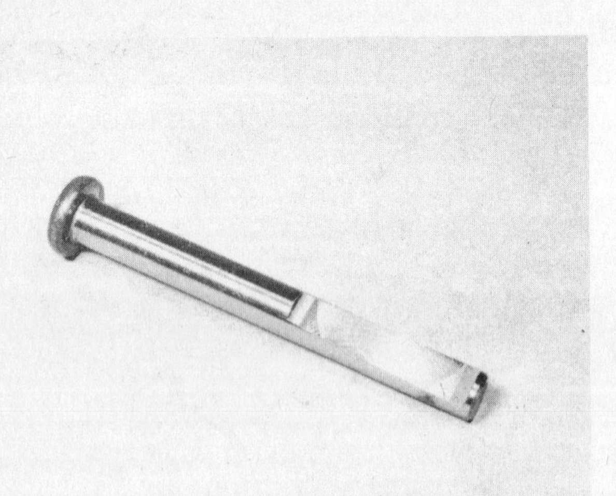

24.6 Tensioner pushrod flat surface must be clean and unmarked, as shown

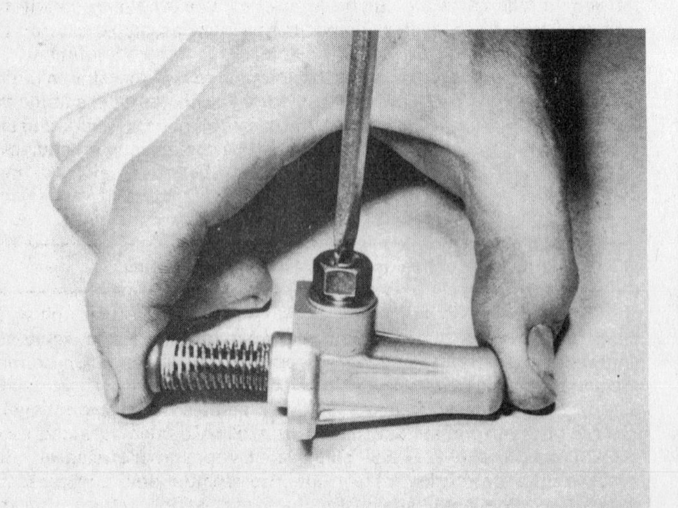

24.7 Reassembling tensioner assembly – do not overtighten adjuster screw

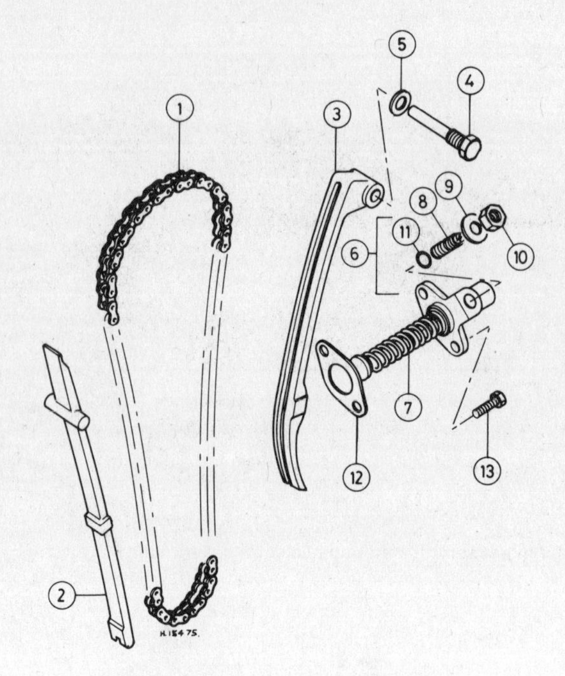

Fig. 1.12 Camchain and tensioner

1	*Camchain*	8	*Adjuster screw*
2	*Guide blade*	9	*Washer*
3	*Tensioner blade*	10	*Locknut*
4	*Mounting bolt*	11	*O-ring*
5	*Washer*	12	*Gasket*
6	*Tensioner adjuster*	13	*Bolt*
7	*Tensioner assembly*		

25 Examination and renovation: starter clutch and drive components – GS125 ES only

1 To check whether the starter clutch is operating correctly, fit the driven gear to the rear of the clutch and rotor. When rotated in a clockwise direction, as viewed from the gear side, the clutch should lock immediately, allowing power to be transmitted from the gear to the crankshaft. When rotated anti-clockwise, the gear should be free to run smoothly. If the movement is unsatisfactory, remove the gear from the clutch. The gear boss should be smooth; scoring or damage to the surface indicates that the rollers are similarly marked and require further inspection.

2 The rollers, springs and plungers may be removed for examination with the clutch still attached to the rear of the rotor. Using a small, flat-bladed screwdriver, carefully push each plunger back against spring pressure until the roller can be removed. Then remove the springs and plungers. Signs of wear will be obvious and will necessitate renewal of the worn or damaged parts.

3 To dismantle the clutch further, the three Allen bolts must be removed. The clutch outer unit and the shim can now be removed from the rear of the rotor. Examine the clutch outer unit for wear in the form of elongation or scoring of the roller housing. If badly worn, the clutch outer unit must be renewed. Check the condition of the roller bearing in the centre of the gear boss by refitting the gear to the crankshaft and feeling for free play; if any is felt, the roller bearing must be renewed. Similarly check the fit of the idler gear on its shaft, and the fit of the shaft in the crankcase. Check also that the teeth of the drive gear, idler gear and driven gear are undamaged and in good condition.

4 On reassembly, refit the shim and clutch outer unit to the rear of the rotor, ensuring that each is located correctly. Thoroughly degrease the threads of the three Allen bolts, apply a few drops of thread locking compound to each, then refit the screws, tightening them to a torque setting of 1.5 – 2.0 kgf m (11 – 14.5 lbf ft). Refit the springs, plungers and rollers, using the reverse of the dismantling procedure, then refit the driven gear to the rear of the rotor, rotating it as necessary to engage the rollers on the gear boss. Keep the rotor/clutch assembly and gear together to avoid losing the springs, plungers and rollers.

25.4a Insert springs into bores in clutch outer unit ...

25.4b ... followed by the plungers

25.4c Press plunger back with small screwdriver while roller is removed or refitted

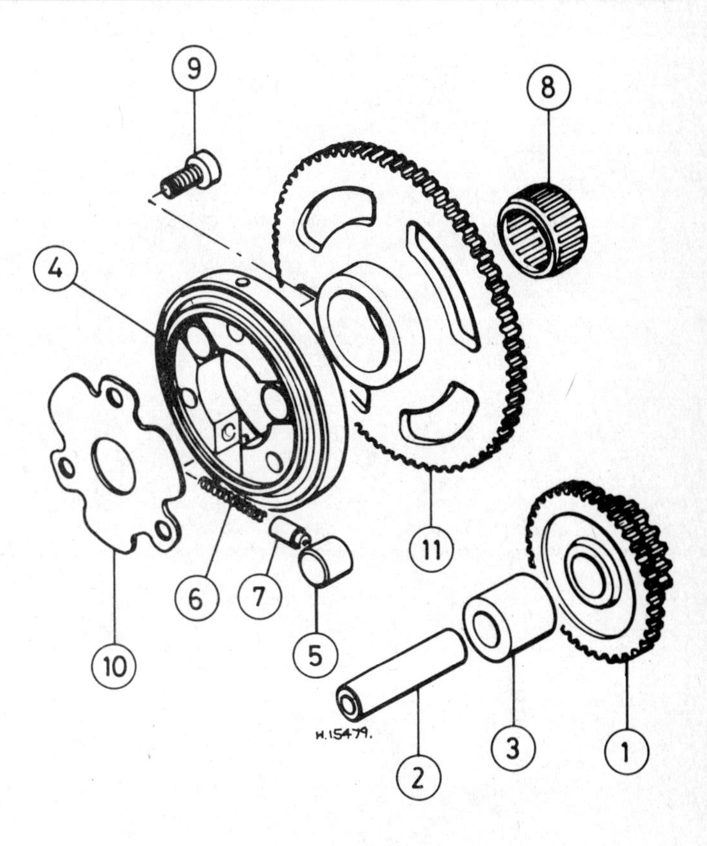

Fig. 1.13 Starter clutch

1	*Idler gear*	7	*Plunger – 3 off*
2	*Idler gear shaft*	8	*Needle roller bearing*
3	*Spacer*	9	*Allen bolt – 3 off*
4	*Clutch outer unit*	10	*Shim*
5	*Roller – 3 off*	11	*Driven gear*
6	*Spring – 3 off*		

26 Examination and renovation: primary drive gears

1 The primary driven gear pinion is riveted to the rear of the clutch outer drum and incorporates six small damper units to act as a transmission shock absorber. Check the condition of the dampers by attempting to rotate the clutch outer drum while holding steady the primary driven gear pinion; if movement is excessive or sloppy, the complete assembly must be renewed.

2 Examine the primary drive gears (and oil pump driven gear), looking for chipped, broken, or worn teeth. If such damage is found the gears must be renewed, but note that matched pairs of gears must be renewed always as a single unit, to avoid the excessive wear and noise that would result from running a part-worn gear with a new component.

27 Examination and renovation: clutch assembly

1 The plain and friction plates should all be given a thorough wash in a petrol/paraffin mix to remove all traces of friction material debris and oil sludge; follow this by cleaning the inside of the clutch drum in a similar manner.

2 The obvious sign of the clutch friction plates having worn beyond their service limit is the advent of clutch slip. To check the degree of wear present on the friction plates, measure the thickness of each plate across the faces of the bonded linings and compare the measurement obtained with the information given in the Specifications Section of this Chapter. If any plate has worn below the service limit of 2.6 mm (0.10 in), then the plates must be renewed as a complete set.

3 Check the condition of the tongues around the periphery of each

friction plate, at the same time checking the slots in the clutch drum wall. In an extreme case, clutch chatter may have caused the plate tongues to make indentations in the slots; these indentations will trap the plates as they are freed, thereby impairing clutch action. If the damage found is only slight, the indentations can be removed by careful work with a fine file and any burrs removed from the plate tongues in a similar fashion. More extensive damage will necessitate renewal of the parts concerned. Note that there is a definite limit to the amount of material that can be removed from each plate tongue, the minimum tongue width allowable being 11.0 mm (0.43 in).

4 Check the clutch plain plates for any signs of warpage. This can be achieved by laying each plate on a completely flat surface, such as a sheet of plate glass, and attempting to pass a feeler gauge between the plate and the surface. Both the plate and the surface must be cleaned of all contamination. The maximum allowable warpage for each plain plate is 0.1 mm (0.004 in).

5 The plain plates should be free from scoring and any signs of overheating, which will be apparent in the form of blueing. Check the condition of both the tongues in the inner periphery of each plain plate and the slots of the clutch centre. Any slight damage found on either of these components should be removed by using a method similar to that described for the friction plates and clutch drum. The final check for the plain plates is to measure each plate for thickness.

6 Examine the clutch pressure plate, clutch centre, and the clutch outer drum for wear or damage of any sort. The likelihood of major components such as these wearing to any great extent is unlikely, but the possibility must be borne in mind in order not to overlook a fault. The only remedy for damage or wear is the renewal of the part concerned.

7 The clutch springs do not wear, but over an extended mileage will lose tension and settle to a shorter length, thus reducing the pressure on the plates and promoting clutch slip. Measure the free length of each spring. If any has settled to a length of 29.5 mm (1.16 in) or less, the springs must all be renewed as a complete set.

8 The clutch release mechanism components should be examined for signs of wear, particularly if clutch engagement and disengagement feel rough or jerky. Check that the thrust bearing is in good condition, renewing it if there is any damage at all to the rollers or cage. The pushrod must be straight and its ends must be unworn with no signs of excessive heat (blueing), and the clutch release shaft must be unworn and free from corrosion. Renew any component that is worn or damaged, not forgetting the pushrod supporting bush set in the input shaft left-hand end.

28 Examination and renovation: kickstart assembly – GS125 and DR125 S only

1 Wear on the kickstart components is only likely after a high mileage has been covered, a damaged or fatigued return spring being the most likely cause of problems.

2 Lay out all components and inspect each item, looking for worn or chipped gear teeth, worn splines and worn bearing surfaces. Any component found to be damaged or worn must be renewed; if there is any doubt about the condition of the kickstart return spring, it must be renewed.

29 Examination and renovation: gearbox components

1 Examine each of the gear pinions to ensure that there are no chipped or broken teeth and that the dogs on the end of the pinions are not rounded. Gear pinions with these defects must be renewed; there is no satisfactory method of reclaiming them. If damage or wear warrants renewal of any gear pinions the assemblies may be stripped down, displacing the various shims and circlips as necessary.

2 If dismantling of the input shaft assembly proves to be necessary, the 2nd gear pinion will have to be pressed from position by using a hydraulic press; no other method of removal is possible. As it is unlikely that this type of tool will be readily available, it is recommended that the complete shaft assembly be returned to an official Suzuki service agent who will be able to remove the pinion, renew any worn or damaged components and return the shaft assembly complete.

3 If a hydraulic press is available and it is decided to attempt removal of the 2nd gear pinion from the input shaft, it is very important to realise fully the dangers involved when using such a tool. Both the tool and the shaft assembly must be set up so that there is no danger of either item slipping. The tool must be correctly assembled in accordance with the maker's instructions as the force exerted by the tool is considerable and perfectly capable of stripping any threads from holding studs or inflicting other damage upon itself and the mainshaft. Always wear proper eye protection in case a component should fail and shatter before it becomes free, as will happen if the component is flawed. Note that the 2nd gear pinion can be removed and refitted only twice; to ensure reliability the input shaft and pinion must be renewed if it is necessary to dismantle the shaft assembly a third time.

4 The accompanying illustrations show how both clusters of the gearbox are assembled on their respective shafts. It is imperative that the gear clusters, including the thrust washers, are assembled in **exactly** the correct sequence, otherwise constant gear selection problems will occur. In order to eliminate the risk of misplacement, make rough sketches as the clusters are dismantled. Also strip and rebuild as soon as possible to reduce any confusion which might occur at a later date.

5 Examine the selector forks carefully, ensuring that there is no sign of scoring on the bearing surface of either their claw ends, their bores or their selector drum guide pins. Check for any signs of cracking around the edges of the bores or at the base of the fork arms. Refer to the Specifications Section at the beginning of this Chapter and measure the thickness of the fork claw ends; renew the fork if the measurement obtained is less than the limit given.

6 Place each selector fork in its respective pinion groove and using a feeler gauge, measure the claw end to pinion groove clearance. If the measurement obtained exceeds the given service limit of 0.5 mm (0.02 in), then it must be decided whether it is necessary to renew one or both components. The acceptable limits for pinion groove width are given in the Specifications Section of this Chapter.

7 Check each selector fork shaft for straightness by rolling it on a sheet of plate glass and checking for any clearance between the shaft and the glass with feeler gauges. A bent shaft will cause difficulty in selecting gears. There should be no sign of any scoring on the bearing surface of the shaft or any discernible play between each shaft and its selector fork(s).

8 The tracks in the selector drum should not show signs of undue wear or damage. Check also that the selector drum bearing surfaces are unworn; renew the drum if it is worn or damaged.

9 Note that certain pinions have a bush fitted within their centres. If any one of these bushes appears to be over-worn or in any way damaged, then the pinion should be returned to an official Suzuki service agent, who will be able to advise on which course of action to take as to its renewal.

10 Finally, carefully inspect the splines on both shafts and pinions for any signs of wear, hairline cracks or breaking down of the hardened surface finish. If any one of these defects is apparent, then the offending component must be renewed. It should be noted that damage and wear rarely occur in a gearbox which has been properly used and correctly lubricated, unless very high mileages have been covered.

11 Clean the gearbox sprocket thoroughly and examine it closely, paying particular attention to the condition of the teeth. The sprocket should be renewed if the teeth are hooked, chipped, broken or badly worn. It is considered bad practice to renew one sprocket on its own; both drive sprockets should be renewed as a pair, preferably with a new final drive chain. Examine the splined centre of the sprocket for signs of wear. If any wear is found, renew the sprocket as slight wear between the sprocket and shaft will rapidly increase due to the torsional forces involved. Remember that the output shaft will probably wear in unison with the sprocket, it is therefore necessary to carry out a close inspection of the shaft splines.

12 Carefully examine the gear selector components. Any obvious signs of damage, such as cracks, will mean that the part concerned must be renewed. Check that the springs are not weak or damaged and examine all points of contact eg selector ratchet assembly or neutral detent plunger and selector stop arm/selector drum, for signs of excessive wear. If doubt arises about the condition of any component it must be compared with a new part to assess the amount of wear that has taken place, and renewed if found to be damaged or excessively worn. Pay careful attention to the springs, plungers and pawls of the selector ratchet assembly.

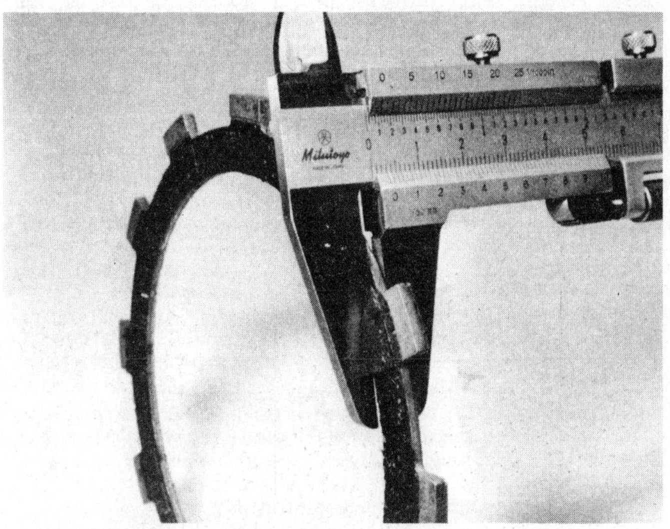

27.2 Measuring friction plate thickness

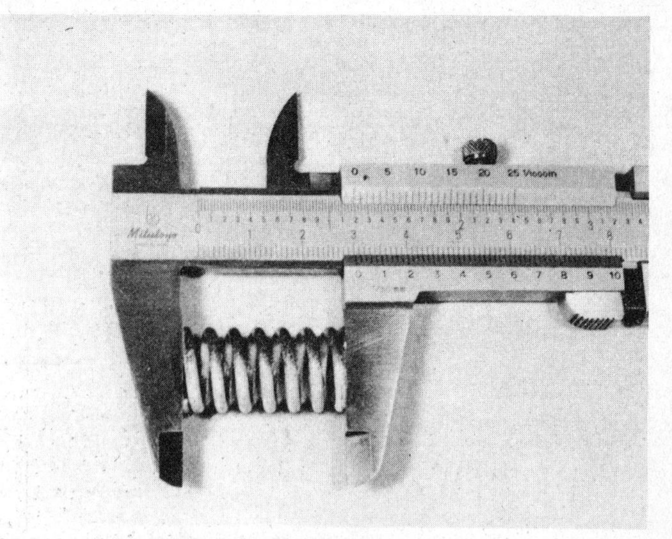

27.7 Measuring clutch spring free length

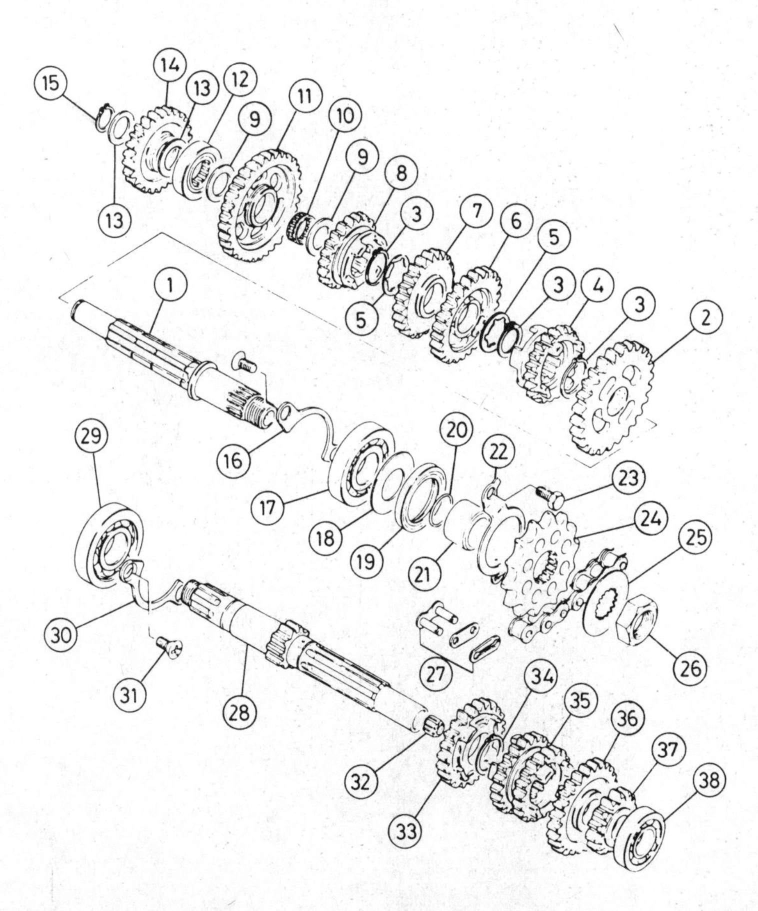

Fig. 1.14 Gearbox shafts

1 *Output shaft*
2 *Output shaft 2nd gear pinion*
3 *Circlip – 3 off*
4 *Output shaft 6th gear pinion (DR model), sliding dog (GS models)*
5 *Splined thrust washer – 2 off*
6 *Output shaft 3rd gear pinion*
7 *Output shaft 4th gear pinion*
8 *Output shaft 5th gear pinion*
9 *Thrust washer – 2 off*
10 *Needle roller bearing*
11 *Output shaft 1st gear pinion*
12 *Output shaft right-hand bearing*
13 *Thrust washer*
14 *Kickstarter idler gear*
15 *Circlip*
16 *Bearing retaining plate*
17 *Output shaft left-hand bearing*
18 *Washer*
19 *Oil seal*
20 *O-ring*
21 *Spacer*
22 *Retaining plate*
23 *Bolt – 2 off*
24 *Final drive sprocket*
25 *Lock washer*
26 *Nut*
27 *Final drive chain master link*
28 *Input shaft and 1st gear pinion*
29 *Input shaft right-hand bearing*
30 *Bearing retaining plate*
31 *Screw – 2 off*
32 *Bush*
33 *Input shaft 5th gear pinion*
34 *Circlip*
35 *Input shaft 3rd/4th gear pinion*
36 *Input shaft 6th gear pinion – DR model only*
37 *Input shaft 2nd gear pinion*
38 *Input shaft left-hand bearing*

30 Gearbox input and output shafts: reassembly

1 Having examined and repaired or renewed the gearbox components as necessary, the gear clusters can be built up ready for fitting as a single unit to the crankcase left-hand half

2 Refer to the line drawings and photographs accompanying this text when assembling the various components and ensure that the bearing surfaces of each component are liberally oiled before fitting. Note that where the terms 'left-hand' and 'right-hand' are used, these refer to the left-hand and right-hand of each component as it would be when installed in the machine.

3 Take the bare input shaft with its integral 1st gear pinion, hold the shaft by its threaded right-hand end and place the 5th gear pinion over the shaft left-hand end with its selector dogs facing to the left. Slide the pinion down the length of the shaft until it butts against the 1st gear pinion and secure it with a circlip. The double 3rd/4th gear pinion is fitted with the 21 tooth 4th gear adjacent to the 5th gear pinion. On DR125 S models only, fit the 6th gear pinion with its selector dogs facing to the right. The 2nd gear pinion must now be fitted.

4 Thoroughly degrease both the input shaft left-hand end and the 2nd gear pinion. Apply Suzuki Lock Super 1303B (Part Number 99000-32030) or a similar locking compound, such as Loctite Bearing Fit, to the internal surface of the pinion and push it as far as possible over the shaft end and down the length of the shaft. The pinion must be fitted so that the protruding shoulder faces to the right. A means

must now be found of pressing the pinion on to the shaft to a precise position. While this can be achieved using a hammer and a long tubular drift to tap the pinion carefully into place, the method employed in practice was as follows. A vice with a jaw opening of at least 8 inches must be found. Place a soft alloy or wooden cover over one of the vice jaws and place the threaded right-hand end of the shaft against this cover. Slip a socket or tubular drift of suitable length and diameter over the left-hand end of the shaft and position this against the other vice jaw. With the socket or tubular drift bearing only on the surface of the pinion, tighten the vice until the outer faces of the 1st and 2nd gear pinions are precisely 87.8 – 88.0 mm (3.457 – 3.465 in) apart on GS125 and GS125 ES models and 88.0 – 88.1 mm (3.465 – 3.469 in) apart on DR125 S models. This measurement must be checked frequently using a micrometer or vernier gauge as shown in the accompanying photograph.

5 On DR125 S models only, ensure that no locking compound is allowed into contact with the 6th gear pinion and spin the pinion by hand to check that it is quite free to rotate. If the pinion does become locked, the 2nd gear pinion must be removed again for the components to be cleaned and refitted correctly. Remember that the 2nd gear pinion can be removed and refitted only twice before the input shaft and pinion must be renewed.

6 Hold the bare output shaft by its threaded left-hand end, place the 2nd gear pinion over the right-hand end with the pinion flat surface facing to the left, and slide the pinion down the shaft to rest against the shaft shoulder. Secure the 2nd gear pinion with a circlip. The next component is the 6th gear pinion (DR125 S models) or the sliding dog (GS125, GS125 ES models), which is fitted against the 2nd gear pinion with its selector fork groove facing to the right. Place a circlip in the groove nearest to the 6th gear pinion/sliding dog and slide a splined thrust washer down the shaft to rest against the circlip.

7 The 3rd gear pinion is fitted with its deeply-recessed face to the left and is followed by the 4th gear pinion which is fitted with its deeply-recessed face to the right. These latter two pinions are located by a splined thrust washer and retained by a circlip. The 5th gear pinion is then fitted with its selector fork groove facing to the left. The large, plain thrust washer is then fitted against the shaft splined shoulder, followed by the needle roller bearing, over which is fitted the 1st gear pinion.

8 A second large, plain thrust washer is fitted against the 1st gear pinion and is followed by the bush which forms the centre of the output shaft right hand (needle roller) bearing. Tighten a rubber band over the shaft right-hand end to secure the components.

9 Where applicable, check that all pinions are free to rotate or to slide easily, check that all bearing surfaces have been well lubricated. and put the completed gear cluster to one side to await reassembly.

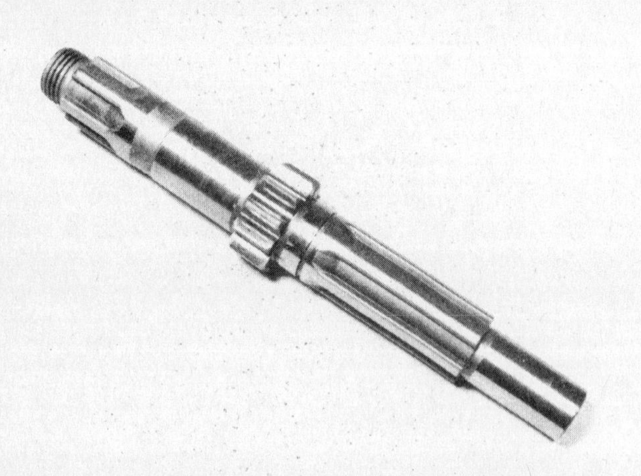

30.3a Input shaft is identified by integral 1st gear pinion

30.3b 5th gear pinion is fitted in direction shown ...

30.3c ... and is secured by a circlip

30.3d 3rd/4th gear pinion is fitted with 4th gear to the right

30.4a Apply locking compound to bore of 2nd gear pinion before refitting

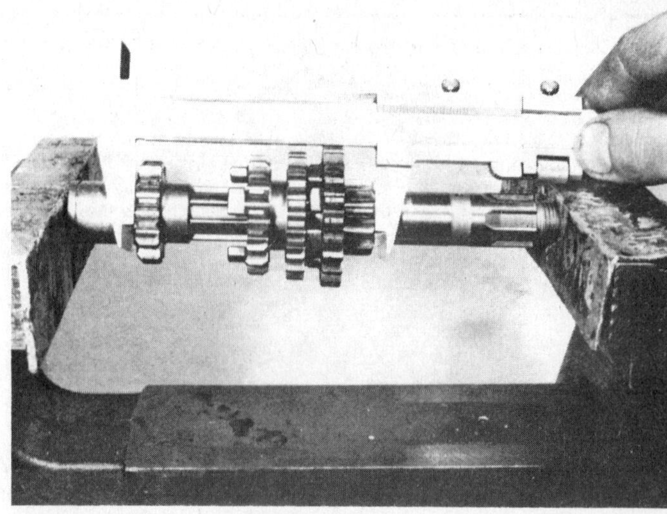

30.4b Use vice as shown to press pinion to correct fitted distance from 1st gear ...

30.4c ... using suitable size socket to bear on pinion – note soft alloy jaw covers

30.6a The output shaft

30.6b 2nd gear pinion is fitted as shown ...

30.6c ... and is secured by a circlip

30.6d Sliding dog (GS125 models only) is fitted as shown (DR125 S gear pinion similar)

30.6e Fit circlip in groove as shown ...

30.6f ... and slide a splined thrust washer against it

30.7a Fit 3rd gear pinion in direction shown ...

30.7b ... followed by 4th gear pinion

30.7c 3rd and 4th gear pinions are located by a second splined thrust washer ...

30.7d ... and secured by a circlip

30.7e 5th gear pinion is fitted as shown

30.7f First large thrust washer fits against shaft splined shoulder

30.7g Oil liberally needle roller bearing before refitting

30.7h 1st gear pinion fits over roller bearing

30.8a Fit second large thrust washer against 1st gear pinion

30.8b Do not omit output shaft right-hand bearing centre bush

31 Engine reassembly: general

1 Before reassembly of the engine/gearbox unit is commenced, the various component parts should be cleaned thoroughly and placed on a sheet of clean paper, close to the working area.
2 Make sure all traces of old gaskets have been removed and that the mating surfaces are clean and undamaged. Great care should be taken when removing old gasket compound not to damage the mating surface. Most gasket compounds can be softened using a suitable solvent such as methylated spirits, acetone or cellulose thinner. The type of solvent required will depend on the type of compound used. Gasket compound of the non-hardening type can be removed using a soft brass-wire brush of the type used for cleaning suede shoes. A considerable amount of scrubbing can take place without fear of harming the mating surfaces. Some difficulty may be encountered when attempting to remove gaskets of the self-vulcanising type, the use of which is becoming widespread, particularly as cylinder head and base gaskets. The gasket should be pared from the mating surface using a scalpel or a small chisel with a finely honed edge. Do not, however, resort to scraping with a sharp instrument unless necessary.
3 Gather together all the necessary tools and have available an oil can filled with clean engine oil. Make sure that all new gaskets and oil seals are to hand, also all replacement parts required. Nothing is more frustrating than having to stop in the middle of a reassembly sequence because a vital gasket or replacement has been overlooked. As a general rule each moving engine component should be lubricated thoroughly as it is fitted into position.
4 Make sure that the reassembly area is clean and that there is adequate working space. Refer to the torque and clearance setting wherever they are given. Many of the smaller bolts are easily sheared if overtightened. Always use the correct size screwdriver bit for the cross-head screws and never an ordinary screwdriver or punch. If the existing screws show evidence of maltreatment in the past, it is advisable to renew them as a complete set.

32 Reassembling the engine/gearbox unit: preparing the crankcases

1 At this stage the crankcase castings should be absolutely clean and dry with any damage, such as worn threads, repaired. If any bearings are to be refitted, the crankcase casting must be heated first, as described in Section 15.
2 Place the heated casting on a wooden surface, ensuring that it is fully supported around the bearing housing. When refitting the output shaft left-hand bearing, remember to insert first the large plain thrust washer into the bearing housing. Position the bearing on the casting, ensuring that it is absolutely square to its housing then tap it fully into place using a hammer and a tubular drift such as a socket spanner which bears only on the bearing outer race. Be careful to ensure that the bearing is kept absolutely square to its housing at all times.
3 Oil seals are fitted into a cold casing in a similar manner. Apply a thin smear of grease to the seal circumference to aid the task, then tap the seal into its housing using a hammer and a tubular drift which bears only on the hard outer edge of the seal, thus avoiding any risk of the seal's being distorted. Tap each seal into place until its flat outer surface is just flush with the surrounding crankcase.
4 Where retaining plates are employed to secure bearings or oil seals, thoroughly degrease the threads of the mounting screws, apply a few drops of thread locking compound to them, and tighten them securely.
5 When all bearings and oil seals have been fitted and secured, lightly lubricate the bearings with clean engine oil and apply a thin smear of grease to the sealing lips of each seal.
6 Support the crankcase left-hand half on two wooden blocks placed on the work surface; there must be sufficient clearance to permit the crankshaft and output shaft to be fitted.

32.4 Do not forget to refit retaining plates (where fitted)

33 Reassembling the engine/gearbox unit: refitting the crankshaft and gearbox components

1 Refit temporarily the primary drive gear retaining nut to the crankshaft right-hand end, remembering that a left-hand thread is employed, and thread the nut down until it is flush with the crankshaft end to prevent damage or distortion during crankshaft fitting. Push the crankshaft as far as possible into the left-hand main bearing, using a smear of oil to ease the task, check that it is absolutely square to the crankcase, and tap it fully into place using only a soft-faced mallet. Remove the protecting nut.
2 Excessive force will not be required to fit the crankshaft; a few firm taps should be sufficient. If any difficulty is encountered do not risk damaging the crankshaft by resorting to excessive force, take the components to a Suzuki dealer for the crankshaft to be drawn into place using the correct service tool.
3 Fit the assembled gear clusters together, ensuring that all pinions are correctly meshed, and lower the assembly as a single unit into the crankcase half. Use a soft-faced mallet to tap the output shaft into its bearing, noting that force will not be necessary. Check that both shafts are fully in place and free to rotate.
4 Engage the selector fork on the input shaft double 3rd/4th gear pinion, using the marks or notes made on dismantling to identify the fork and placing it so that its longer, webbed, boss faces to the left.
5 Refit the plain washer to the selector drum left-hand end, insert the key into the selector drum keyway and place the selector stop arm camplate on the drum so that the camplate plain face faces outwards. Align the camplate keyway with the key and push the camplate into

position against the plain washer. Insert the selector drum into its bearing in the crankcase left-hand half and rotate it so that the recess machined to accept the neutral detent plunger faces the oil filter screen housing (see accompanying photographs).

6 Slide the selector fork across until its guide pin engages with the appropriate selector drum track, then lubricate the shorter selector fork shaft and press the shaft through the selector fork bore and into its crankcase housing.

7 Using the marks made on dismantling, identify the two remaining selector forks and engage each in turn on its correct respective output shaft pinion and selector drum track; position each fork so that its longer, webbed, boss faces to the left. Lubricate the longer selector fork shaft and press it firmly through both fork bores and into its crankcase housing.

8 Engage the stop arm return spring shorter end on the stop arm then position the plain washer on the crankcase boss next to the selector drum. Ensuring that its roller engages correctly on the stop arm camplate, place the stop arm over the plain washer, apply a few drops of thread locking compound to the threads of the stop arm shouldered pivot bolt and screw the bolt into the crankcase wall to secure the stop arm and washer. Tighten securely the pivot bolt then check that the stop arm is aligned correctly with the camplate and that it is free to move. Hook the return spring longer end over the post set in the crankcase wall.

9 Make a final check that all components have been correctly fitted, that all bearing surfaces are well lubricated, and that all shafts are free to rotate. Rotate the selector drum to check that all gears can be selected, then return the drum to the neutral position.

10 The kickstart assembly fitted to DR125 S and GS125 models only must now be refitted to the crankcase right-hand half. Referring to the accompanying illustration, check that the kickstart stopper plate and ratchet guide plate are correctly placed on the crankcase half, apply a few drops of thread locking compound to the threads of each of the plate mounting bolts, then refit and tighten securely the two bolts. Place the ratchet over the kickstart shaft plain left-hand end so that the ratchet teeth face to the right, then slide the ratchet as far as possible down the shaft splines ensuring that the punch mark on the ratchet left-hand face is aligned with the punch mark on the shaft splines.

11 Refit first the light coil spring then the plain washer over the shaft left-hand end, then insert the shaft into the crankcase half, pushing it through from right to left until the ratchet arm is engaged under the ratchet guide plate. Holding the shaft in position, invert the crankcase half, fit the kickstart return spring around the shaft left-hand end and engage the spring outer end in its recess in the crankcase. Using a suitable pair of pliers to grasp firmly the spring inner end, tension the spring by rotating clockwise its inner end through approximately half a turn until the inner end can be inserted into the hole in the shaft. Refit the return spring guide ensuring that the guide slot aligns with the return spring inner end, then secure the complete assembly by fitting a circlip to the groove in the shaft left-hand end.

34 Reassembling the engine/gearbox unit: joining the crankcase halves

1 Apply a thin film of sealing compound to the gasket surface of the crankcase left-hand half, then press firmly the two locating dowels into their recesses in the crankcase mating surface. Make a final check that all components are in position and that all bearings and bearing surfaces are lubricated, and remove the rubber band from the output shaft end.

2 Lower the crankcase right-hand half into position, using firm hand pressure only to push it home. It may be necessary to give a few gentle taps with a soft-faced mallet to drive the casing fully into place. Do not use excessive force, instead be careful to check that all shafts and dowels are correctly fitted and accurately aligned, and that the crankcase halves are exactly square to each other. If necessary, pull away the crankcase right-hand half to rectify the problem before starting again.

3 When the two halves have joined correctly and without strain, refit the crankcase retaining screws using the cardboard template to position correctly each screw. Working in a diagonal sequence from the centre outwards, progressively tighten the screws until all are securely and evenly fastened.

4 Wipe away any excess sealing compound from around the joint area, then check the free running and operation of the crankshaft and gearbox components. If a particular shaft is stiff to rotate, a smart tap on each end using only a soft-faced mallet, will centralise the shaft in its bearings. If this does not work, or if any other problem is encountered, the crankcases must be separated again to find and rectify the fault.

5 On GS125 ES models only refit the thrust washer over the output shaft right-hand end and secure it with a circlip.

6 Refit the oil filter gauze, tightening securely the two retaining screws. Fit a new O-ring to the triangular filter cover, using a smear of grease to stick it in place, then refit the filter cover ensuring that the cast arrow mark points to the front of the crankcase. Refit and tighten securely the three cover retaining bolts. Check the condition of the drain plug sealing washer, renewing it if necessary, then refit the drain plug, tightening it to a torque setting of 1.8 – 2.0 kgf m (13 – 14.5 lbf ft).

7 Fit a new O-ring to the bore of the gearbox sprocket spacer, then smear liberally the O-ring and output shaft splines with grease. Carefully refit the spacer, taking care not to damage either the O-ring as it passes over the shaft splines or the output shaft oil seal as the spacer is inserted. Push the spacer firmly into place.

33.1 Crankshaft must be absolutely square to crankcase

33.3 Insert gear clusters as a single unit

33.4 Insert first selector fork as shown

33.5 Rotate selector drum so that recess (arrowed) is in position shown

33.6 Engage selector fork in drum track and insert fork shaft

33.7a Insert second and third selector forks as shown

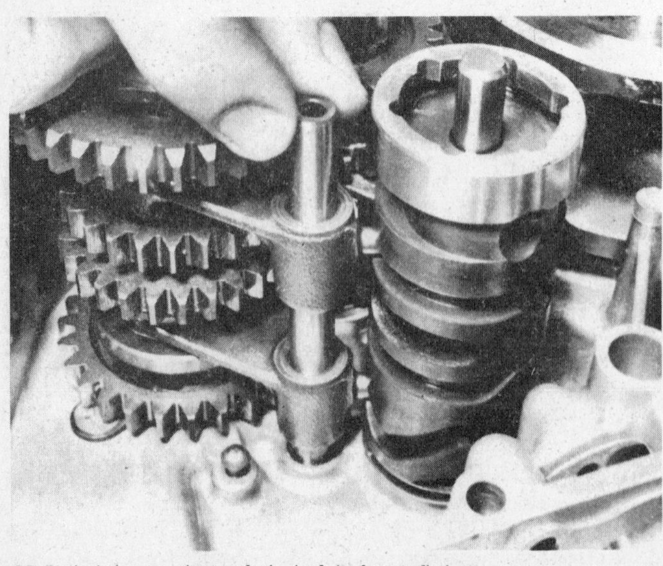

33.7b Lubricate selector fork shaft before refitting

33.8 Selector stop arm assembly in position

34.1 Apply sealing compound to gasket surface – note two locating dowels (arrowed)

34.5a GS125 ES models only – check that output shaft right-hand bearing centre bush is in place, then refit thrust washer ...

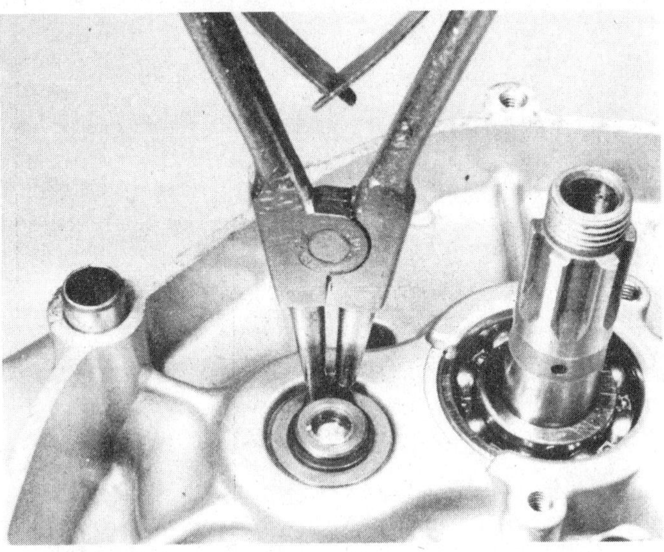

34.5b ... and secure with circlip

34.6 Tighten securely filter gauze retaining screws

34.7a Always renew O-ring in gearbox sprocket spacer ...

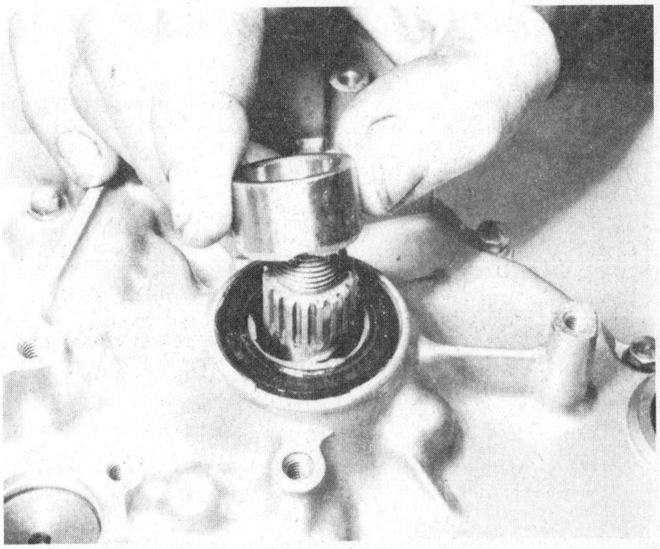

34.7b ... before refitting to output shaft – note liberal use of grease to prevent oil seal damage

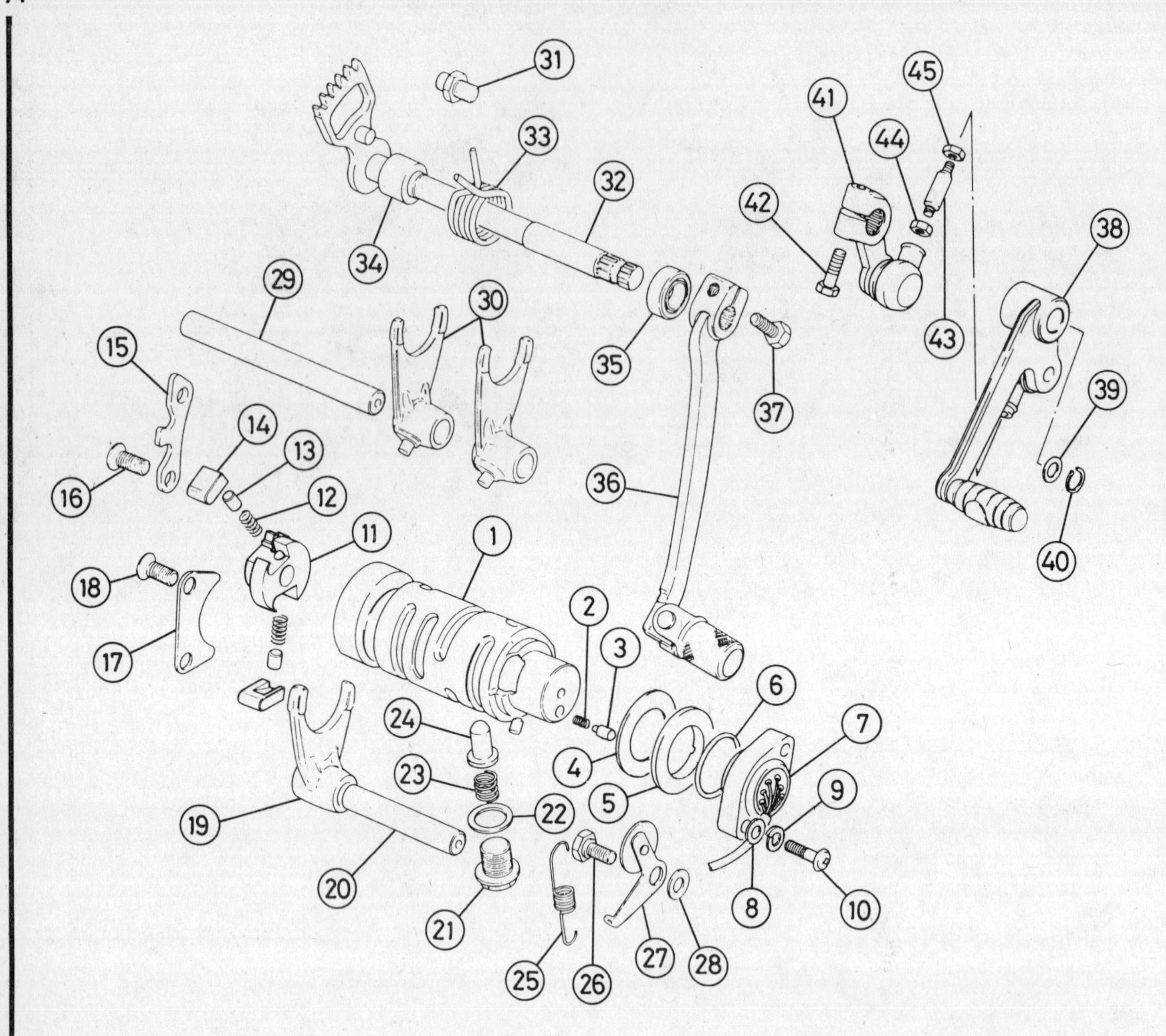

Fig. 1.15 Gearchange mechanism

1 Selector drum
2 Spring
3 Neutral contact pin
4 Washer
5 Camplate
6 O-ring
7 Neutral indicator switch
8 Washer – 2 off
9 Spring washer – 2 off
10 Screw – 2 off
11 Ratchet
12 Spring – 2 off
13 Pin – 2 off
14 Pawl – 2 off
15 Ratchet locating plate
16 Screw – 2 off
17 Pawl guide plate
18 Screw – 2 off
19 Input shaft selector fork
20 Selector fork shaft
21 Plunger housing cap
22 Washer
23 Spring
24 Neutral detent plunger
25 Spring
26 Pivot bolt
27 Selector stop arm
28 Washer
29 Selector fork shaft
30 Output shaft selector forks
31 Spring anchor
32 Gearchange shaft
33 Return spring
34 Bush
35 Oil seal
36 Gearchange lever – DR models
37 Bolt
38 Gearchange lever – GS models
39 Washer
40 Circlip
41 Linkage front arm
42 Bolt
43 Linkage rod
44 Locknut
45 Locknut

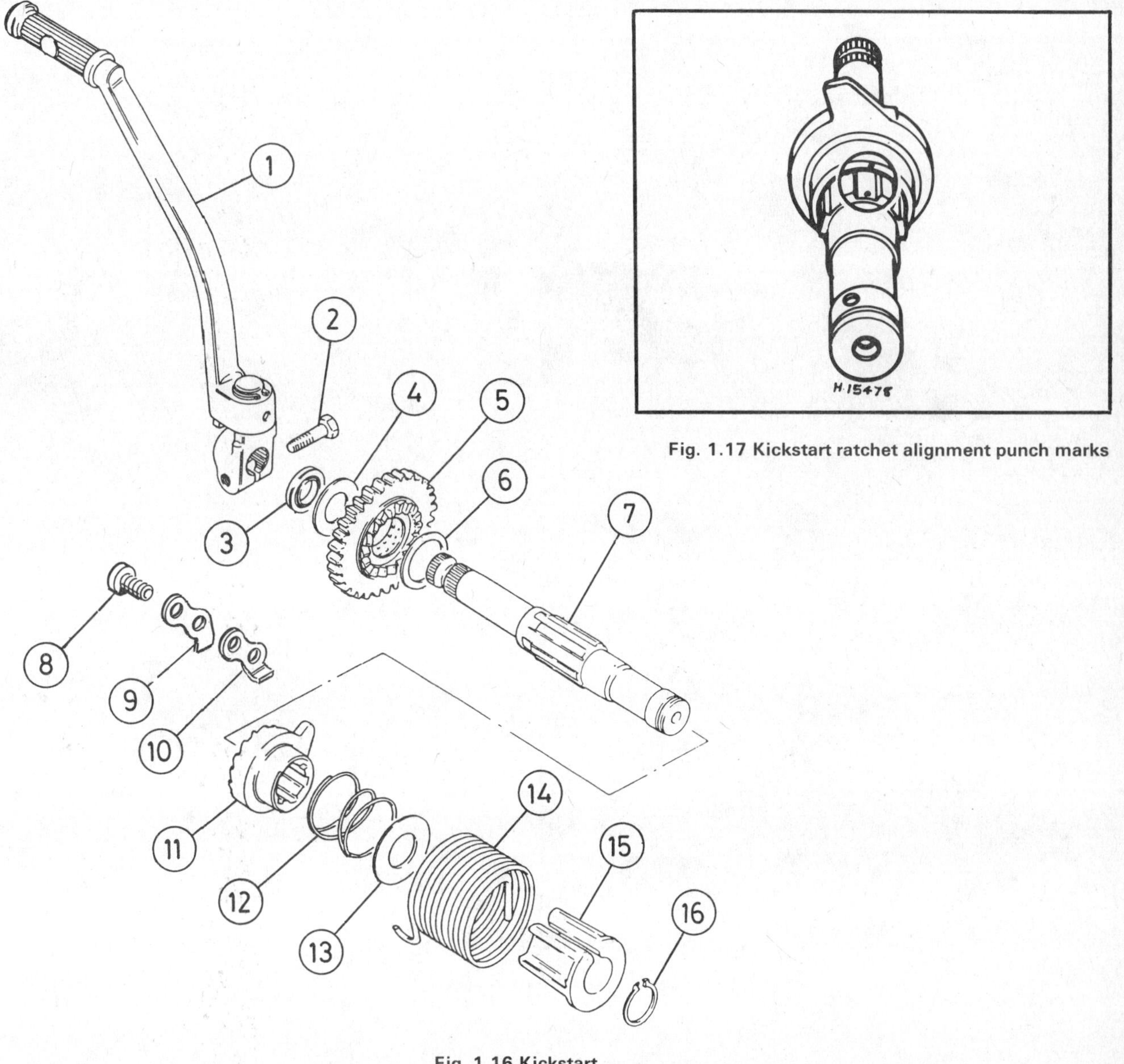

Fig. 1.17 Kickstart ratchet alignment punch marks

Fig. 1.16 Kickstart

1 Kickstart lever
2 Bolt
3 Oil seal
4 Thrust washer
5 Drive pinion
6 Thrust washer
7 Kickstart shaft
8 Bolt
9 Ratchet guide plate
10 Stopper plate
11 Ratchet
12 Light coil spring
13 Washer
14 Return spring
15 Return spring guide
16 Circlip

35 Reassembling the engine/gearbox unit: refitting the gear selector external components

1 Insert the coil spring into the neutral detent plunger and insert the two together into the crankcase to engage with the selector drum. Check the condition of the plunger cap sealing washer, renewing it if damaged or distorted, then refit and tighten securely the plunger cap.
2 Insert each coil spring into its recess in the body to the selector ratchet assembly, followed by the plunger. The pawls are refitted with the wider shoulder on the outside, as shown in the accompanying photographs. Restraining the two pawls so that they cannot fly out, refit the selector ratchet assembly to the selector drum so that the teeth face towards the gearchange shaft bore.
3 Position the ratchet assembly locating plate and the pawl guide plate on the crankcase, using the accompanying photographs to position them correctly. Apply a few drops of thread locking compound to the threads of each of the retaining screws, then refit the screws and tighten them securely using an impact driver.
4 Refit the gearchange shaft return spring to the shaft, ensuring that the spring ends are correctly engaged as shown in the accompanying photograph, then refit the shaft spacer, sliding it along the shaft to fit inside the spring coils. To protect the sealing lips of the oil seal as the shaft is refitted, wrap a thin layer of insulating tape around the shaft splines or smear a liberal quantity of grease over them. Insert the shaft into the crankcase nose and push it carefully into place. The return spring ends must engage on each side of the crankcase stop, and the central tooth on the shaft selector plate must engage with the central tooth of the selector ratchet assembly, as shown in the accompanying photograph.
5 Temporarily refit the gearchange lever to the shaft splines and check that all gears can be selected, rotating the input shaft to assist selection. While a certain amount of difficulty is inevitable, it should be possible to select all gears with relative ease, whether changing up or down. If any undue difficulty is encountered, the cause must be found and rectified before reassembly work can proceed.

35.1a Insert neutral detent plunger ...

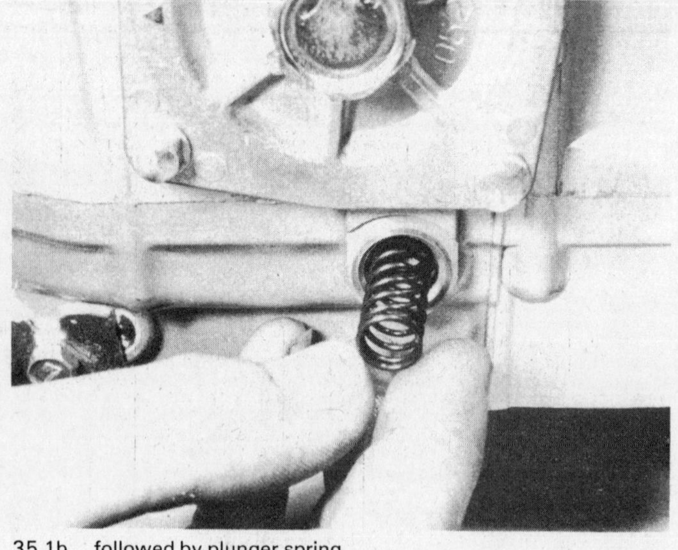

35.1b ... followed by plunger spring

35.1c Tighten securely plunger cap

35.2a Insert spring, followed by plunger, into body of ratchet

35.2b Note difference in width of pawl shoulders (arrowed) – wider shoulder must be on outside

35.3a Use thread locking compound and impact driver to fasten securely the selector ratchet plate retaining screws

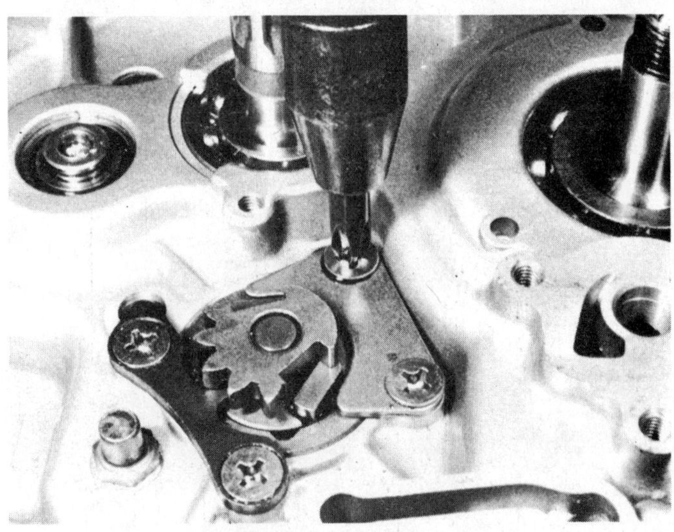
35.3b Pawl guide plate is positioned as shown

35.4a Gearchange return spring and spacer correctly fitted on gearchange shaft

35.4b Selector plate and ratchet assembly must align as shown

36 Reassembling the engine/gearbox unit: refitting the oil pump, primary drive gear and kickstart drive gears

1 On GS125 and DR125 S models only, fit a thrust washer over the output shaft right-hand end, then refit the kickstart idler pinion with its recessed face to the left. Refit the second thrust washer and secure all three components with a circlip. The kickstart pinion is refitted over the kickstart shaft with its toothed surface to the left, and is followed by the large plain thrust washer.
2 Temporarily refit the kickstart lever and check the operation of the kickstart assembly.
3 Insert the Woodruff key into the crankshaft keyway then refit the primary drive gear, aligning its keyway with the key and ensuring that the gear tapered surface faces to the left. Tap the gear firmly into place.
4 Liberally oil the oil pump, ensuring that all the bearing surfaces of the pump are well lubricated, then refit the pump to the crankcase wall. Thoroughly degrease the oil pump retaining screws, apply a few drops of thread locking compound to the threads of each, then refit and tighten securely the three screws. Insert the small pin into the pump shaft and refit the pump driven gear with its recessed face to the left, aligning the cutouts in the centre of the gear with the small pin. Secure the driven gear by refitting the retaining circlip and check that the pump rotates easily.
5 Refit the oil pump drive gear to the crankshaft, aligning its keyway with the Woodruff key and ensuring that the gear tapered surface faces to the left, as shown in the accompanying illustration. The retaining nut lock washer should be renewed if all its locking tabs have been used previously or if it is otherwise damaged. Refit the lock washer with its raised inner tab inserted into the oil pump drive gear keyway, then refit the drive gear retaining nut. The retaining nut employs a left-hand thread. Lock the crankshaft by the method used on dismantling and tighten the nut to a torque setting of 4.0 - 6.0 kgf m (29 - 43 lbf ft). Bend up an unused portion of the lock washer against one of the flats of the nut.

37 Reassembling the engine/gearbox unit: refitting the clutch assembly

1 Smear grease over the entire surface of the clutch release shaft and over the sealing lips and outer circumference of the new release shaft oil seal. Place the locating washer and oil seal over the release shaft splined upper end, taking care not to damage the oil seal lips, then insert the shaft into the crankcase, rotating it so that the operating face is to the right. Press the oil seal squarely into its housing by hand, then use a hammer and a tubular drift to tap down the seal until it is flush with the surrounding crankcase. Check that the shaft is free to rotate, then apply a few drops of thread locking compound to the threads of the seal retaining screw and refit the screw, tightening it securely.
2 The clutch pushrod ends are smaller in diameter, the thinner parts being of different lengths. Oil the whole length of the pushrod and insert it into the input shaft centre with the longer small diameter length to the left. Refit temporarily the clutch release thrust piece and apply light finger pressure while rotating slightly the clutch release shaft to check that the release shaft and pushrod are correctly engaged. As the release shaft is rotated clockwise through $\frac{1}{4} - \frac{1}{2}$ a turn the pushrod and thrust piece should move out, returning under finger pressure as the shaft is released. Withdraw the thrust piece.
3 If the clutch components are to be correctly located on the input shaft, the two thrust washers must be identified by comparison and refitted in their correct positions; although the two are apparently identical, one is 2.0 mm (0.08 in) thick and the other is 1.5 mm (0.06 in) thick. Refit the 2.0 mm thick thrust washer to the input shaft, sliding it down to rest against the shaft right-hand bearing. On 'Z' (1982) models only, oil liberally and refit the separate bush.
4 Lubricate the bush at the centre of the clutch outer drum and refit the drum, rotating it as necessary to assist the engagement of the teeth on the primary drive gears, and kickstart drive gears (where applicable). Refit the 1.5 mm thick thrust washer and then the clutch centre. The clutch centre retaining nut lock washer should be renewed if all its locking tabs have been used previously or if it is otherwise

damaged. Refit the lock washer with its raised outer tab engaging with the squared edge of the clutch centre boss. Refit the clutch centre retaining nut, lock the clutch centre by the method used on dismantling and tighten the nut to a torque setting of 3.0 - 5.0 kgf m (22 - 36 lbf ft). Bend up an unused portion of the lock washer against one of the flats of the nut. Check that the clutch components are free to rotate independently of each other.

5 If new clutch friction plates are to be fitted, they should be coated first with a film of oil. Starting with a friction plate followed by a plain plate, build up alternately the plates to finish with a friction plate.

6 Lubricate the clutch release thrust bearing and place it over the head of the thrust piece, followed by the thrust washer. Insert the thrust piece assembly into the input shaft, ensuring that it engages correctly with the clutch pushrod. Apply a spanner to the flats of the thrust piece to prevent it from rotating, slacken the adjuster locknut and unscrew fully the adjuster screw. Refit the clutch pressure plate, followed by the clutch springs and their retaining bolts and washers. Working progressively and in a diagonal sequence, tighten securely the clutch spring bolts.

7 Screw in the adjuster screw until light resistance is encountered, then turn back the screw by $\frac{1}{4} - \frac{1}{2}$ turn. Prevent the thrust piece from rotating and tighten securely the adjuster locknut.

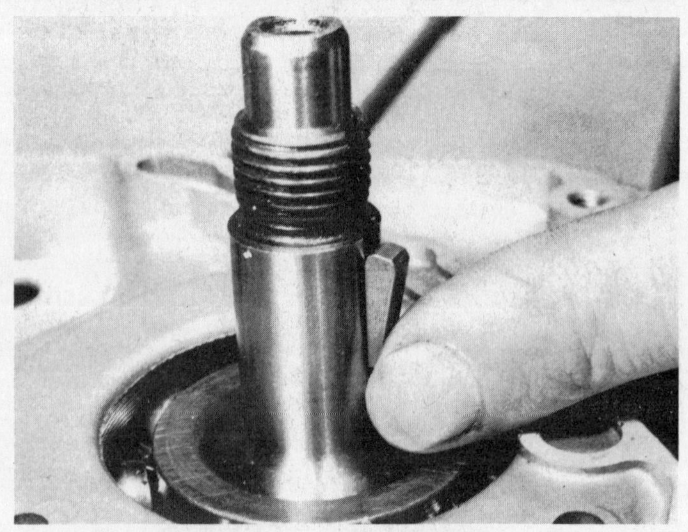

36.3a Insert Woodruff key into crankshaft keyway

36.3b Primary drive gear recessed surface faces inwards

36.4a Liberally oil oil pump before refitting

36.4b Apply thread locking compound to pump retaining screws – tightening securely

36.4c Cutouts in centre of drive gear must align with pin (arrowed)

36.5a Pump drive gear tapered surface inwards – align keyway with key

36.5b Lock washer inner tab (arrowed) fits into drive gear keyway

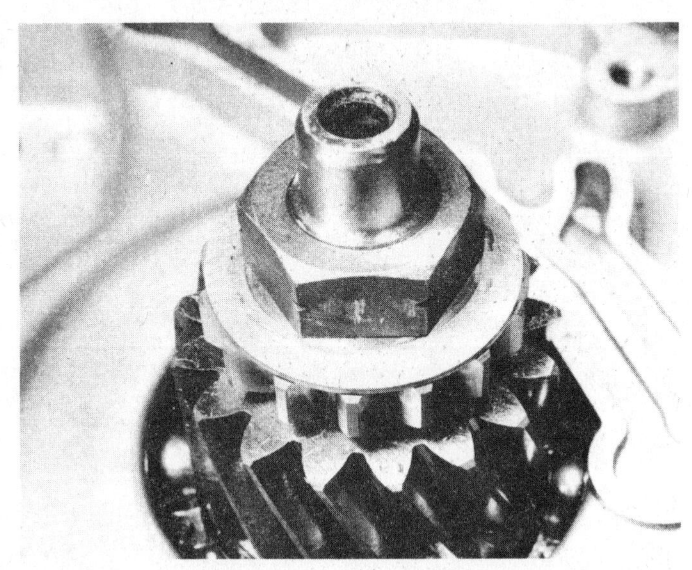

36.5c Drive gear retaining nut employs left-hand thread – tighten anti-clockwise

36.5d Use punch to secure nut with lock washer

37.3a Identify correct thrust washer by measurement before refitting

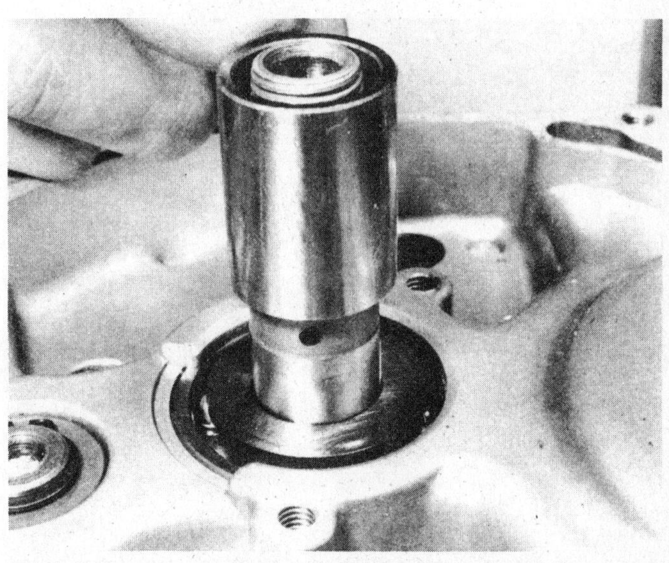

37.3b Separate bush is fitted to Z (1982) models only – oil liberally before refitting

37.4a Rotate clutch outer drum as it is refitted to engage gears

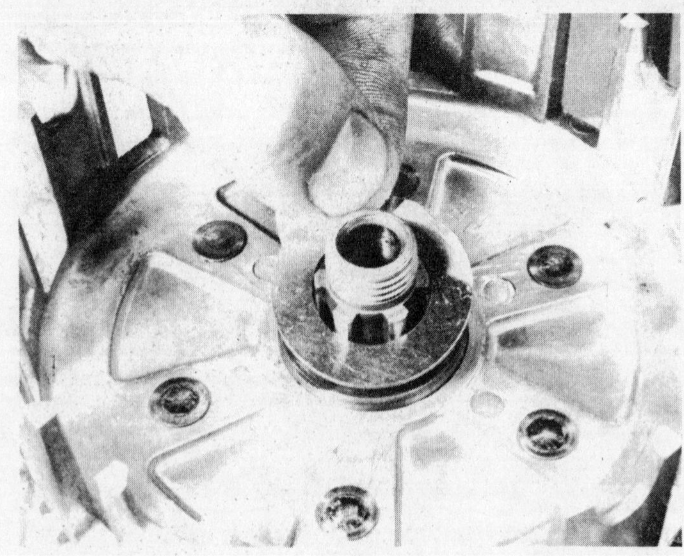

37.4b Correct thickness thrust washer fits between clutch outer drum ...

37.4c ... and clutch centre – note flat (arrowed)

37.4d Lock washer tab engages as shown with flat on clutch centre

37.4e Secure retaining nut with lock washer as shown

37.5a Refit clutch pushrod in direction shown – longer thin length

37.5b Clutch thrust piece assembly is built up as shown ...

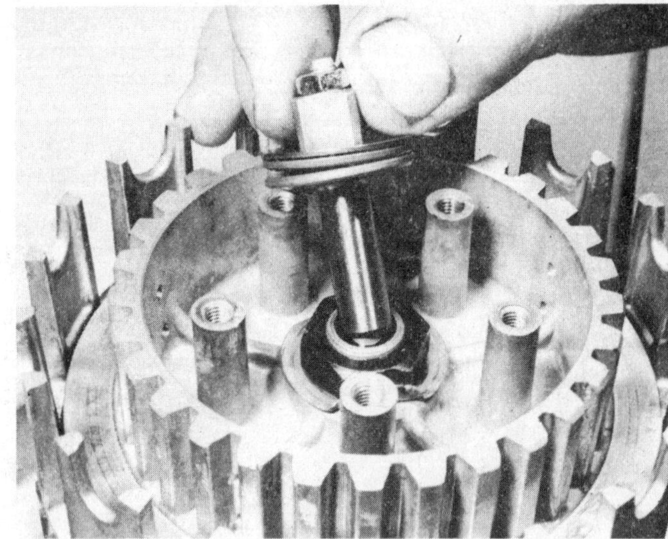

37.5c ... and inserted into input shaft centre to engage with clutch pushrod

37.5d Oil friction plates before refitting ...

37.5e ... alternately with clutch plain plates

37.6a Slacken fully adjuster screw before refitting pressure plate ...

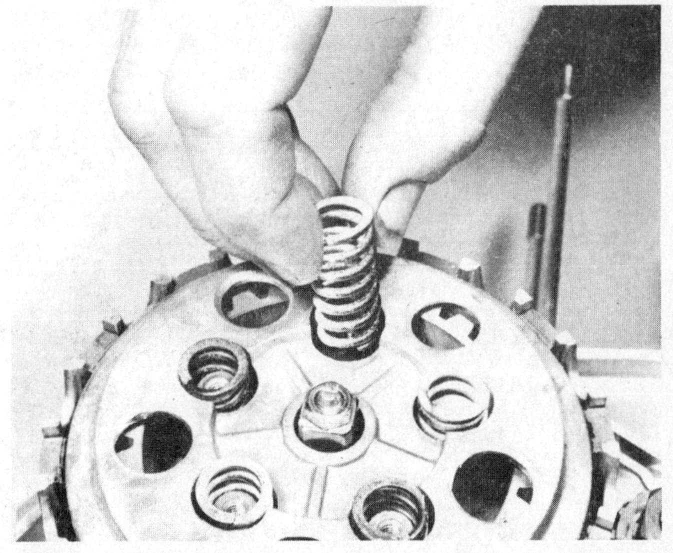

37.6b ... refit clutch springs ...

37.6c ... tighten spring bolts and adjust release mechanism (see text)

38 Reassembling the engine/gearbox unit: refitting the crankcase right-hand cover

1 Check that all components are correctly fitted and secured, and that all bearing surfaces are lubricated. Refit the two locating dowels into their recesses in the crankcase gasket surface and fit a new gasket, using a smear of grease to stick it in place.
2 If the crankshaft oil seal and (where applicable) the kickstart shaft oil seal were removed, they must now be refitted. Smear grease over the sealing lips of each seal.
3 Lower the cover into position, using only firm hand pressure to press it home and ensuring that it is aligned correctly on the crankshaft and on the locating dowels. It may be necessary to give a few taps with a soft-faced mallet to seat the cover, but do not force it into position if resistance is encountered, remove instead the cover and rectify the problem before starting again.
4 Refit the cover retaining bolts, using the cardboard template to position correctly each screw and not forgetting to renew the sealing washer under the head of the bolt above and to the right of the filler plug orifice. Working progressively and in a diagonal sequence from the centre outwards, tighten securely the cover retaining bolts.
5 Renew the two O-rings fitted in the oil filter chamber and cap, using a smear of grease to stick them in place. Insert the filter into the chamber, refit the chamber cap and spring, then refit and tighten securely the cap retaining screws or nuts.

39 Reassembling the engine/gearbox unit: refitting the alternator and ignition components, the starter motor and drive components, the cam chain and the neutral/gear position indicator switch

1 Loop the cam chain over the crankshaft left-hand end, engage it on the drive sprocket and feed the remainder up through the crankcase tunnel. Secure the chain to prevent it from dropping into the crankcase.

GS125 and DR125 S models

2 Insert the Woodruff key into the crankshaft keyway, then refit the stator plate. The mounting screw slot nearest to the ignition pulser coil is stamped with a single line which acts as a timing index mark; rotate the plate to align exactly the index mark with the centre of the screw hole. Apply thread locking compound to the threads of the screws, then refit and tighten securely the three screws.
3 Refit the rotor, aligning its keyway with the Woodruff key, then gently tap the rotor centre with a soft-faced mallet to seat it on the crankshaft taper. Apply a few drops of thread locking compound to the crankshaft thread, then refit the plain washer, the lock washer and the retaining nut. Use the method employed on dismantling to lock the crankshaft and tighten the rotor retaining nut to a torque setting of 3.0 - 4.0 kgf m (22 - 29 lbf ft).

GS125 ES models

4 Insert the Woodruff key into the crankshaft keyway. Check that the starter clutch assembly is correctly fitted to the rear of the rotor and functioning correctly, then refit the rotor/starter clutch assembly as a single unit to the crankshaft. Apply a few drops of thread locking compound to the crankshaft thread and refit the rotor retaining nut. Use the method employed on dismantling to lock the crankshaft and tighten the nut to a torque setting of 3.0 - 4.0 kgf m (22 - 29 lbf ft).
5 Renew the O-ring fitted around the starter motor boss and lightly smear it with grease. Refit the starter motor and tighten securely its two mounting bolts. Place the starter idler gear on the crankcase boss, insert the gear shaft and place the thick spacer over the shaft end.

All models

6 Insert the switch spring, followed by the switch contact, into the drilling in the selector drum left-hand end. Renew the sealing O-ring and refit the neutral/gear position indicator switch ensuring that the wiring projects upwards. Refit and tighten securely the switch retaining screws, but be careful not to overtighten the screws or the switch may crack.
7 Apply a thin smear of grease to a new gasket and stick the gasket to the crankcase surface, not forgetting the single locating dowel fitted to GS125 ES models only. On GS125 and DR125 S models only, pass the alternator and ignition wiring through the gasket.
8 On GS125 ES models only, if the alternator stator and ignition pulser coil were removed, they must now be refitted. Refit first the alternator stator, then the pulser coil, ensuring that the wiring of each component is correctly routed and cannot be trapped or pinched, then press the sealing grommet into its recess in the cover wall. Apply a few drops of thread locking compound to the threads of each of the screws, then refit and tighten securely the screws.
9 Lower the crankcase left-hand cover into position, noting that on GS125 and DR125 S models only, the sealing grommet fitted around the alternator and ignition wiring must be pressed into the recess in the cover wall. Check that the gasket and cover are correctly positioned, then refit and tighten securely the cover retaining screws, using the cardboard template to position correctly each screw and tightening the screws progressively and evenly working in a diagonal sequence from the centre outwards.

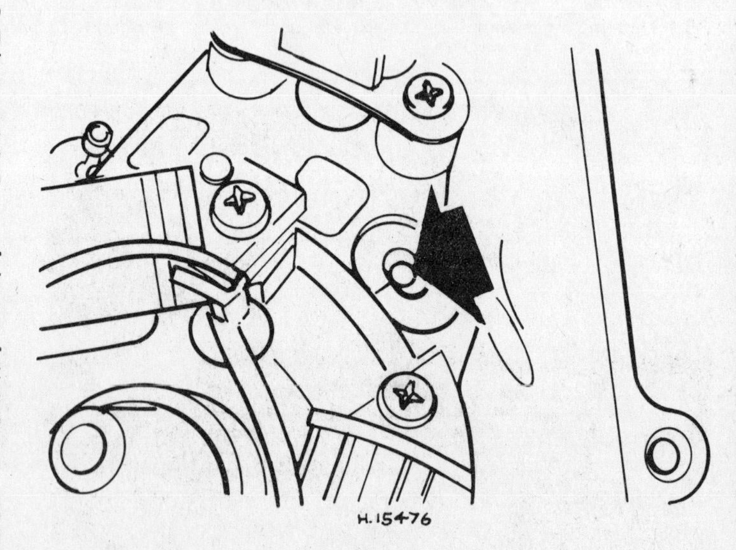

Fig. 1.18 Alternator stator plate positioning marks

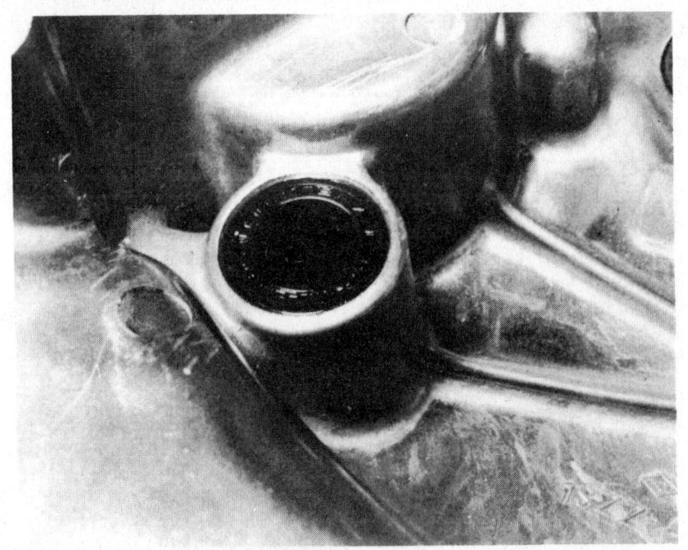
38.2 Always renew crankshaft oil seal whenever it is disturbed as it is essential to maintain correct oil pressure – apply grease on refitting

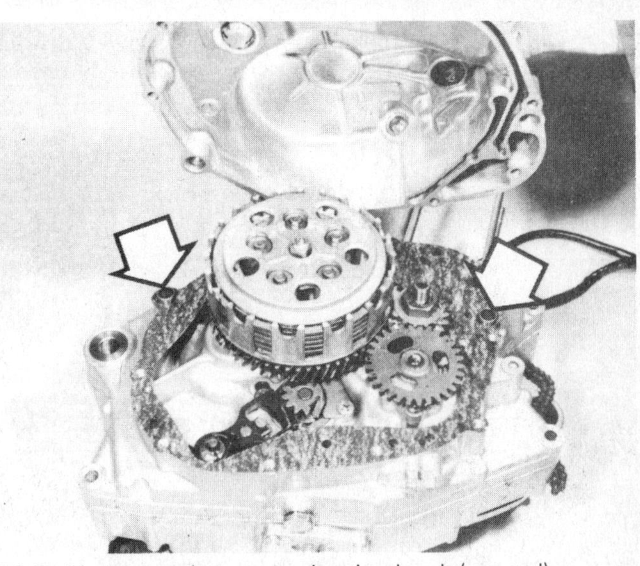
38.3 Position new gasket over two locating dowels (arrowed)

39.1 Engage cam chain on sprocket and feed into crankcase tunnel

39.4 Hold rotor with spanner while tightening nut to torque setting – GS125 ES only

39.5a GS125 ES model only – first insert starter idler gear ...

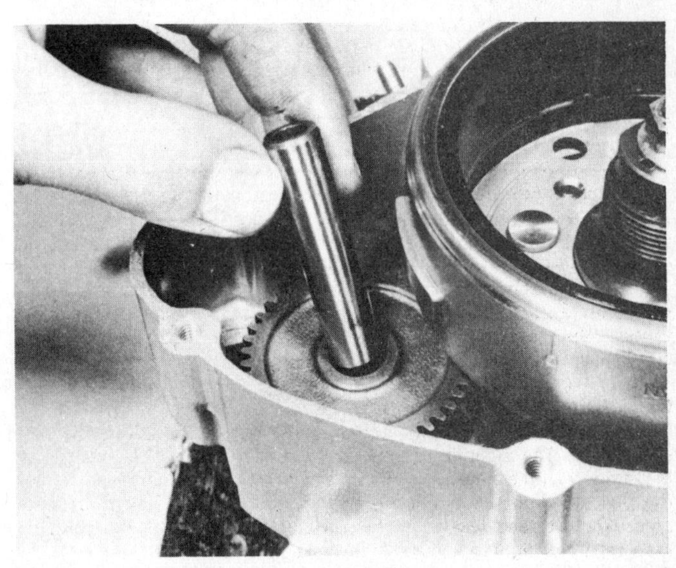
39.5b ... then lubricate and refit gear shaft ...

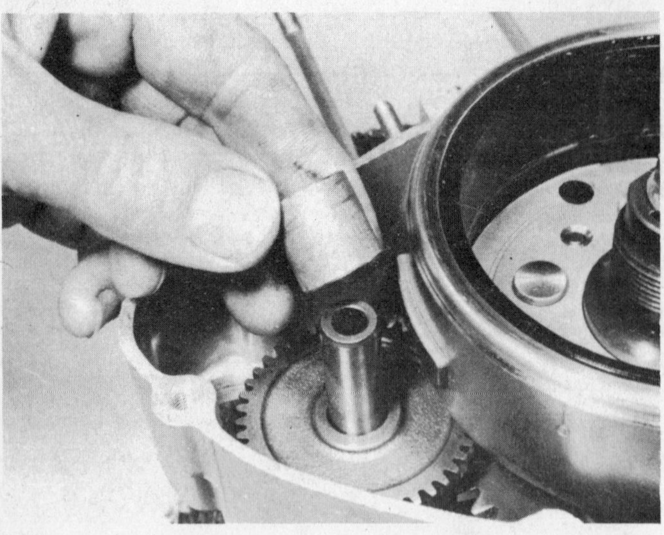

39.5c ... and refit spacer

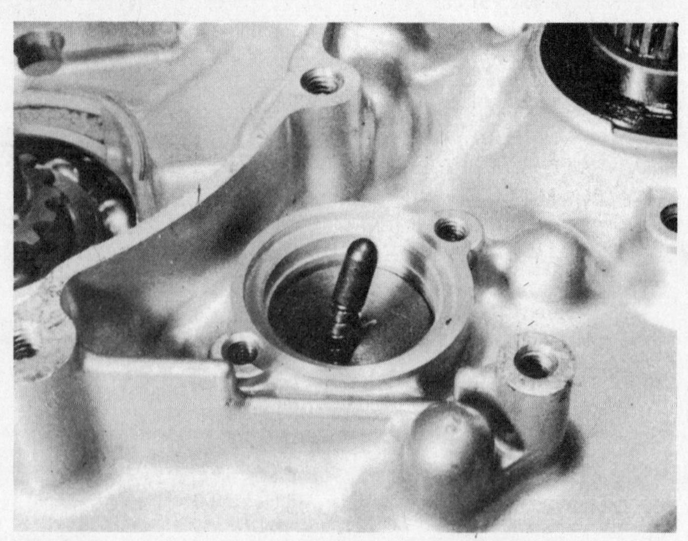

39.6a Insert indicator switch spring and contact into selector drum drilling ...

39.6b ... renew O-ring if necessary – use grease to stick in place while indicator switch body is refitted

39.7 Cover gasket is located by single dowel (arrowed) on GS125 ES model only

40 Reassembling the engine/gearbox unit: refitting the piston and cylinder barrel

1 Refit the piston rings to the piston, using the method employed on removal. The first component to be fitted is the oil control ring central spacer, followed by its two side rails which are fitted one on each side of the spacer; while none of these components have a specific top or bottom surface when new, if part-worn components are being reused the wear marks will reveal which way up they were originally fitted. Always refit part-worn components in their exact original locations, to prevent excessive wear.

2 Refit the thin expander ring to the bottom compression ring groove. The top surface of both compression rings is marked by the letters 'RN' stamped or etched into the metal. Refer to the accompanying illustration to identify the two compression rings and note that as the top ring is chrome-plated it will appear lighter in colour than the bottom ring. Ensure that the compression rings are fitted the correct way up and in the correct groove.

3 Fit a new circlip to the groove in either of the piston bosses, then insert the gudgeon pin from the opposite side of the piston. If the gudgeon pin is a very tight fit, heat the piston as described in Section 7 of this Chapter. Do not allow the gudgeon pin to project beyond the inside of the boss.

4 Bring the connecting rod to the top of its stroke and lubricate the small-end eye and gudgeon pin. Pack the crankcase mouth and cam chain tunnel with clean rag. Place the piston over the connecting rod ensuring that the arrow embossed on the piston crown is facing to the front of the engine unit, then push through the gudgeon pin. Secure the gudgeon pin by fitting the second circlip and ensure that both circlips are correctly located in their grooves. Never re-use old circlips; new ones should be used always.

5 Refit both locating dowels in their recesses around the left-hand pair of larger diameter cylinder retaining studs, then fit a new cylinder base gasket, using a smear of grease to stick it in place. Rotate the piston rings until their end gaps are in the positions shown in the accompanying illustration, then oil liberally the piston rings, piston skirt and cylinder bore. Place the cylinder barrel over the studs and lower it to meet the piston.

6 If a piston ring compressor is not available, insert the piston crown into the cylinder bore, ensuring that the piston is absolutely square to the bore, and compress by hand each ring in turn while pressing the barrel down with a gentle twisting motion to help the rings enter the bore. While a generous lead-in is provided to help this task, great care must be taken or the piston rings will break.

7 Pull the barrel down until the entire piston is in the cylinder bore, then hold the barrel in that position while the rag is removed and the cam chain is passed up through the tunnel in the barrel casting. Keeping taut the cam chain, slide the barrel down to rest on the gasket surface, then refit and tighten, by hand only the barrel retaining nuts. Secure the cam chain to prevent it from dropping into the crankcase.

40.2 Stamped letters identify piston top surface

40.4a Arrow mark stamped on piston crown faces exhaust (front)

40.4b Ensure circlip is correctly engaged in groove

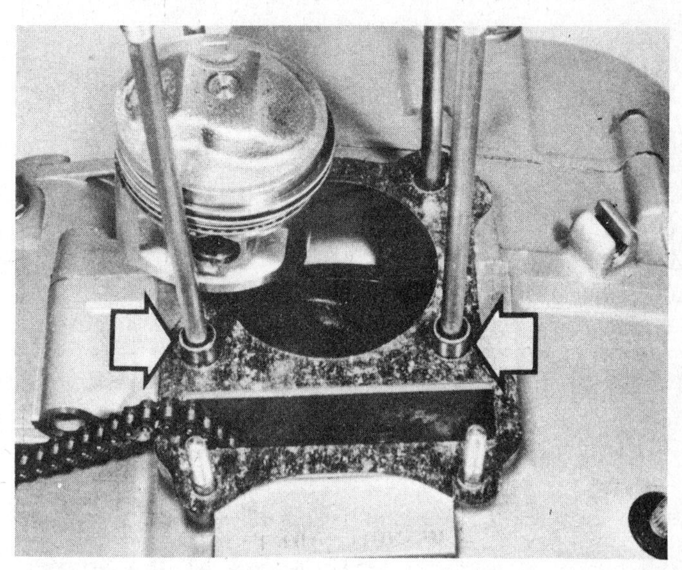

40.5 Always use new cylinder base gasket – note two locating dowels (arrowed)

40.7 Tighten cylinder barrel nuts by just enough to retain barrel

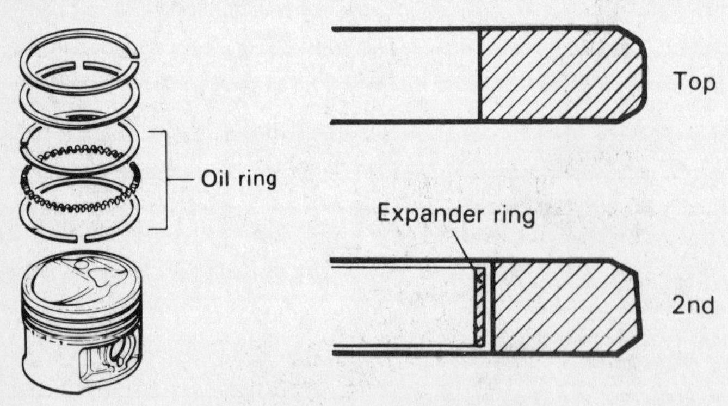

Fig. 1.19 Piston ring identification

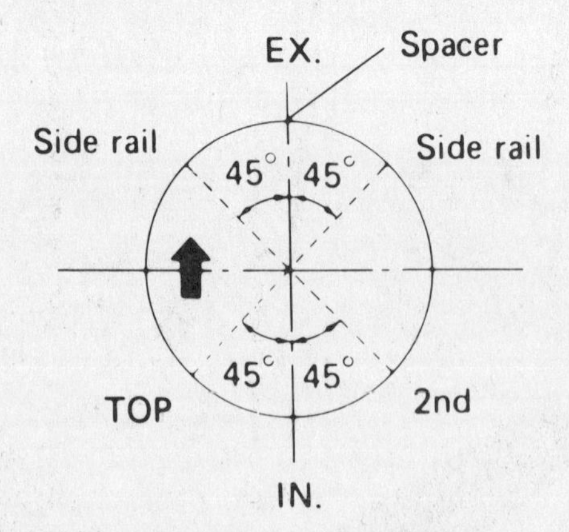

Fig. 1.20 Piston ring end gap positions

41 Reassembling the engine/gearbox unit: refitting the cylinder head, camshaft and the cylinder head cover

1 Insert the cam chain guide blade into the front of the cam chain tunnel, using a torch to ensure that the blade bottom end locates correctly in its recess in the crankcase wall. Fit the blade upper locating projections into the recesses in the cylinder barrel gasket surface. Refit the two locating dowels to their recesses around the cylinder studs and position a new cylinder head gasket on the gasket surface.

2 Fit the cam chain tensioner blade to the rear of the cylinder head cam chain tunnel and secure it by refitting the blade mounting bolt. Do not forget to renew if necessary the mounting bolt sealing washer, and check that the blade is free to pivot about the mounting bolt. The tensioner blade should be refitted at this stage only if the engine is removed from the frame; if this is not the case, it should be refitted only after the cylinder head has been placed on the cylinder barrel.

3 Keeping taut the cam chain, hook a length of stiff wire to its upper end. Pass the wire up through the cylinder head cam chain tunnel, lower the cylinder head over the wire and cam chain to rest on the cylinder head gasket. Secure the cam chain.

4 Check that the cylinder head is seated correctly and refit the cam chain tensioner blade as described above, if this has not already been done. Referring to the accompanying illustration, refit the cylinder head retaining nuts and their washers. Working in a diagonal sequence, tighten progressively the four 8 mm nuts to a torque setting of 1.5 – 2.0 kgf m (11 – 14.5 lbf ft). The two 6 mm retaining nuts are tightened next, followed by the two cylinder barrel retaining nuts; these last four nuts are tightened to a torque setting of 0.7 – 1.1 kgf m (5 – 8 lbf ft).

5 Keeping taut the cam chain, rotate the engine via the alternator rotor retaining nut until the piston is at TDC. This is indicated when the index mark (a straight line with the letter 'T' or 'O') stamped on the rotor rim and visible via the aperture in the top surface of the crankcase left-hand cover is aligned exactly with the arrow cast on the crankcase cover (all GS125 models) or is exactly in the middle of the aperture (DR125S model). Do not disturb the crankshaft from this position and keep checking it as the camshaft is refitted.

6 Insert the camshaft locating half ring into its groove in the left-hand bearing surface and insert the sprocket locating pin into the camshaft flange, ensuring that the pin is firmly fixed with grease. Coat the camshaft and cylinder head bearing surfaces with a high quality molybdenum disulphide-based lubricant, or failing this, a copious supply of clean engine oil. Place the sprocket over the camshaft left-hand end and position the camshaft and sprocket on the cylinder head, aligning the camshaft wth its locating half-ring and looping the chain over the camshaft left-hand end. Rotate the camshaft until the line stamped on its left-hand end is exactly parallel with the cylinder head/cylinder head cover mating surface when the sprocket locating pin is in approximately the 11 o'clock position (see accompanying photographs).

7 Engage the sprocket on the chain so that the sprocket will fit exactly on the camshaft when the chain front run is taut. Lift the sprocket on to the camshaft shoulder and press it into place over the locating pin. Check that the crankshaft and the camshaft timing marks are aligned exactly when the chain front run is taut. If the marks do not align as described, do not move either the crankshaft or the camshaft but move the sprocket by one tooth at a time until the correct position is achieved.

8 Fit the sprocket mounting bolt lock washer, ensuring that it covers the sprocket locating pin, then screw in temporarily one of the mounting bolts, tightening it by hand only to retain the sprocket and lock washer. Keeping taut the cam chain front run, rotate the engine anti-clockwise until the other mounting bolt hole is exposed.

9 Apply a few drops of thread locking compound to the threads of the second sprocket mounting bolt, then refit the bolt, tightening it to a torque setting of 1.0 – 1.3 kgf m (7 – 9 lbf ft). Bend up an unused portion of the lock washer against one of the flats of the bolt head. Keeping taut the cam chain front run, rotate slowly the engine until the first sprocket mounting bolt is again exposed. Unscrew the bolt so that thread locking compound can be applied to its threads, then refit the bolt, tightening it to the correct torque setting and securing it with the lock washer as described. It is essential that the lock washer is placed to cover the sprocket locating pin to prevent the pin from dropping out. Rotate the engine to the TDC position.

10 Apply a thin film of sealing compound to the cylinder head/cylinder head cover gasket surface, being careful to keep surplus sealing compound away from the camshaft right-hand bearing, then insert the two locating dowels and the rubber camshaft end cap into their recesses in the gasket surface. There have been instances of leakage at this joint area, most of which have been attributed to the use of the wrong type of jointing compound. Suzuki recommend the use of one type of compound only; Suzuki Bond 1215. This can be obtained from Suzuki dealers as Part Number 99000-31110. Slacken fully the valve adjusters and oil liberally the rockers, the valve assemblies and the camshaft, filling the pocket beneath the cam lobes with oil.

11 Refit the cylinder head cover, pressing it firmly on to the cylinder head and ensuring that the camshaft end cap is correctly located. Refit the ten cover retaining bolts, noting that a new sealing washer must be fitted under the head of the bolt marked in the accompanying illustration. Following the sequence shown in the illustration tighten progressively the bolts to a torque setting of 0.9 – 1.0 kgf m (6.5 – 7 lbf ft).

12 Placing a new gasket on the cylinder barrel mating surface, refit the cam chain tensioner assembly noting that the adjuster screw must face to the left. Tighten securely the tensioner mounting bolts and slacken the adjuster locknut. Unscrew by one turn the adjuster screw; if the cam chain front run was kept taut as described the tensioner will move into place with an audible click, taking up all the chain free play. Tighten carefully the adjuster screw, being careful not to overtighten it, then tighten securely the locknut.

13 Following the procedure described in the Routine Maintenance Chapter of this Manual, adjust the valve clearances. Refit the two valve adjuster inspection caps.

14 Insert the tachometer driven gear into its housing, renew the housing sealing O-ring and apply a smear of grease to the housing spigot. Insert the assembly into the cylinder head cover, ensuring that the teeth on the camshaft and driven gear mesh correctly. Secure the housing by tightening securely the retaining screw.

15 Check that the spark plug is clean and correctly gapped and refit it, tightening it by hand only at first until it is correctly seated, then apply a suitable box spanner and tighten the plug by a further $\frac{1}{4}$ turn. Refit the two inspection caps set in the crankcase left-hand cover.

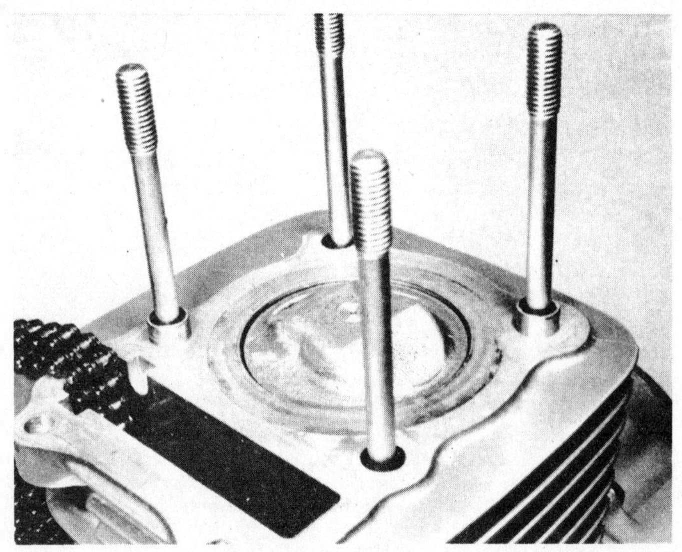
41.1a Insert two locating dowels in gasket surface recesses ...

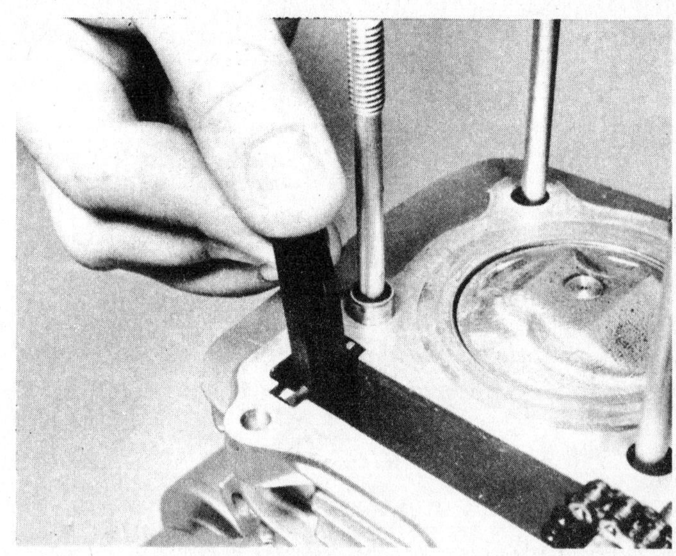
41.1b ... then refit cam chain guide blade ...

41.1c ... and position correctly a new head gasket

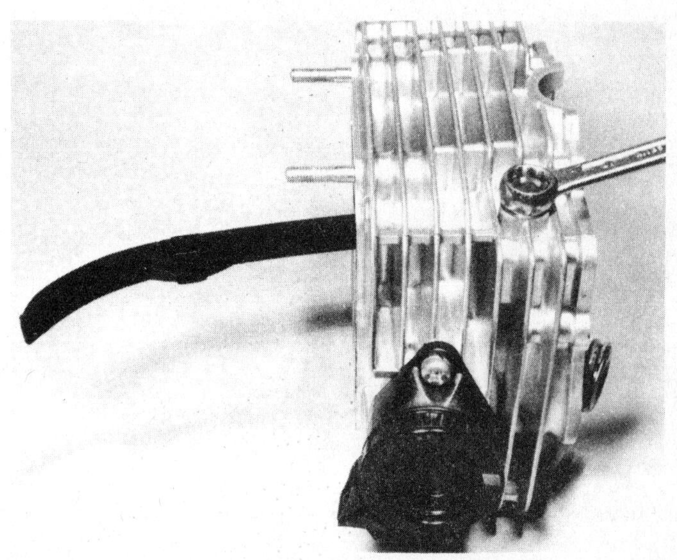
41.2 Cam chain tensioner blade is secured by a single bolt

41.3 Keep cam chain taut while refitting cylinder head

41.4a Do not forget the two nuts underneath the head ...

41.4b ... and the cylinder barrel retaining nuts

41.6a Insert camshaft locating half ring into groove

41.6b Sprocket locating pin (arrowed) must be fitted into camshaft flange

41.6c Fit camshaft, cam chain and sprocket as shown ...

41.6d ... then align camshaft timing marks with mating surface – note position of sprocket locating pin (arrowed)

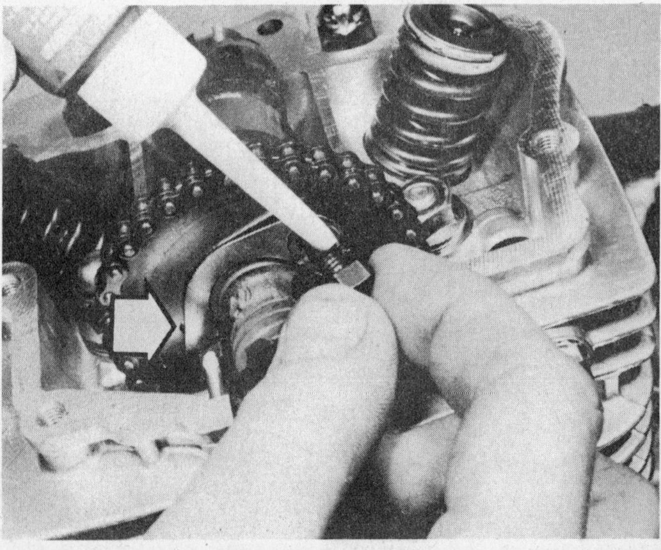

41.9a Apply thread locking compound to sprocket bolt threads – note sprocket locating pin (arrowed) is covered by lock washer

41.9b Tighten sprocket bolt to recommended torque setting ...

41.9c ... and secure by bending up lock washer tab – repeat for second bolt

41.10a Liberally oil camshaft bearing surfaces and fill pocket with oil

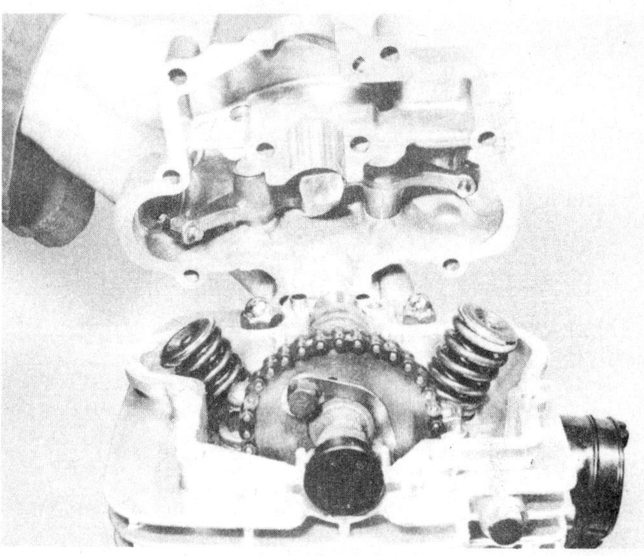
41.11 Refit camshaft end cap and cylinder head cover – do not omit locating dowels

41.12 Use a new gasket when refitting tensioner assembly

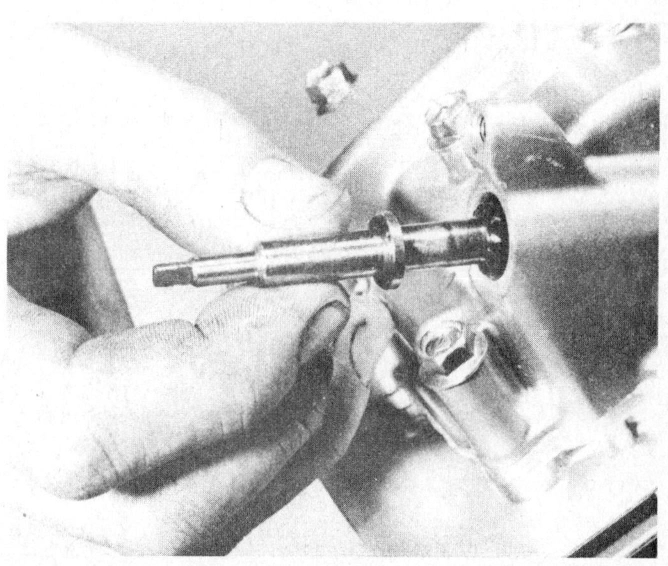
41.14a Tachometer drive assembly must be fitted after cylinder head cover is secured

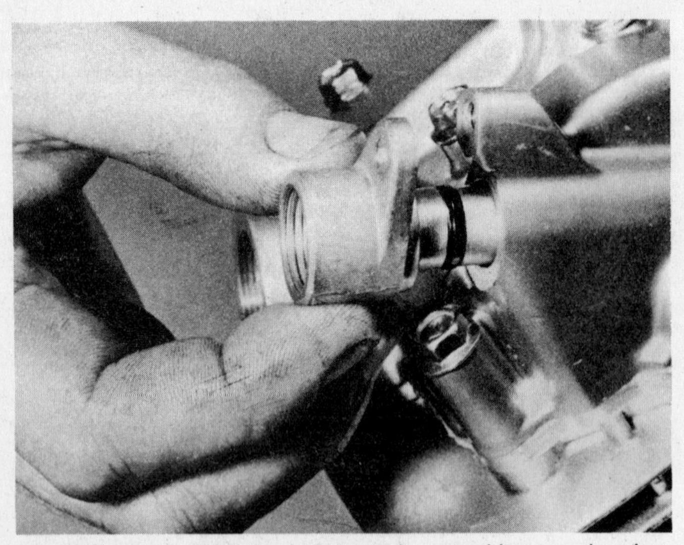

41.14b Renew O-ring before refitting tachometer driven gear housing

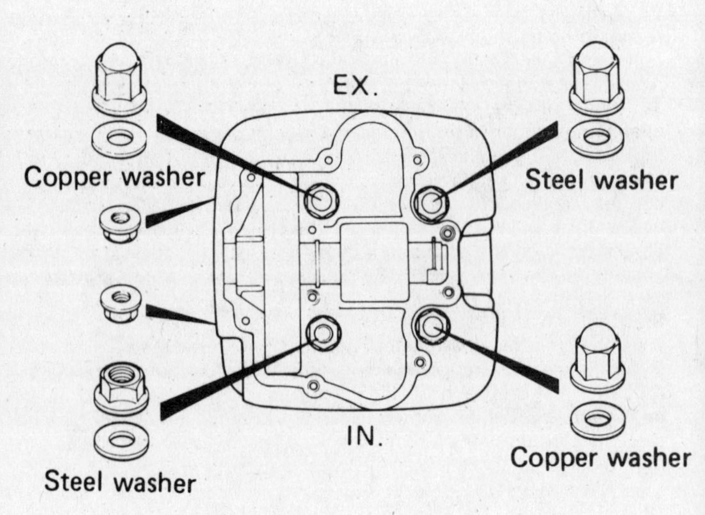

Fig. 1.21 Cylinder head nut and washer positions

Fig. 1.22 Positioning the piston at TDC

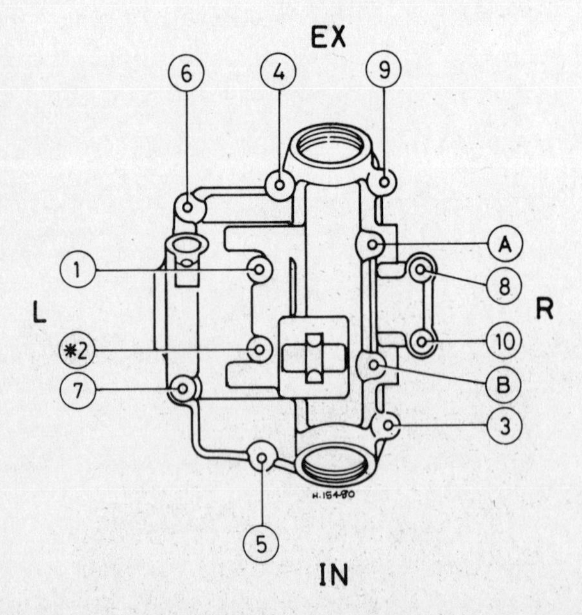

Fig. 1.23 Cylinder head cover retaining bolt tightening sequence

A and B – do not disturb
** – sealing washer under this bolt*

42 Refitting the engine/gearbox unit in the frame

1 Lift the engine/gearbox unit into the frame and hold it while inserting the left-hand plate and engine mounting bolts of the engine top mounting/cylinder head steady assembly. Refit the right-hand plate and retaining nuts of the assembly, tightening the nuts by hand only. Check that the engine/gearbox unit is aligned correctly on its mountings and that no components are trapped.

2 Pack grease into the crankcase rear lug and into both swinging arm pivot lugs, smear grease along the exposed length of the swinging arm pivot bolt, then tap the pivot bolt into place, ensuring that it passes through both sides of the frame, through both swinging arm pivots, and through the crankcase rear lug. Refit the retaining nut and washer and wipe off any surplus grease.

3 Refit the remaining engine mounting bolts and mounting plate. Smear grease over each bolt as it is refitted, to prevent corrosion. Push the bolts through from left to right and refit their retaining nuts. Note that the nuts are of the self-locking type which should be renewed, if the locking medium no longer functions. Tighten the nuts by hand only at first.

4 Check that the engine is aligned correctly and is held securely and without strain in its mountings, then use a torque wrench to tighten securely the nuts. First the swinging arm pivot bolt retaining nut is tightened to a torque setting of 5.0 – 8.0 kgf m (36 – 58 lbf ft), then the retaining nuts of the two 80 mm (3.15 in) long bolts of the engine front mounting assembly (the two bolts which pass through the mounting plate and crankcase) are tightened to 3.7 – 4.5 kgf m (27 – 33 lbf ft). The retaining nuts of all the remaining bolts are tightened to a torque setting of 2.8 – 3.4 kgf m (20 – 25 lbf ft).

5 Route carefully the electrical leads up the frame tubes and connect them to the main wiring loom at their multi-pin block connectors or individual snap connectors. Use the colour-coding of the wires in conjunction with the wiring diagram at the back of this Manual to connect correctly the wiring, and fasten the leads to the frame so that they are out of the way using the clamps or other fasteners provided for this task. The crankcase retaining bolt above and to the right of the output shaft is used to retain the engine earth lead; remove the bolt, refit the lead terminal under the bolt head, then refit the bolt, aligning correctly the lead and tightening securely the bolt.

6 Engage the gearbox sprocket on the chain and refit the sprocket over the output shaft splines. Check the condition of the lock washer, renewing it if it is damaged or if all of its locking tabs have been used, then refit the washer. Apply a few drops of thread locking compound to the output shaft thread and refit the sprocket retaining nut. Apply hard the rear brake and tighten the nut to a torque setting of 8.0 – 10.0 kgf m (58 – 72 lbf ft). Bend up an unused portion of the lock washer against one of the flats of the nut. Referring to Routine

Maintenance, check, and adjust if necessary, the chain tension and the rear brake adjustment.
7 Refit the gearbox sprocket cover, tightening securely the three retaining bolts, then refit the gearchange lever on DR125S models, tightening securely the pinch bolt. On all GS125 models, smear grease over the gearchange linkage frame pivot point and offer up the linkage as shown in the photographs accompanying the text; bring the gearchange pedal to the height required by rotating the linkage front arm about the gearchange shaft before pushing the front arm on to the shaft splines and tightening securely the pinch bolt. Fine adjustment can be achieved by slackening the two locknuts and rotating as necessary the link rod, but note that the linkage will function most efficiently when the angles formed by the linkage front and rear arms and the link rod are at 90°. Secure the linkage by refitting the plain washer and circlip.
8 Refit the kickstart lever (where fitted) and, on DR125S models only, the crankcase bashplate, tightening securely the retaining bolts.
9 Refit the carburettor, manoeuvring it carefully into place. Check that the inlet stub and air filter hose are correctly engaged on the carburettor body, then tighten securely the two clamp screws. Check that the throttle cable is correctly adjusted as described in Routine Maintenance, and that the drain hose is correctly routed.
10 Fit a new gasket to the exhaust port, using a smear of grease to stick it in place. On DR125S models check the condition of the exhaust pipe/silencer gasket, renewing it if it is damaged or worn. Refit the exhaust system or exhaust pipe (as applicable), aligning it on its mountings with the fasteners only loosely secured and tighten first the front mountings, then the rear; all exhaust system mounting nuts or bolts are tightened to a torque setting of 0.9 – 1.2 kgf m (6.5 – 8.5 lbf ft), except for the pillion footrest mounting bolt on all GS125 models which is tightened to 2.7 – 3.6 kgf m (19.5 – 26 lbf ft).
11 Refit the breather hose to its crankcase stub and secure it with its spring clip. On GS125 ES models only, connect the starter motor lead to its terminal, tightenin° securely the retaining nut and lock washers, then refit the rubber terminal cover. Insert the tachometer cable into the driven gear housing, rotating the engine to engage the driven gear and cable, then use a suitable pair of pliers to tighten securely the cable retaining ring. Refit the spark plug cap to the spark plug.
12 Slacken the adjuster lock nuts and screw fully in both adjusters to gain the maximum free play in the clutch cable. Insert the cable inner wire into the retaining lug set on the crankcase top surface and on all GS125 models only, screw in fully the cable lower end adjuster. Refit the clutch operating lever to the release shaft splines, aligning the slit in the operating lever with the line scribed on the top of the shaft as shown in the accompanying illustration. This may prove difficult to achieve if a new cable has been fitted; check that both adjusters are screwed in as far as possible and that the release shaft is rotated fully clockwise. Tighten securely the pinch bolt. Using only the lower cable adjuster, adjust the clutch cable until there is 4 mm (0.16 in) free play, measured between the butt end of the lever and the handlebar clamp. The handlebar clamp adjuster may be used if necessary, but should be reserved for minor or quick roadside adjustments. Tighten the adjuster locknut and replace the rubber adjuster covers, where fitted. Use the clamp provided to secure the cable clear of the exhaust.
13 Remove all traces of corrosion from the battery terminals and connect the battery, using the wire colour-coding to ensure correct polarity. Smear petroleum jelly over the terminals and check that the battery breather pipe is correctly routed and free from kinks or blockages.
14 Refit the fuel tank, securing the fuel pipe with its spring clip, then switch the tap to the 'Res' position and check for fuel leaks. Refit the seat and sidepanels, tightening all mounting bolts.
15 Remove the filler plug set in the crankcase right-hand cover and check that the oil drain plug, both filters and both filter caps have been correctly refitted and securely fastened, then pour in 1300 cc (2.3 pint) of SAE 10W/40 SE or SF engine oil. Refit the filler plug, but remember to check the oil level after the machine has been run for the first time.

42.1 Hang engine/gearbox unit from top mounting then refit ...

42.2 ... swinging arm pivot bolt – pack pivot bearings with grease

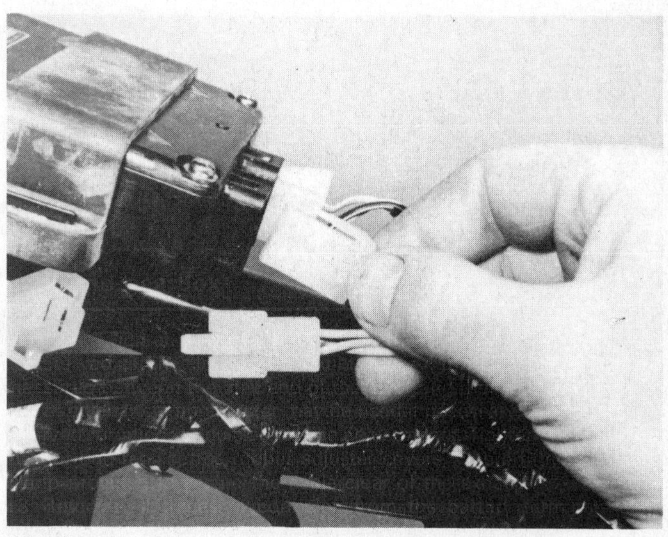
42.5 Connect all leads and secure wiring to prevent damage

42.6a Refit gearbox sprocket and lock washer over output shaft end

42.6b Refit retaining nut as shown – apply thread locking compound

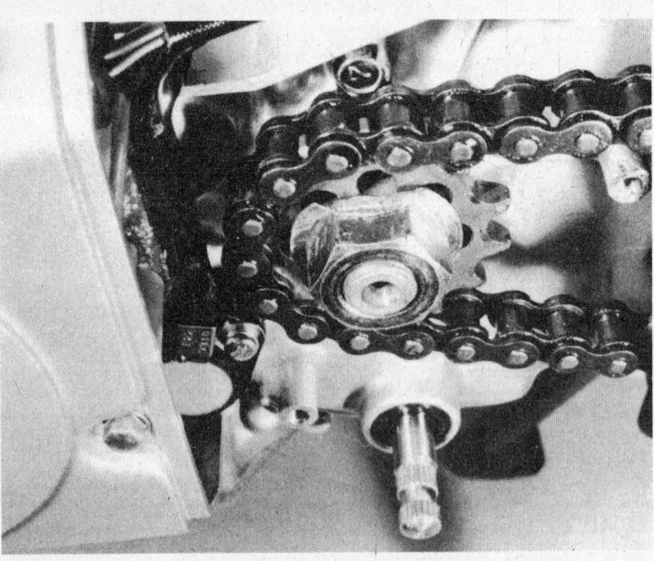

42.6c Tighten and secure nut with lock washer

42.7a GS125 models only – gearchange linkage must be aligned as shown

42.10a Position exhaust pipe and retaining flange ...

42.10b ... tighten exhaust front mountings ...

42.10c ... before the rear mountings

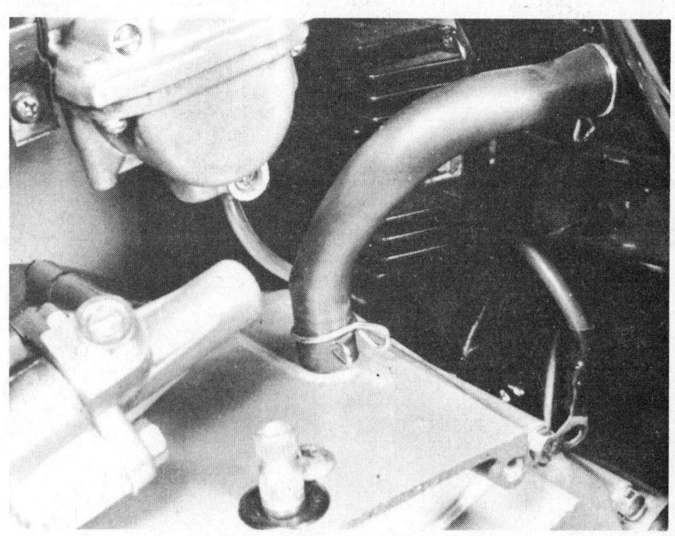
42.11 Do not forget to refit crankcase breather hose

42.12a Adjust clutch cable at lower adjuster (GS125 shown) ...

42.12b ... and secure cable clear of exhaust pipe

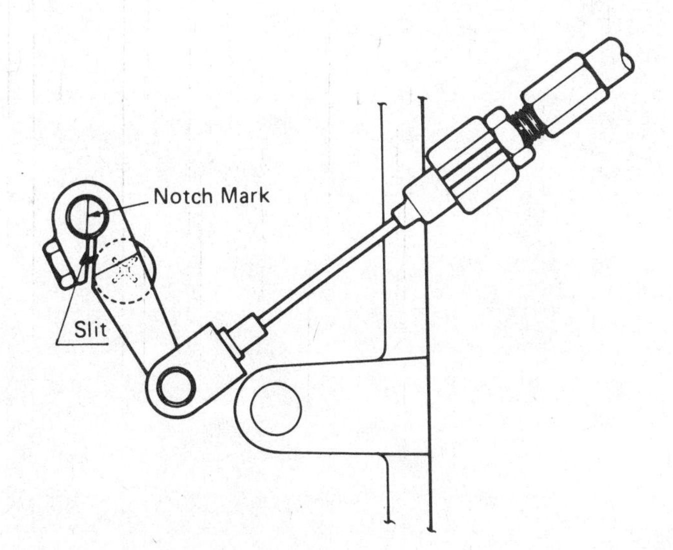

Fig. 1.24 Clutch operating lever and release shaft alignment

43 Starting and running the rebuilt engine

1 Attempt to start the engine using the usual procedure adopted for a cold engine. Do not be disillusioned if there is no sign of life initially. A certain amount of perseverance may prove necessary to coax the engine into activity even if new parts have not been fitted. Should the engine persist in not starting, check that the spark plug has not become fouled by the oil used during re-assembly. Failing this go through the fault finding charts and work out what the problem is methodically.

2 When the engine does start, keep it running as slowly as possible to allow the oil to circulate. Open the choke as soon as the engine will run without it. During the initial running, a certain amount of smoke may be in evidence due to the oil used in the reassembly sequence being burnt away. The resulting smoke should gradually subside.

3 Check the engine for blowing gaskets and oil leaks. Before using the machine on the road, check that all the gears select properly, and that the controls function correctly.

4 As soon as the engine has warmed up to its normal operating temperature, allow it to idle and check the carburettor idle speed adjustment as described in Chapter 2 of this Manual. If any repair work has been carried out and new components have been fitted, carburettor adjustment will almost certainly be required. Once the

engine is ticking over smoothly at its correct speed, check and reset if necessary the throttle cable adjustment.

5 Stop the engine and allow the oil level to settle for one or two minutes, then check the oil level. With the machine standing upright on level ground the oil level should be between the 'F' (full) and 'L' (low) marks adjacent to the level window set in the crankcase right-hand cover. Remove the filler plug and add oil if necessary to bring the level up to the 'F' mark.

44 Taking the rebuilt machine on the road

1 Any rebuilt machine will need time to settle down, even if parts have been replaced in their original order. For this reason it is highly advisable to treat the machine gently for the first few miles to ensure oil has circulated throughout the lubrication system and that any parts fitted have begun to bed down.

2 Even greater care is necessary if the engine has been rebored or if a new crankshaft has been fitted. In the case of a rebore, the engine will have to be run-in again, as if the machine were new. This means greater use of the gearbox and a restraining hand on the throttle until at least 500 miles have been covered. There is no point in keeping to any set speed limit; the main requirement is to keep a light loading on the engine and to gradually work up performance until the 500 mile mark is reached. These recommendations can be lessened to an extent when only a new crankshaft is fitted. Experience is the best guide since it is easy to tell when an engine is running freely.

3 If at any time a lubrication failure is suspected, stop the engine immediately, and investigate the cause. If any engine is run without oil, even for a short period, irreparable engine damage is inevitable.

4 When the engine has cooled down completely after the initial run, recheck the various settings, especially the valve clearances. During the run most of the engine components will have settled into their normal working locations. Check the various oil levels, particularly that of the engine as it may have dropped slightly now that the various passages and recesses have filled.

Chapter 2 Fuel system and lubrication

Refer to Chapter 7 for information relating to DR125 SF, SH, SJ 'Raider', GN125 and GZ125 'Marauder' models

Contents

Specifications

Fuel tank capacity	**GS125 models**	**DR125 S model**
Overall	11.0 lit (2.4 gal)	8.5 lit (1.9 gal)
Reserve	1.6 lit (2.8 pint)	2.0 lit (3.5 pint)

Fuel grade

Manufacturer's recommendation	Minimum octane rating 91 RON, unleaded or low-lead

Carburettor	**GS125 with slide carburettor**	**GS125 with CV carburettor**	**DR125 S model**
Make	Mikuni	Mikuni	Teikei
Type	VM24SS	BS26SS	S22P
ID number	05320	05311	05210 (05260 D, E models)
Bore	24 mm	26 mm	22 mm
Main jet	95	102.5	100 (98 D, E models)
Main air jet	1.4 mm	1.6 mm	1.1 mm
Jet needle	4JT38	4DZ35	5C71 (5C72 D, E models)
Clip position – grooves from top	3rd	2nd	3rd
Needle jet	0-4	P-2	2.595 or V95
Throttle valve cutaway	1.5	N/Av	2.5
Pilot jet	20	35	38 (40 D, E models)
By-pass	0.8 mm	0.8 mm	1.0 mm
Pilot outlet	0.7 mm	0.9 mm	0.8 mm
Valve seat	2.0 mm	2.0 mm	2.0 mm
Starter jet	42.5	25	60 (both)
Pilot mixture screw – turns out	$^3/_4$	1 $^5/_8$	3 ± $^1/_2$ (1$^1/_2$ E model)
Pilot air jet	1.2 mm	180	1.2 mm (1.1 mm D, E models)
Float height	24.5 ± 1.0 mm (0.97 ± 0.04 in)	21.4 ± 1.0 mm (0.84 ± 0.04 in)	26.5 ± 0.5 mm (1.04 ± 0.02 in)
Idle speed	1450 ± 50 rpm	1450 ± 50 rpm	1450 ± 50 rpm

Lubrication system

Type	Pressure fed, wet sump
Oil pump type	Trochoid
Oil capacity:	
Oil change	850 cc (1.5 pint)
Oil and filter change	950 cc (1.7 pint)
Engine overhaul	1300 cc (2.3 pint)
Oil type	SAE 10W/40 SE or SF engine oil

1 General description

The fuel system comprises a tank from which fuel is fed by gravity to the carburettor float chamber via a tap which has 'Off', 'On' and 'Reserve' positions. The latter position provides a reserve supply of fuel when the level in the tank is running low. Air is drawn into the carburettor via an oil-impregnated polyurethane foam filter element which is contained in a housing to the rear of the cylinder barrel.

On all DR models and all GS models except the ESM, the carburettor is of the slide type, whereas on the GS125 ESM a constant velocity (CV) carburettor is fitted.

Engine lubrication is of the wet sump type, the oil contained in the reservoir formed by the crankcase castings being drawn through a gauze filter by an oil pump. The pump is of the Eaton trochoid type and is driven by gears from the crankshaft; the oil is forced through a full-flow pleated paper oil filter element and then to the crankshaft, the camshaft and the two gearbox shafts. All components not provided with a direct supply of oil are lubricated by splash, whereupon the oil drains back into the crankcase reservoir. A by-pass valve is fitted to maintain full oil flow if the paper element should become so clogged as to restrict the flow.

The exhaust system fitted to all GS125 models is a single unit comprising the exhaust pipe and silencer welded together. The unit is finished in a heat-resistant matt black coating and does not incorporate baffles that can be removed for cleaning purposes. The DR125 S exhaust system is similar but incorporates a second, separate silencer assembly and is routed upwards over the engine/gearbox unit to pass down the right-hand side of the machine at a high level in keeping with the trail bike styling of the machine.

2 Fuel tank: removal, examination and refitting

1 In the event that it is necessary to remove the fuel tank for repairs the following points should be noted. Fuel tank repair, whether necessitated by accident damage or by fuel leaks, is a task for the professional. Welding or brazing is not recommended unless the tank is purged of all fuel vapour; which is a difficult condition to achieve. Resin-based tank sealing compounds are a much more satisfactory method of curing leaks, and are now available through suppliers who advertise regularly in the motorcycle press. Accident damage repairs will inevitably involve re-painting the tank; matching of modern paint finishes, especially metallic ones, is a very difficult task not to be lightly undertaken by the average owner. It is therefore recommended that the tank be removed by the owner, and then taken to a motorcycle dealer or similar expert for professional attention.

2 Carefully detach the sidepanels and remove the dual seat, then turn the petrol tap to the 'Off' position. Use a suitable pair of pliers to release the wire petrol pipe retaining clip to allow the pipe to be pulled off the stub at the rear of the tap. Careful use of a small screwdriver may be necessary to help ease the pipe off the stub. Once the pipe is detached, allow any fuel in the pipe to drain into a small clean container. The tank may now be detached from the frame by unscrewing the retaining bolt(s) at the rear of the tank and pulling the tank up and rearwards off its front mounting rubbers. Place the tank mounting components in a safe place ready for refitting. Inspect the mounting rubbers for signs of damage or deterioration and if necessary renew them before refitting of the tank is due to take place.

3 Store the tank in a safe place whilst it is removed from the machine, well away from any naked lights or flames. It will otherwise represent a considerable fire or explosion hazard. Check that the tap is not leaking and that it cannot be accidentally knocked into the 'On' position. It is well worth taking simple precautions to protect the paint finish of the tank whilst in storage. Placing the tank on a soft protected surface and covering it with a protective cloth or mat may well avoid damage being caused to the finish by dirt, grit, dropped tools, etc.

4 To refit the tank, reverse the procedure adopted for its removal. Move it from side to side before it is fully home, so that the rubber buffers engage with the guide channels correctly. If difficulty is encountered in engaging the front of the tank with the rubber buffers, apply a small amount of petrol to the buffers to ease location. Secure the tank with the retaining bolt(s) whilst ensuring that the mounting components are correctly located and that there is no metal to metal contact between the tank and frame.

5 Finally, always carry out a leak check on the fuel pipe connections after fitting the tank and turning the tap lever to the 'On' position. Any leaks found must be cured; as well as wasting fuel, any petrol dropping onto hot engine castings may well result in a fire or explosion occurring.

3 Fuel tap: removal, examination and refitting

1 The fuel tap assembly is secured to the underside of the tank by two bolts, the joint being sealed by an O-ring.

2 When working on any component in a fuel system, always take careful precautions against the risk of fire; for example, never smoke or use a naked light to illuminate the work surface. Always use a soft-bristled brush such as an old toothbrush when cleaning to avoid damage to the delicate filter gauze. Some form of eye protection is recommended against the drops of fuel which will inevitably be flying around during cleaning operations. If work is being carried out to cure fuel leaks, first ensure that all retaining fasteners are tight but not overtightened. The tap components are delicate and easily distorted; overtightening will merely exacerbate any leaks. If the components are correctly fastened and the leaks persist, the tap must be stripped and the relevant seal renewed.

3 Petrol tap removal must be preceded by draining the petrol remaining in the tank. Remove the petrol feed pipe from the carburettor, switch the tap to the 'Res' position and allow all the fuel to drain into a clean, dry, container of suitable size. Note that this container must be constructed of metal and clearly marked if the petrol is to be stored for any length of time. Alternatively the tank can be removed, as described in the preceding Section, and placed on one side so that the petrol level is below that of the tap. Take care not to damage the tank paintwork if this method is employed, and ensure that petrol does not leak from the filler cap.

4 Unscrew the filter bowl by using a close-fitting ring spanner on the moulded square end of the bowl. Pick out the sealing O-ring. Unscrew the two bolts which retain the tap to the underside of the tank and carefully withdraw the tap assembly. The filter gauze can then be removed for cleaning. The tap lever is retained by a grub screw threaded into the underside of the tap body. Remove the screw, then withdraw the tap lever. It may be necessary to pad the jaws of a suitable pair of pliers with cloth or tape and to pull the lever out as it is sealed by a tight-fitting O-ring. Pick out the small spring and the shaped plug behind it.

5 Thoroughly clean and inspect all the tap components, renewing any parts that are worn or damaged. Again, pay particular attention to the fibre sealing washers beneath the heads of the two tap mounting bolts and to the large O-ring set in the tap mating flange surface. It is recommended that these components are renewed whenever the tap is disturbed as their condition is critical to the prevention of persistent petrol leaks.

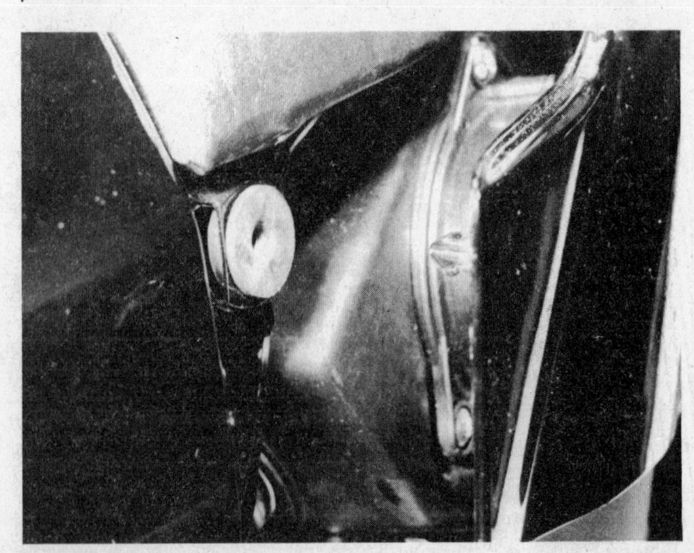

2.2a Sidepanels are retained by prongs pressed into rubber grommets

2.2b Release wire clip to remove fuel pipe from tap spigot

2.2c GS125 fuel tank is retained by two bolts at rear

2.2d Check condition of mounting rubbers – renew if necessary

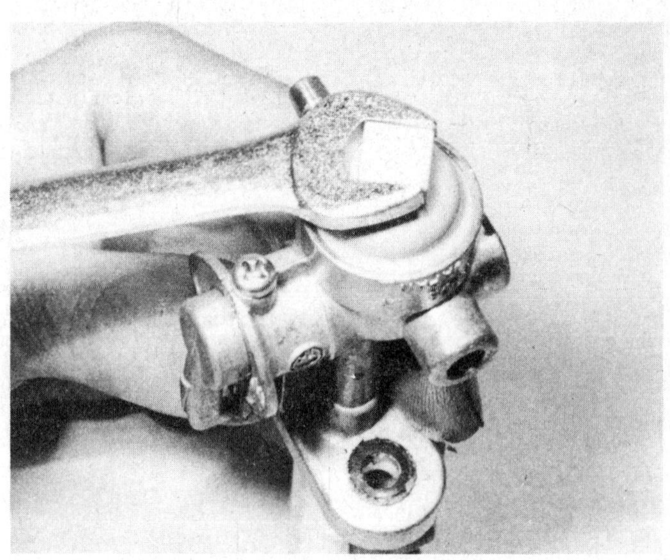

3.4a Use close-fitting spanner to remove or refit tap filter bowl

3.4b Renew O-ring if worn or damaged – remove grub screw (arrowed) to release tap lever components

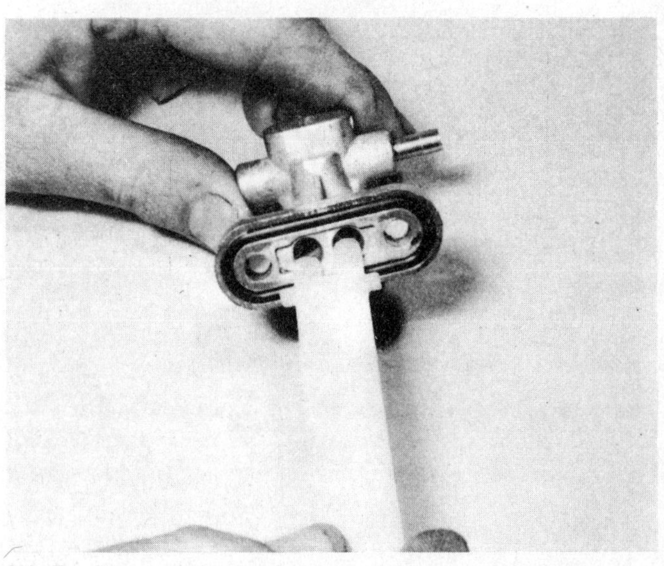

3.4c Tap filter gauze can be displaced after tap is removed from fuel tank

Fig. 2.1 Fuel tap

1 Fuel tap
2 O-ring
3 O-ring
4 Filter bowl
5 Bolt - 2 off
6 Washer - 2 off
7 Tap lever
8 O-ring
9 Spring
10 Shaped plug
11 Screw

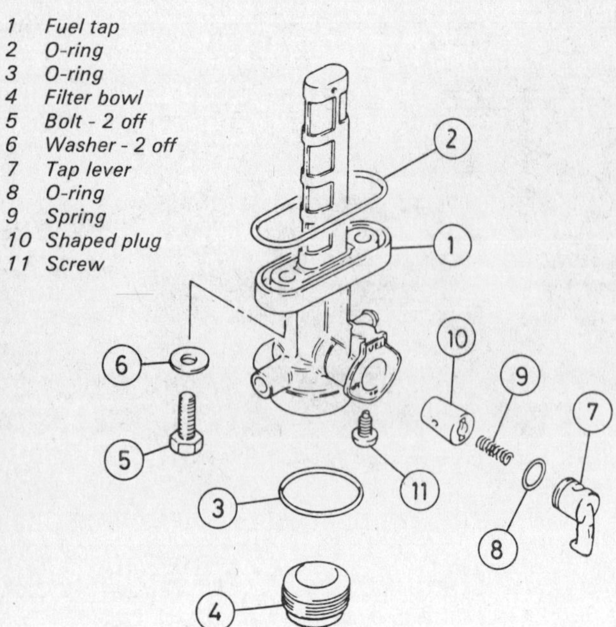

4 Fuel feed pipe: examination

1 The fuel feed pipe is made from thin walled synthetic rubber and is of the push-on type. It is necessary to replace the pipe only if it becomes hard or splits. It is unlikely that the retaining clips will need replacing due to fatigue as the main seal between the pipe and union is effected by an interference fit.
2 If the fuel pipe has been replaced with a transparent plastic type for any reason, look for signs of yellowing which indicate that the pipe is becoming brittle due to the plasticiser being leached out by the petrol. It is a sound precaution to renew a pipe when this occurs, as any subsequent breakage whilst in use will be almost impossible to repair. **Note:** On no account should natural rubber tubing be used to carry petrol, even as a temporary measure. The petrol will dissolve the inner wall, causing blockages in the carburettor jets which will prove very difficult to remove.

5 Carburettor: removal and refitting

1 As a general rule, the carburettor should be left alone unless it is in obvious need of overhaul. Before a decision is made to remove and dismantle, ensure that all other possible sources of trouble have been eliminated. This includes the more obvious candidates such as a fouled spark plug, a dirty air filter element or damaged exhaust system. If a fault has been traced back to the carburettor, proceed as follows.
2 Make sure that the petrol tap is turned off, then prise off the petrol feed pipe at the carburettor union.

Slide carburettor

3 Slacken the screws of the clips which secure the carburettor to its inlet stub and air filter hose so that the carburettor can be twisted free of them and partially removed. This affords access to the threaded carburettor top, which should be unscrewed to allow the throttle valve assembly to be withdrawn. It is not normally necessary to remove this from the cable and it can be left attached and taped clear of the engine. If removal is necessary, however, proceed as follows.
4 Holding the carburettor top, compress the throttle return spring against it and hold it in position against the cap so that the cable can be pushed down and slid out of its locating groove. The various parts can now be removed and should be placed with the carburettor.

CV carburettor

5 Slacken the locknuts of the throttle cable adjuster at the carburettor pulley and disconnect the cable trunnion from the pulley.
6 Slacken the screws of the clamps which secure the carburettor to the inlet stub and air filter hose and manoeuvre it free of the machine.

Both carburettor types

7 The carburettor is refitted by reversing the removal sequence. Note that it is important that the instrument is mounted vertically to ensure that the fuel level in the float bowl is correct. A locating tab is fitted on slide carburettors to provide a good guide to alignment but it is worthwhile checking this for accuracy. Once refitted, check the carburettor adjustments as described later in this Chapter.
8 **Note:** If the carburettor is to be set up from scratch it is important to check jet and float level settings prior to installation. To this end, refer to Sections 7 and 8 before the carburettor is refitted.

6 Carburettor, dismantling, examination and reassembly

1 Invert the carburettor and remove the float chamber by withdrawing the four retaining screws. The float chamber bowl will lift away, exposing the float assembly, hinge and float needle. There is a gasket between the float chamber bowl and the carburettor body which need not be disturbed unless it is leaking.
2 With a pair of thin-nose pliers, withdraw the float pivot pin to free the floats and the float needle. Check that neither of the floats has punctured and that the float needle and seating are both clean and in good condition. If the needle has a ridge, it should be renewed in conjunction with its seating.
3 The two floats are made of plastic or brass, and are connected by a brass bridge and pivot piece. If either float is leaking, it will produce the wrong fuel level in the float chamber, leading to flooding and an over-rich mixture. The floats cannot be repaired successfully, and renewal will be required.
4 On all slide carburettor GS125 models, unscrew the main jet and then the hexagon-headed jet holder. Pick out the O-ring, invert the carburettor and push out the needle jet from above, using a soft wooden drift. Be careful to use only a soft drift, or the jet will be damaged, and note that moderate pressure will be quite enough to dislodge the jet from its seating. On all DR125 S models and CV carburettor GS125 models, unscrew the main jet, withdraw the plain washer, and push

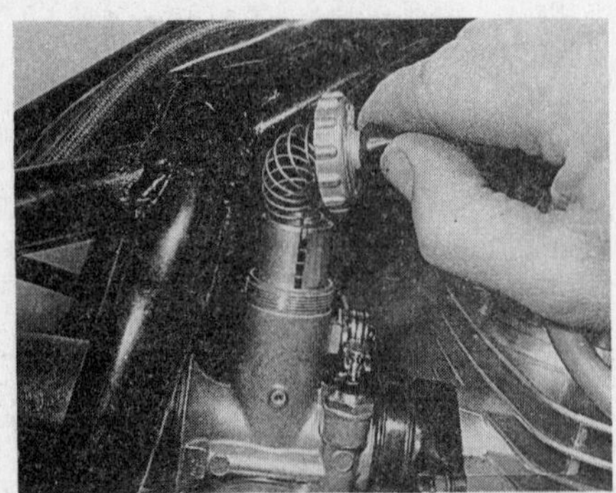
5.3 On slide carburettors, unscrew carburettor top and withdraw throttle valve ...

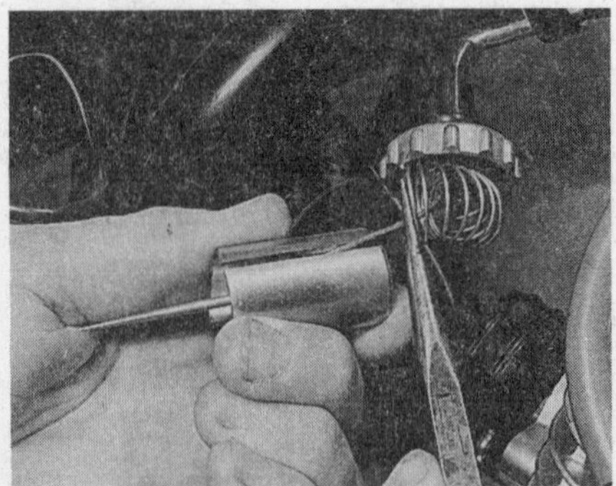
5.4a ... compress spring with pliers and slide cable out of slot in valve

5.4b Jet needle retaining plate – GS125 slide carburettor

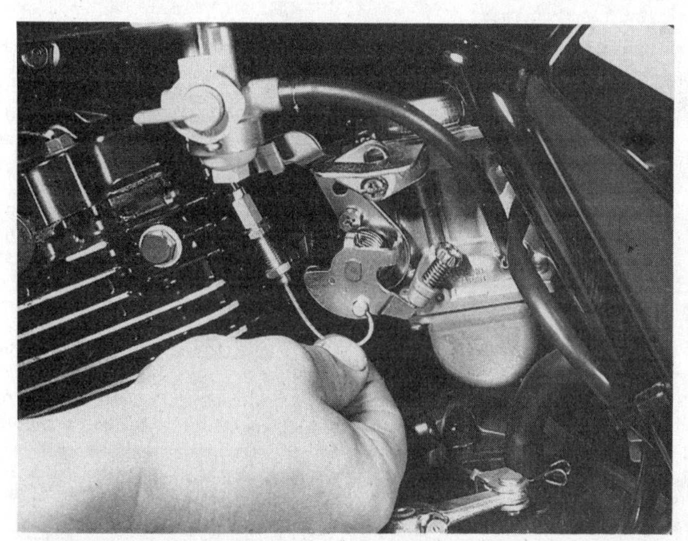

5.5 On CV carburettors, once adjuster has been freed, cable can be disconnected from pulley

5.7 Align tab and inlet stub slot (arrows) to correctly align slide type carburettor

out from above the combined main jet holder and needle jet, using a soft wooden drift as previously described. Turn the carburettor upside down again and use a small electrical screwdriver to unscrew the pilot jet from its seating adjacent to the main/needle jet holder; note that a plug may obscure the pilot jet head. On slide carburettors note the location of the main air jet on the air filter side of the venturi; it is positioned at the base of the venturi, to the right of the pilot air jet.

5 The float needle valve seat is retained by a separate plate on all GS125 models and is screwed into place on all DR125 S models; remove the valve seat and inspect both the float needle and the valve seat for signs of wear, which normally manifests itself in the form of a slight ridge or groove around the seating taper of the float needle. As previously mentioned, any such wear will allow leakage and raise the fuel level, thus richening the air/fuel mixture, and must be rectified by renewing both components together. On the CV carburettor, clean the valve filter carefully. Lastly, check the condition of the valve seat washer or O-ring and renew it if it is at all worn or damaged.

6 To gain access to the jet needle on slide carburettors, the throttle cable must first be disconnected from the valve as described in Section 5. Release the retaining plate (GS125) or spring seat and holder (DR125 S) to free the jet needle from the valve. On CV carburettors, remove the carburettor top (4 screws) and the large return spring. Carefully peel the rubber diaphragm out of its groove and lift out the combined diaphragm/throttle valve. Remove the two screws from inside the valve to release the retaining plate and free the jet needle. On all models, check that the needle is straight by rolling it on a flat surface such as a sheet of plate glass and then examine both the jet needle and the needle jet for signs of wear or damage. Any damage to either component will mean that the two must be renewed together. Do not attempt to straighten a bent needle as they are easily broken; also, if the machine has been running for any length of time with a bent needle, the jet needle and needle jet must be renewed anyway to rectify the uneven wear which will have occurred. On CV carburettors, note that when refitting the needle retaining plate its cutout must align with the hole in the valve.

7 The jet needle is suspended from the valve, where it is retained by a circlip in the groove specified at the front of this Chapter, but other grooves are provided as a means of adjustment so that the mixture strength can be either increased or decreased by raising or lowering the needle. Care is necessary when replacing the carburettor top (slide carburettor) or throttle valve/diaphragm (CV carburettor) because the needle is easily bent if it does not fit inside the needle jet.

8 After an extended period of service the throttle valve will wear and may produce a clicking sound within the carburettor body. Wear will be evident from inspection, usually at the base of the slide and in the locating groove (slide carburettor). On CV carburettors inspect the diaphragm carefully; hold it up to a good light source and check that there are no cracks or holes. A damaged diaphragm must be renewed together with the throttle valve. When refitting the diaphragm, make sure its tab aligns with the cutout in the body. The throttle valve butterfly on CV carburettors should not require attention during the life of the machine. It is secured to the throttle shaft by two screws.

9 The manually operated choke is unlikely to require attention during the normal service life of the machine. To gain access to the choke plunger on GS125 models, first disconnect its operating lever (slide carburettor) or operating shaft and fork (CV carburettor); on DR models the operating knob is removed with the plunger. Unscrew the plunger from the carburettor body and check that it is unworn, that its seating is clean and undamaged and that it operates smoothly. Any fault will mean that the complete assembly must be renewed as repairs are not possible and very few parts can be purchased separately. Check that the plunger housing and the various passages in the carburettor body are clean and free from any particles of foreign matter.

10 If removal of the throttle stop and pilot mixture adjustment screw is required, screw each one in carefully until it seats lightly, counting and recording the number of turns required, and then unscrew each one, complete with its spring, washer and O-ring (where fitted). Remove any dirt or corrosion and check for signs of wear or damage, renewing any component which needs it. When refitting the adjustment screws, ensure that, where fitted, the sealing O-rings are fitted first, then the screw retaining springs, and the screw itself. Tighten the screws carefully until they seat lightly, then unscrew each one by the number of turns counted on removal to return it to its original position; this will serve as a basis for subsequent adjustments.

11 On slide carburettors the pilot air jet is located to the left of the main air jet in the air filter side of the carburettor venturi. On CV carburettors, it is located in the upper part of the carburettor body. In both cases, the jet can be unscrewed for cleaning.

12 Before the carburettor is reassembled, using the reversed dismantling procedure, it should be cleaned out thoroughly, preferably by the use of compressed air. Avoid using a rag because there is always a risk of fine particles of lint obstructing the internal air passages or the jet orifices. Check carefully the condition of the carburettor body and float chamber, looking for distorted or damaged mating surfaces or any other signs of wear. If severe damage or wear is found, the carburettor assembly will have to be renewed. Check the condition of all O-rings and gaskets, renewing any that are worn or distorted.

13 Never use a piece of wire or sharp metal object to clear a blocked jet. It is only too easy to enlarge the jet under these circumstances and increase the rate of petrol consumption. Always use compressed air to clear a blockage; a tyre pump makes an admirable substitute when a compressed air line is not available.

14 Do not use excessive force when reassembling the carburettor because it is quite easy to shear the small jets or some of the smaller screws. Before attaching the air cleaner hose, check that the throttle slide rises smoothly when the throttle is opened. Check the float height before fitting the float chamber and adjust the throttle cable after the carburettor has been refitted. If the pilot mixture screw and throttle stop screw settings have been disturbed, carry out adjustment as described in Section 8.

6.2a Remove the four retaining screws and lift off the float chamber

6.2b Withdraw pivot pin to release floats

6.2c Lift float needle out of its seating and examine needle tip

6.4a GS125 slide carburettor – unscrew main jet (arrow) ...

6.4b ... followed by jet holder and O-ring

6.4c GS125 CV carburettor – unscrew main jet and washer ...

6.4d ... and remove needle jet from above

6.4e When fitting needle jet, align cutout with peg (arrow) in body (CV carburettor)

6.4f Pilot jet location – slide carburettor

6.4g Pilot jet location – CV carburettor

6.5a Release retaining plate on GS125 models ...

6.5b ... to release float needle seating

6.5c Detach gauze filter for cleaning on CV carburettors

6.6a On CV carburettors remove carburettor top and return spring ...

6.6b ... carefully peel diaphragm off body (note alignment tab arrowed), and lift throttle valve out of carburettor ...

6.6c ... two screws retain jet needle plate

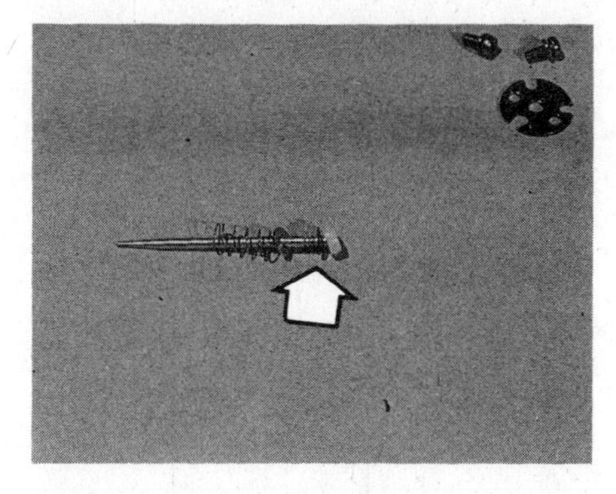

6.7 Jet needle assembly on CV carburettor, clip arrowed

6.8 Two screws retain butterfly valve to throttle shaft on CV carburettor

6.9a Choke lever is retained by single screw on GS125 slide carburettor

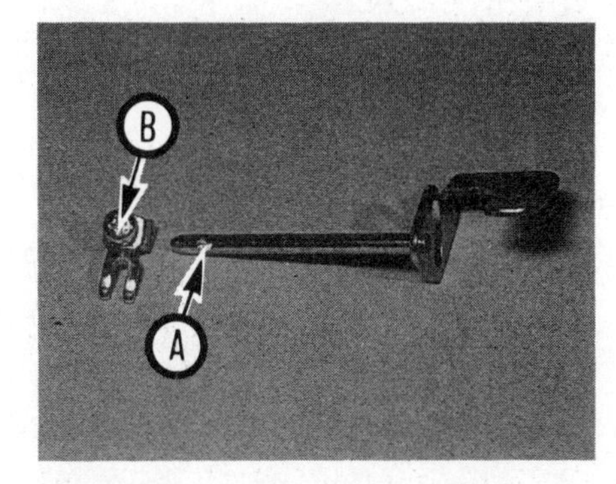

6.9b Detent on choke lever shaft (A) must align with screw of fork (B) on CV carburettor

6.9c Unscrew choke plunger from body for examination

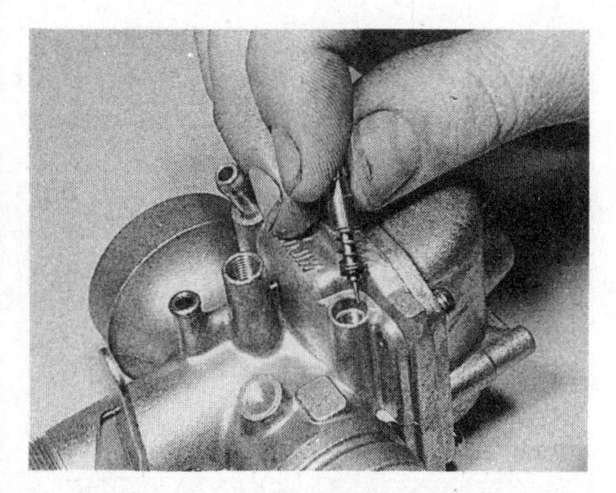

6.10a Pilot mixture screw position on GS125 slide carburettor

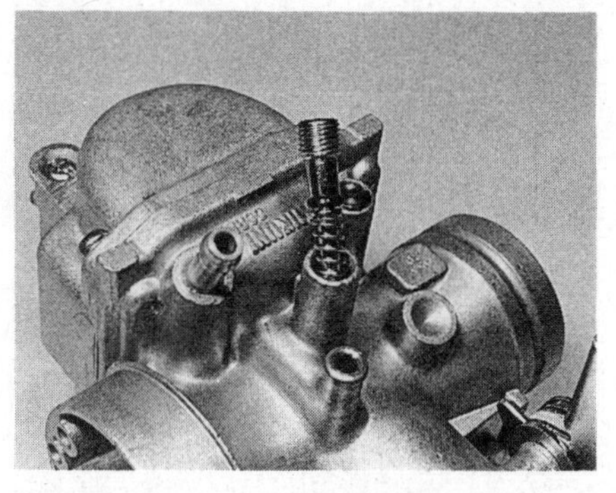

6.10b Throttle stop screw position on GS125 slide carburettor

6.11 Pilot air jet (arrow) on CV carburettor

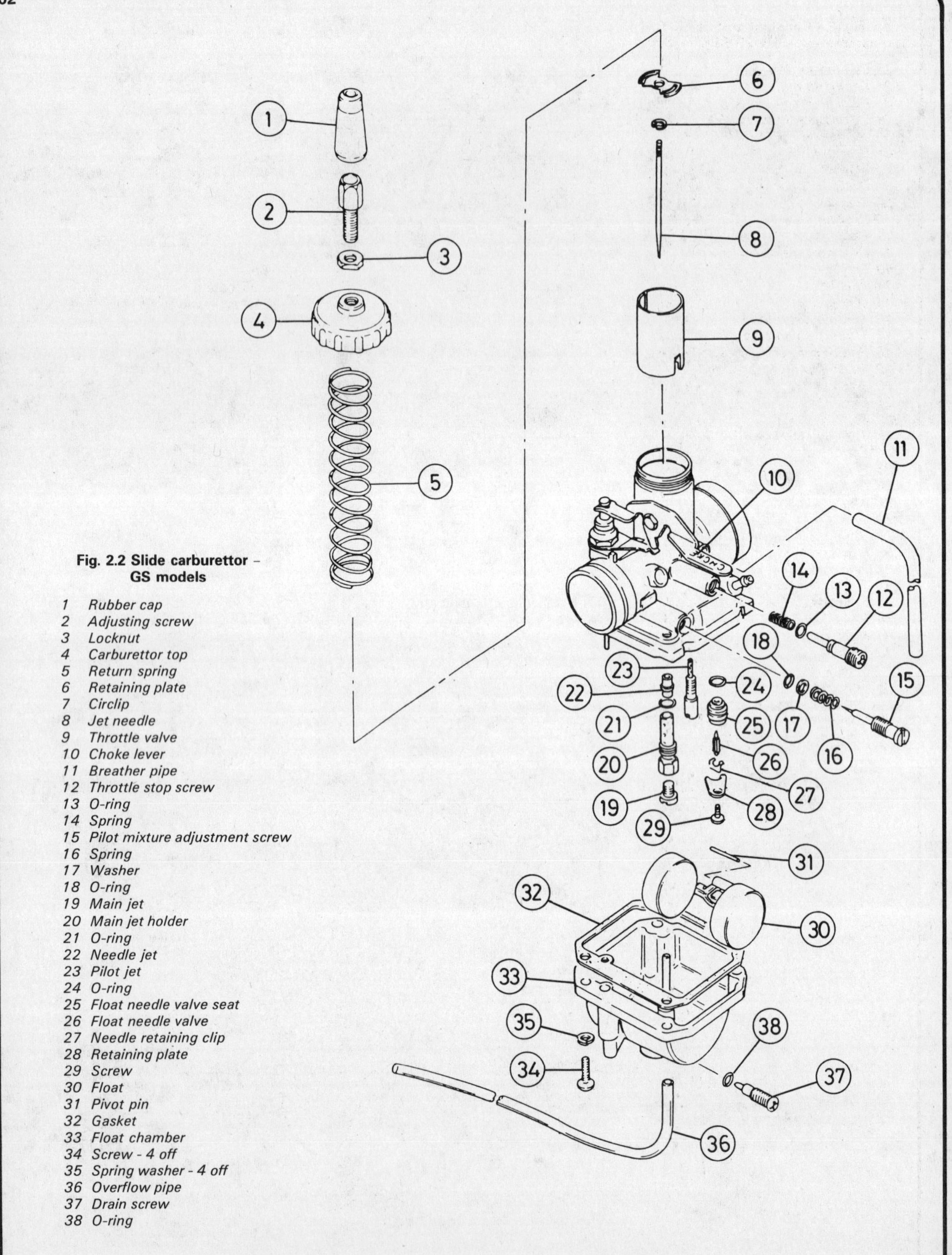

Fig. 2.2 Slide carburettor – GS models

1 Rubber cap
2 Adjusting screw
3 Locknut
4 Carburettor top
5 Return spring
6 Retaining plate
7 Circlip
8 Jet needle
9 Throttle valve
10 Choke lever
11 Breather pipe
12 Throttle stop screw
13 O-ring
14 Spring
15 Pilot mixture adjustment screw
16 Spring
17 Washer
18 O-ring
19 Main jet
20 Main jet holder
21 O-ring
22 Needle jet
23 Pilot jet
24 O-ring
25 Float needle valve seat
26 Float needle valve
27 Needle retaining clip
28 Retaining plate
29 Screw
30 Float
31 Pivot pin
32 Gasket
33 Float chamber
34 Screw - 4 off
35 Spring washer - 4 off
36 Overflow pipe
37 Drain screw
38 O-ring

Fig. 2.3 Constant velocity (CV) carburettor – GS125 ESM models

1 Carburettor top
2 Return spring
3 Retaining plate
4 Diaphragm/throttle valve
5 Jet needle
6 Spring
7 Washer
8 Circlip
9 Ring
10 Pilot air jet
11 Plug
12 Pilot mixture screw
13 Spring
14 Washer
15 O-ring
16 Throttle butterfly
17 Throttle spindle and cable pulley
18 Return spring
19 Sealing ring
20 Washer
21 Circlip
22 Cap
23 Throttle stop screw
24 Spring
25 Plug
26 Pilot jet
27 Float chamber
28 Gasket
29 Drain screw and washer
30 Floats
31 Pivot pin
32 Float valve, O-ring and seat
33 Gauze filter
34 Retaining plate
35 Main jet
36 Washer
37 Needle jet
38 Choke lever
39 Choke plunger
40 Choke link

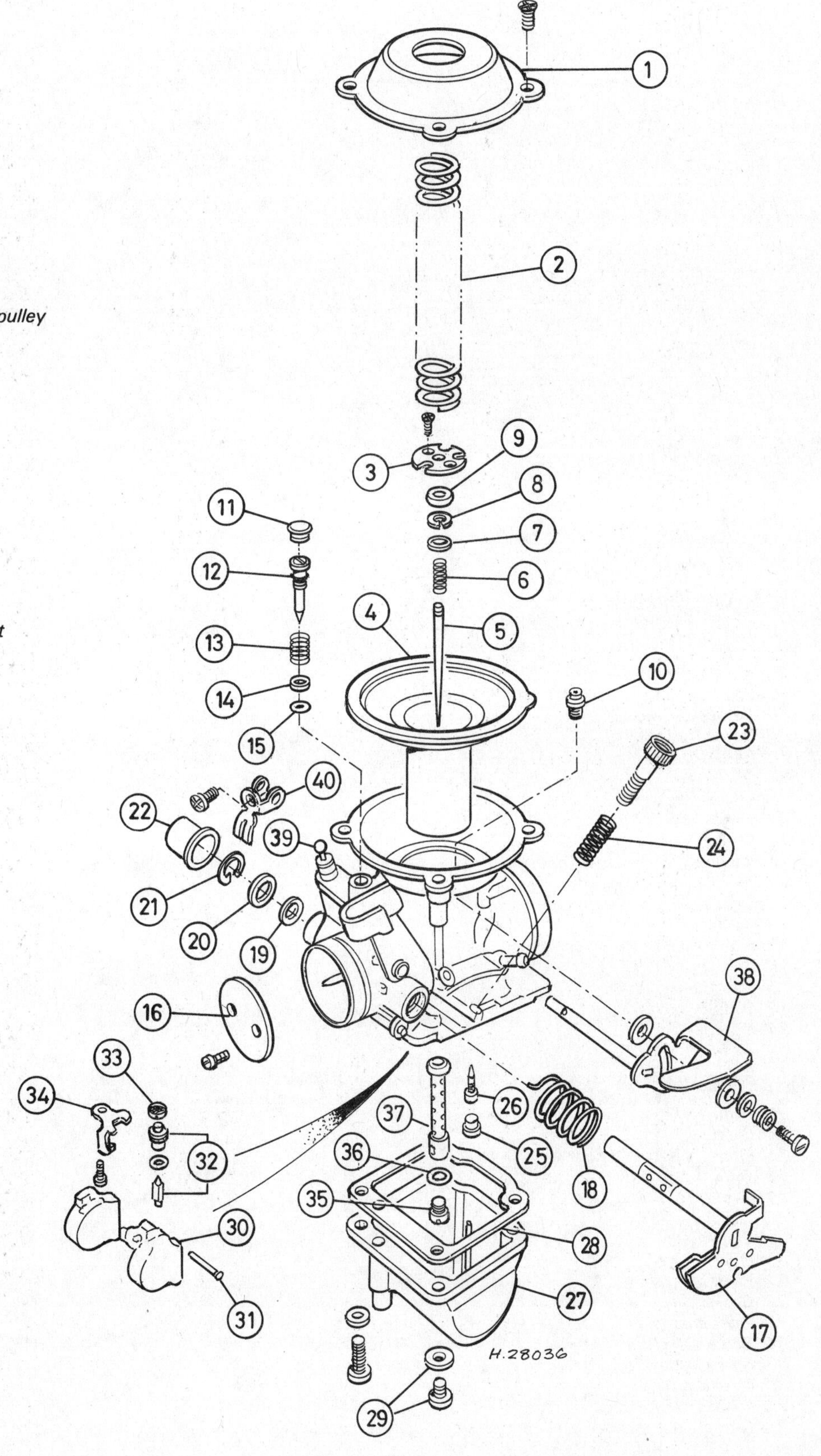

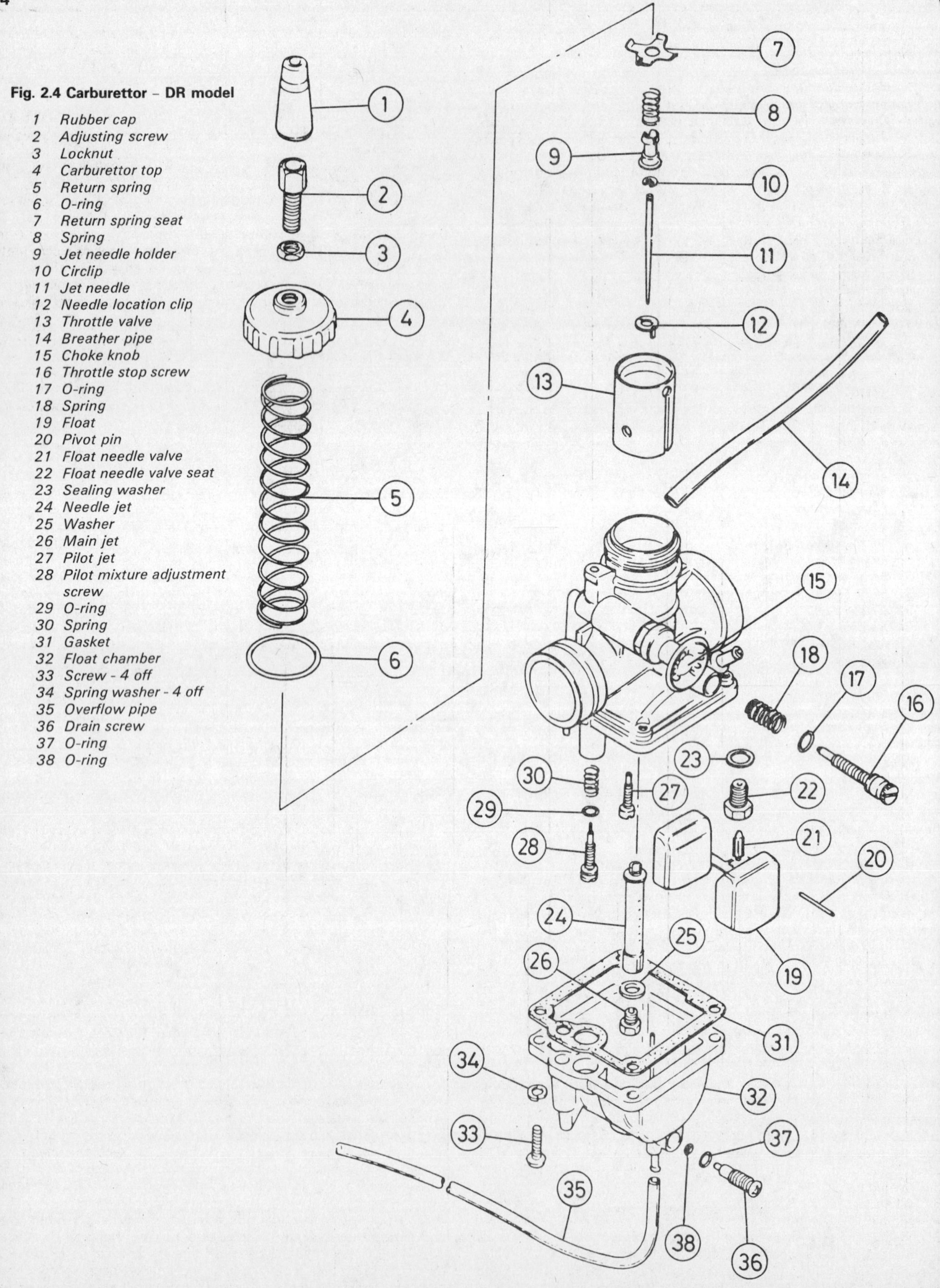

Fig. 2.4 Carburettor – DR model

1 *Rubber cap*
2 *Adjusting screw*
3 *Locknut*
4 *Carburettor top*
5 *Return spring*
6 *O-ring*
7 *Return spring seat*
8 *Spring*
9 *Jet needle holder*
10 *Circlip*
11 *Jet needle*
12 *Needle location clip*
13 *Throttle valve*
14 *Breather pipe*
15 *Choke knob*
16 *Throttle stop screw*
17 *O-ring*
18 *Spring*
19 *Float*
20 *Pivot pin*
21 *Float needle valve*
22 *Float needle valve seat*
23 *Sealing washer*
24 *Needle jet*
25 *Washer*
26 *Main jet*
27 *Pilot jet*
28 *Pilot mixture adjustment screw*
29 *O-ring*
30 *Spring*
31 *Gasket*
32 *Float chamber*
33 *Screw - 4 off*
34 *Spring washer - 4 off*
35 *Overflow pipe*
36 *Drain screw*
37 *O-ring*
38 *O-ring*

7 Carburettor: checking the settings

1 The various jet sizes, throttle valve cutaway and needle position are predetermined by the manufacturer and should not require modification. Check with the Specifications list at the beginning of this Chapter if there is any doubt about the types fitted. If a change appears necessary it can often be attributed to a developing engine fault unconnected with the carburettor. Although carburettors do wear in service, this process occurs slowly over an extended length of time and hence wear of the carburettor is unlikely to cause sudden or extreme malfunction. If a fault does occur check first other main systems, in which a fault may give similar symptoms, before proceeding with carburettor examination or modification.
2 Where non-standard items, such as exhaust systems, air filters or camshafts have been fitted to a machine, some alterations to carburation may be required. Arriving at the correct settings often requires trial and error, a method which demands skill borne of previous experience. In many cases the manufacturer of the non-standard equipment will be able to advise on correct carburation changes.
3 If alterations to the carburation must be made, always err on the side of a slightly rich mixture. A weak mixture will cause the engine to overheat which may cause engine seizure. Reference to Routine Maintenance and the spark plug condition colour page, will show how, after some experience has been gained, the condition of the spark plug electrodes can be interpreted as a reliable guide to mixture strength.

8 Carburettor: adjustment

1 Before any dismantling or adjustment is undertaken, eliminate all other possible causes of running problems, checking in particular the spark plug, ignition timing, air cleaner, valve clearances and the exhaust. Checking and cleaning these items as appropriate will often resolve a mysterious flat spot or misfire.
2 The first step in carburettor adjustment is to ensure that the jet sizes, needle position and float height are correct, which will require the removal and dismantling of the carburettors as described in Sections 5 and 6 of this Chapter.
3 If the carburettor has been removed for the purpose of checking jet sizes, the float height should be measured at the same time. It is unlikely that once this is set up correctly there will be a significant amount of variation, unless the float needle or seat have worn. These should be checked and renewed, if necessary, as described in Section 6.
4 Remove the float bowl from the carburettor body, if this has not already been done, and very carefully peel away the float chamber gasket. Check that the gasket surface of the carburettor body is clean and smooth once the gasket is removed. Hold the carburettor body so that the venturi is now vertical with the air filter side upwards and the floats are hanging from their pivot pin. Carefully tilt the carburettor to an angle of about 60° – 70° from the vertical so that the tang of the float pivot is resting firmly on the float needle and the float valve is therefore closed, but also so that the spring-loaded pin set in the float needle itself is not compressed. Measure the distance between the gasket face and the bottom of one float with an accurate ruler or a vernier caliper and compare the reading with that given in the Specifications.
5 If adjustment is required, remove the float assembly and bend by a very small amount the small tang which acts on the float needle pin. Reassemble the float and measure the height again. Repeat the process until the measurement is correct, then check that the other float is exactly the same height as the first. Bend the pivot very carefully and gently if any difference is found between the heights of the two floats.
6 When the jet sizes have been checked and reset as necessary, reassemble the carburettor and refit it to the machine as described in Sections 5 and 6 of this Chapter.
7 Start the engine and allow it to warm up to normal operating temperature, preferably by taking the machine on a short journey. Stop the engine and screw the pilot mixture screw in until it seats lightly, then unscrew it by the number of turns shown in the Specifications Section for the particular model. Start the engine and set the machine to its specified idle speed by rotating the throttle stop screw as necessary. Try turning the pilot mixture screw inwards by about $\frac{1}{4}$ turn at a time, noting its effect on the idling speed, then repeat the process, this time turning the screw outwards.
8 The pilot mixture screw should be set in the position which gives the fastest consistent tickover. The tickover speed may be reduced further, if necessary, by unscrewing the throttle stop screw the required amount. Check that the engine does not falter and stop after the throttle twistgrip has been opened and closed a few times.
9 Throttle cable adjustment should be checked at regular intervals and after any work is done to the carburettor or to the cable itself. The amount of free play specified for the throttle cable is 0.5 - 1.0 mm (0.02 – 0.04 in); use the adjuster on the carburettor top (slide carburettor) or at the pulley (CV carburettor) followed if necessary by the adjuster below the twistgrip, to achieve the correct setting. Tighten securely the adjuster locknuts and replace the rubber sleeves over the adjusters.

9 Air filter: general

The air filter will require no attention other than the regular cleaning and re-oiling described in the Routine Maintenance Chapter of this Manual.

10 Exhaust system: general

1 Exhaust systems fitted to four-stroke machines do not become clogged with carbon deposits as a general rule. The most common cause of problems will be corrosion, which will attack the components from inside as well as outside.
2 Check at regular intervals that the exhaust is securely fastened and that there are no leaks at any of the joints; leaks caused by damaged gaskets must be cured by the renewal of the gasket. Check that the system components are intact and have not been rotted away by corrosion; the only cure is the renewal of the affected component.
3 The matt-black painted finish employed in pursuit of current styling trends is cheaper to renovate but less durable than a conventional chrome-plated system. It is inevitable that the original finish will deteriorate to the point where the system must be removed from the machine and repainted. Reference to the advertisements in the national motorcycle press, or to a local Suzuki dealer and to the owners of machines with similarly-finished exhausts will help in selecting the most effective finish. The best are those which require the paint to be baked on, although some aerosol sprays are almost as effective.
4 On all GS125 models the exhaust system is retained by two Allen bolts to the cylinder head and by a single bolt to the pillion footrest mounting; removal and refitting is a straightforward task requiring no preliminary dismantling. On DR125 S models, the system is fastened to the cylinder head by two nuts and to the rear of the frame by three bolts, the two parts of the system being secured by a clamp bolt. Heatshields are screwed both to the exhaust pipe and to the silencer tailpipe; these can be removed to permit thorough cleaning and repainting.
5 To remove the system on DR125 S models, detach carefully the right-hand sidepanel, remove the two nuts securing the exhaust pipe to the cylinder head then slacken fully the clamp bolt and tap forwards the pipe using a soft-faced mallet. Remove the three mounting bolts which secure the silencer to the frame and manoeuvre the silencer clear of the machine.
6 Refit the exhaust system by reversing the removal sequence. Always renew the exhaust gaskets to prevent leaks, and refit the system tightening the fasteners only by hand at first. When the system is correctly aligned on its mounting tighten securely the mounting nuts or bolts working from the front backwards. The recommended torque setting for all exhaust system fasteners is 0.9 - 1.2 kgf m (6.5 - 8.5 lbf ft), except for the pillion footrest mounting bolt on all GS125 models which is tightened to 2.7 - 3.6 kgf m (19.5 - 26 lbf ft).
7 Do not at any time attempt to modify the exhaust system in any way. The exhaust system is designed to give the maximum power possible consistent with legal requirements and yet to produce the minimum noise level possible. If an aftermarket accessory system is being considered, check very carefully that it will maintain or increase performance when compared with the standard system.

8.8 Adjust tickover speed with the throttle stop screw - CV carburettor type shown

8.9 Throttle cable adjuster on carburettor top - slide carburettor type

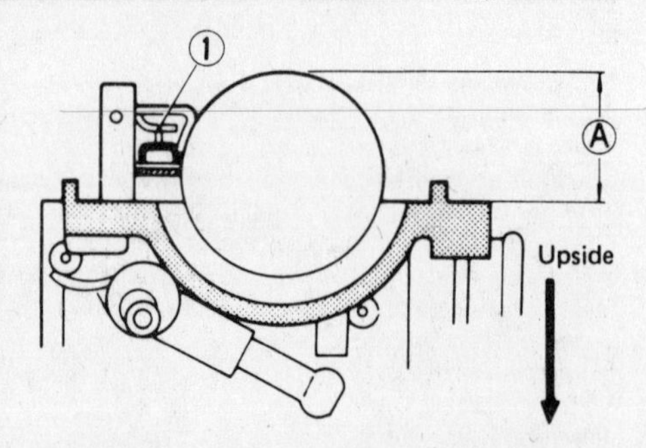

Fig. 2.5 Measuring the float height

A Float height 1 Adjustment tang

10.4a DR125 S only - silencer is retained by front . . .

10.4b . . . bottom . . .

10.4c . . . and top rear mounting bolts

11 Engine lubrication system: general

If the oil and paper oil filter element are changed and the gauze filter cleaned regularly as described in the Routine Maintenance Section of this Manual, the lubrication system will require no further attention.

If doubts arise about the efficiency of the system, it can be checked only by checking the oil pressure as described in the next Section of this Chapter.

12 Checking the oil pressure

1 The operation of checking the oil pressure is used primarily to check the efficiency of the oil pump itself, but can be used to check the efficiency of the system as a whole. The Suzuki service tool Part Number 09915 - 74510 comprises a pressure gauge with the necessary attachments and must be used if the test results are to be accurate. If the tool is not available, take the machine to a local Suzuki dealer for the work to be carried out.

2 To eliminate as many variables as possible, follow the instructions given in Routine Maintenance to change the engine oil and paper element and to clean the filter gauze. Ensure that the engine is filled to exactly the correct level with clean, good quality engine oil. This is vital to ensure that the correct conclusions are drawn from the results of the test.

3 Remove the blanking plug screwed into the crankcase right-hand cover, immediately below the oil filter chamber cap. Screw in the adaptor of the pressure gauge assembly, tighten securely all connections and hang the gauge on a convenient point on the frame. Start the engine and allow it to warm up to normal operating temperature; this is achieved by running the engine at a steady 2000 rpm for 10 minutes (summer conditions) or 20 minutes (winter conditions).

4 With the engine fully warmed up, increase speed to a steady 3000 rpm; the gauge should show a reading of 0.1 - 0.3 kg/cm^2 (1.42 - 4.27 psi). Stop the engine, remove the gauge and refit the blanking plug, ensuring that its sealing washer is renewed if damaged or worn. Tighten securely the blanking plug and check the oil level, topping it up if necessary.

5 If a figure was recorded lower than that given the causes may be as follows: oil level too low, oil of too thin a grade or of poor quality, blocked filter gauze, worn oil pump, worn crankcase cover seal, incorrectly seated oil filter element, or worn engine components. By changing the oil and filter element and by cleaning the gauze filter before conducting the pressure test the majority of the above causes can be eliminated. It will be necessary to remove the crankcase right-hand cover as described in Section 9 of Chapter 1 to gain access to the seal and to the oil pump. The seal is set in the crankcase cover and must fit correctly over the crankshaft right-hand end. Because its efficiency is of critical importance in maintaining the correct oil pressure, renew the seal if in any doubt about its condition. The oil pump must be regarded as a sealed unit; renew it as described in Sections 11 and 26 of Chapter 1.

6 If the figure recorded was higher than that given, the causes may be as follows: oil of too thick a grade, oil filter element clogged, incorrectly fitted or wrongly seated, or a blockage in one or more of the internal oilways. The first two possibilities will be eliminated by the correct execution of the pressure check as described above, the last cause can be cured only by dismantling the engine until the blockage is found.

Chapter 3 Ignition system

Refer to Chapter 7 for information relating to DR125 SF, SH, SJ 'Raider', GN125 and GZ125 'Marauder' models

Contents

Specifications

Ignition system		
Type:		
GS125 ES	Electronic, transistor controlled	
GS125, DR125 S	Capacitor discharge ignition (CDI)	
Ignition timing:		
Up to 1950 rpm	13° BTDC	
Above 3800 rpm	38° BTDC	
Flywheel generator		
Pulser coil resistance:		
GS125 ES	90 – 120 ohm	
GS125, DR125 S	100 – 120 ohm	
Source coil resistance – GS125, DR125 S only:		
Red – Black/red	3 – 6 ohm	
Black/red – White	280 – 340 ohm	
Ignition HT coil		
Spark gap	At least 8 mm (0.32 in)	
Primary winding resistance:		
GS125 ES	2 – 5 ohm	
GS125, DR125 S	0.5 – 1.5 ohm	
Secondary winding resistance	15 – 25 K ohm	
Spark plug		
Make	**NGK**	**ND**
Type	DR8ES-L	X24ESR-U
Gap	0.6 – 0.7 mm (0.024 – 0.028 in)	

1 General description

Since the ignition system is fully electronic there are no mechanical components which can wear out, and no need, therefore, for regular checking and adjustment.

Apart from the ignition and engine kill switches, the HT coil and spark plug and the relevant wiring, the transistor-controlled system of the GS125 ES model comprises only a pulser or pickup coil mounted on the crankcase left-hand cover and the ignitor unit mounted on the frame top tube, while the capacitor discharge system fitted to the GS125 and DR125 S models comprises a source coil and a pulser coil mounted on the generator stator and the CDI unit mounted on the frame top tube or frame front downtube, as appropriate.

2 Ignition system: fault diagnosis

1 As no means of adjustment is available, any failure of the system can be traced to the failure of a system component or a simple wiring fault. Of the two possibilities, the latter is by far the most likely. In the event of failure, check the system in a logical fashion, as described below.

2 Remove the spark plug, giving it a quick visual check, noting any obvious signs of flooding or oiling. Fit the plug into the plug cap and rest it on the cylinder head so that the metal body of the plug is in good contact with the cylinder head metal. The electrode end of the plug should be positioned so that sparking can be checked as the engine is spun over.

3 *Important note.* The energy levels in electronic systems can be very high. On no account should the ignition be switched on whilst the plug or plug cap is being held. Shocks from the HT circuit can be most unpleasant. Secondly, it is vital that the plug is in position and soundly earthed when the system is checked for sparking. The ignition components can be seriously damaged if the HT circuit becomes isolated.

4 Having observed the above precautions, turn the ignition and engine kill switches to 'On' and kick the engine over. If the system is in good condition a regular, fat blue spark should be evident at the plug electrodes. If the spark appears thin or yellowish, or is non-existent, further investigation will be necessary. Before proceeding further, turn the ignition off and remove the key as a safety measure.

5 Ignition faults can be divided into two categories, namely those where the ignition system has failed completely and those which are due to a partial failure. The likely faults are listed below, starting with the most probable source of failure. Work through the list systematically, referring to the subsequent sections for full details of the necessary checks and tests.

a) Defective spark plug or spark plug cap
b) Loose, corroded or damaged wiring connections, broken or shorted wiring between any of the component parts of the ignition system
c) Faulty ignition switch or engine kill switch
d) Faulty ignition coil
e) Faulty ignitor unit (GS125 ES model) or CDI unit (GS125, DR125 S models)
f) Faulty source coil (GS125, DR125 S only) or pulser coil (all models)

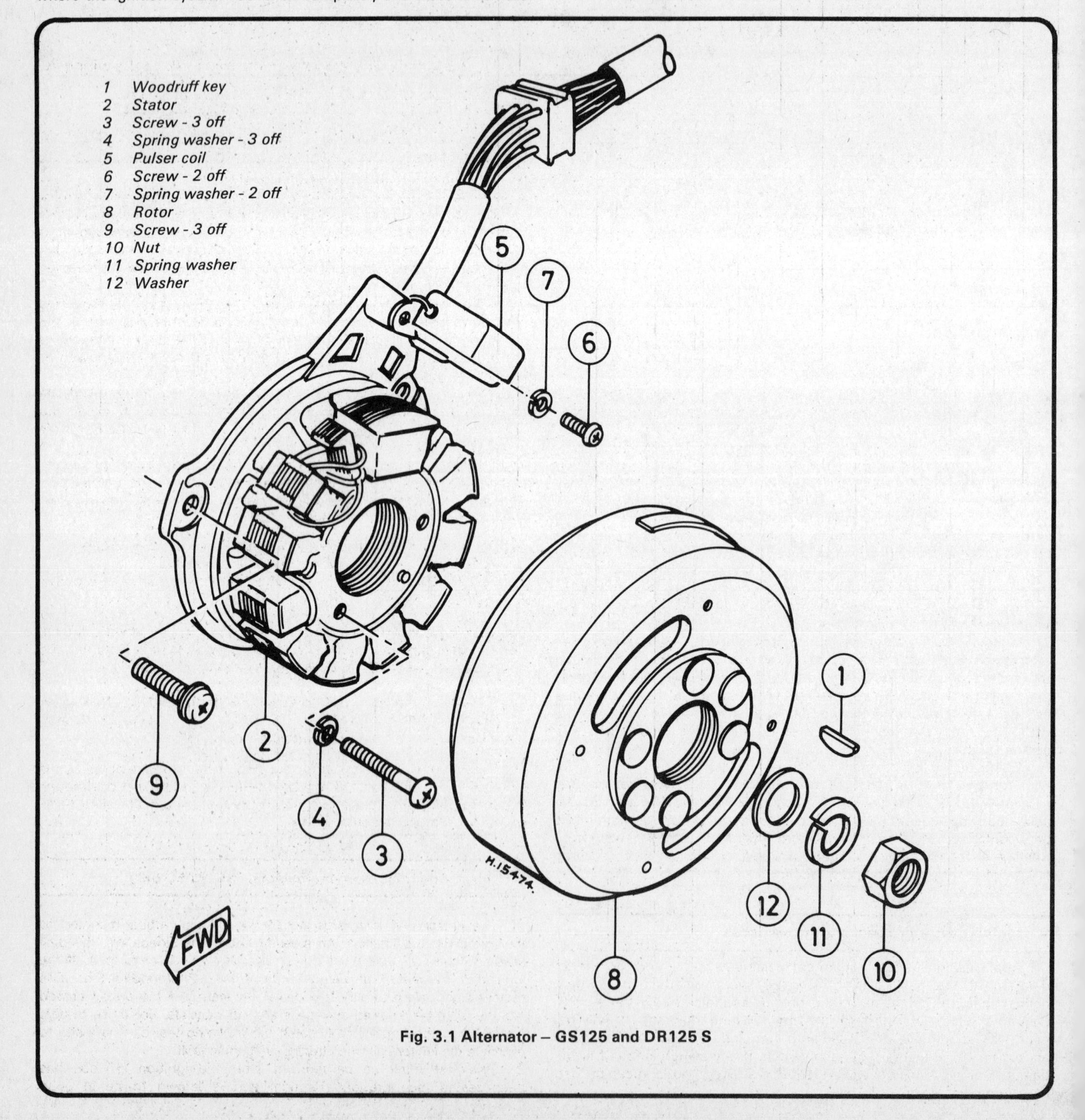

Fig. 3.1 Alternator – GS125 and DR125 S

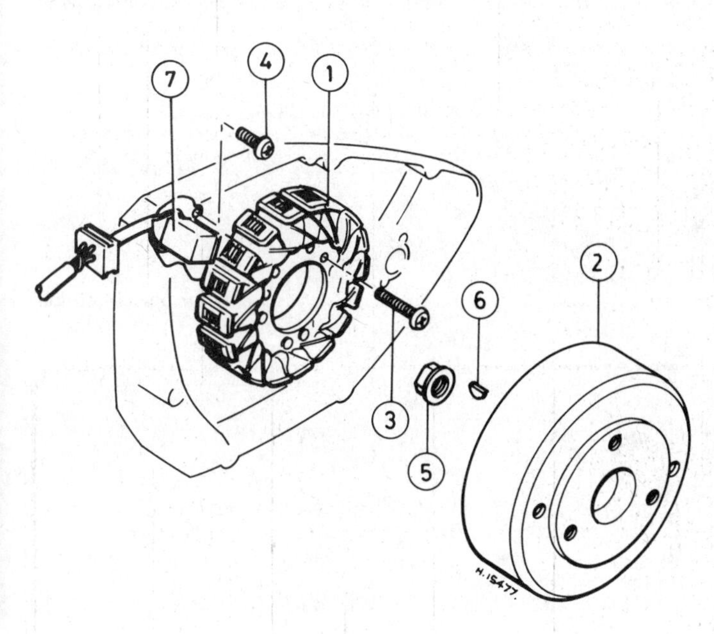

Fig. 3.2 Alternator – GS125 ES

1 *Stator*
2 *Rotor*
3 *Screw - 3 off*
4 *Screw - 2 off*
5 *Nut*
6 *Woodruff key*
7 *Pulser coil*

3 Ignition system: checking the wiring

1 The wiring should be checked visually, noting any signs of corrosion around the various terminals and connectors. If the fault has developed in wet conditions it follows that water may have entered any of the connectors or switches, causing a short circuit. A temporary cure can be effected by spraying the relevant area with one of the proprietary de-watering aerosols, such as WD40 or similar. A more permanent solution is to dismantle the switch or connector and coat the exposed parts with silicone grease to prevent the ingress of water. The exposed backs of connectors can be sealed off using a silicone rubber sealant.
2 Light corrosion can normally be cured by scraping or sanding the affected area, though in serious cases it may prove necessary to renew the switch or connector affected. Check the wiring for chafing or breakage, particularly where it passes close to part of the frame or its fittings. As a temporary measure, damaged insulation can be repaired with PVC tape, but the wire concerned should be renewed at the earliest opportunity.
3 Using the wiring diagram at the end of the manual, check each wire for breakage or short circuits using a multimeter set on the resistance scale or a dry battery and bulb wired as shown in the accompanying illustration. In each case, there should be continuity between the ends of each wire. Note, that a simple test is given in Section 6 for GS125 ES models only which will eliminate several stages of an otherwise lengthy procedure.

4 Ignition system: checking the switches

Again using the test equipment described above and the wiring diagram at the back of this Manual, check whether the switch connections are being made and broken as indicated by the switch diagrams which are part of the main wiring diagram. It will be necessary to remove the headlamp rim and reflector unit to gain access to the connectors of each switch. If any switch is found to be defective usually renewal will be required although nothing can be lost by attempting repair.

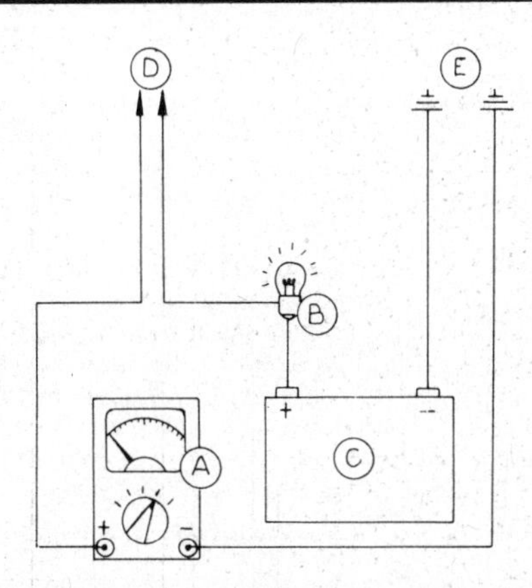

Fig. 3.3 Simple testing arrangement for checking the wiring

A *Multimeter*
B *Bulb*
C *Battery*
D *Positive probe*
E *Negative probe*

5 Ignition HT coil: location and testing

1 The ignition HT coil is a small black or grey plastic sealed unit that is readily identified by the HT lead protruding from one end. It is mounted underneath the frame top tube to the rear of the steering head; the sidepanels, the dual seat and the fuel tank must be removed to gain access to it.
2 Maintenance is restricted to ensuring that the coil mounting bracket/frame earth points are clean and free from corrosion, that the mounting bolts are securely tightened and that the wiring connections are in good order. If the coil proves defective it must be renewed as repairs are not possible.
3 If the coil is suspected of being faulty it should be removed and taken to a competent Suzuki dealer or auto-electrical expert for checking on a spark gap tester. A sound coil will produce a reliable spark across a minimum gap of 8 mm (0.32 in).
4 If a spark gap tester is not available a basic test can be conducted using a multimeter with ohm and kilo ohm scales. The coil primary windings are tested by making the meter connections as shown below and noting the readings obtained:

GS125 ES model	Orange/white to White leads	2 – 5 ohm
GS125, DR125 S models	Black lead to earth (mounting bracket)	0.5 – 1.5 ohm

The coil secondary windings are tested by making the meter connections as shown below and noting the readings obtained. Note that the resistance value of the standard suppressor cap is included.

GS125 ES model	HT to Orange/white leads	15 – 25 K ohm
GS125, DR125 S models	HT lead to earth (mounting bracket)	15 – 25 K ohm

5 If either of the readings obtained is appreciably above or below the set figure, the coil should be considered faulty and renewed, although since the set figures are approximate the coil should be taken to an expert for a more thorough check.

6 Ignitor unit: location and testing – GS125 ES only

1 The ignitor unit is a large square sealed unit rubber-mounted to the frame top tube behind the steering head. The sidepanels, the dual seat and the fuel tank must be removed to gain access to it. If the ignitor is suspected of being faulty it is recommended that it is removed and taken to a Suzuki dealer for testing; if the unit is proved faulty it must be renewed as repairs are not possible. For those owners who have a good quality multimeter, the following test can be made to eliminate the ignitor unit when tracing an ignition fault.
2 This test must be carried out after the ignition HT coil has been tested (see previous Section) and is known to be in good

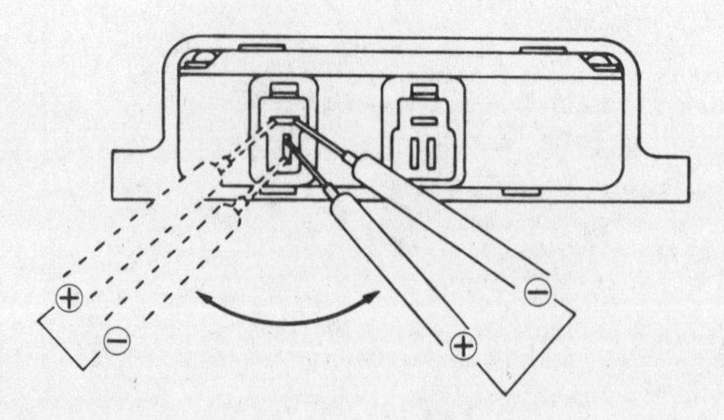

Fig. 3.4 Ignitor unit elimination test

order. Disconnect from the rear of the ignitor unit the two-pin block connector (blue and green wires) thus isolating the pulser coil. Remove the spark plug then refit it to the spark plug cap and lay the plug on the cylinder head finning so that its metal body is in good contact with the metal of the cylinder, then switch on the ignition. With a multimeter set to the x 1 ohm range, connect the meter positive (+) probe to the ignitor blue wire terminal and meter negative (–) probe to the ignitor green wire terminal. A spark should appear at the spark plug electrodes. Note that you may need to reverse the meter probes to obtain the spark, depending on the polarity of the meter.

3 If the results of this test are satisfactory, the fault must lie in the pulser (pick-up) coil or its wiring; if the test is not satisfactory and the wiring, switches, HT coil, and spark plug have been checked and found to be in good order, the ignitor unit could be at fault also and must be checked more thoroughly.

4 Although test details are produced by the manufacturer, testing on home workshop equipment (ie using a tester other than the Suzuki tester) has produced inaccurate readings. Owners are therefore advised to take the ignitor unit to a Suzuki dealer for testing, or to substitute it with an ignitor from an identical model for the purpose of testing.

5.1 Location of ignition HT coil – DR125 S (GS125 models similar)

7 CDI unit: locating and testing – GS125, DR125 S models

1 The CDI unit is a large square sealed unit mounted on the frame top tube behind the steering head on GS125 models, and on the frame front downtube beneath the fuel tank on DR125 S models; it will be necessary to remove the sidepanels, the seat and the fuel tank to gain access to the unit or to its connectors.

2 Although test details are produced by the manufacturer, testing on home workshop equipment (ie using a tester other than the Suzuki tester) has produced inaccurate readings. Owners are therefore advised to take the CDI unit to a Suzuki dealer for testing, or to substitute it with a CDI unit from an identical model for the purpose of testing. If the CDI unit is proved faulty it must be renewed as repairs are not possible.

6.1 Ignitor unit – GS125 ES model (GS125 CDI unit is mounted similarly)

7.1 Location of CDI unit – DR125 S model

8 Source and pulser coils: location and testing – all models

1 The ignition pulser coil is mounted on the crankcase left-hand cover (GS125 ES models) or on the alternator stator plate (GS125, DR125S models), while the ignition source coil (GS125, DR125S models only) is incorporated in the alternator stator assembly. If any of these coils are found through testing to be faulty they must be renewed, although an auto-electrical expert may be found who can attempt a repair: it should be noted that in some cases the coils cannot be purchased as a separate replacement part and can be obtained only as part of the stator assembly.
2 Removal and refitting of the coils themselves are described in Sections 8 and 39 of Chapter 1, but each coil can be tested in situ: it will be necessary to remove the sidepanels, the seat and the fuel tank to gain access to the various connectors, and a good quality multimeter must be available.
3 To test the pulser (or pickup) coil, disconnect the wires from the ignitor or CDI unit at the multi-pin block connector, make the meter connections as shown below, and compare the readings obtained with those given.

Pulser coil resistance values

GS125 ES models	Blue to green leads	90 – 120 ohm
GS125, DR125 S models	Blue/red to red/ green leads	100 – 120 ohm

4 To test the ignition source coil (GS125, DR125 S models only) disconnect the various wires as described above and use the meter in the same way but making the connections shown below.

Source coil resistance values

Red to black/red leads	3 – 6 ohm
Black/red to white leads	280 – 340 ohm

5 If any of the readings obtained do not correspond with those given, the coil must be considered faulty and renewed. It is worthwhile, however, to take the machine to a competent Suzuki dealer for accurate testing.

9 Ignition timing: checking

1 Although there is no provision for the adjustment of the ignition timing since there should be no need for it, it may prove necessary on occasion to check that the ignition timing is correct and that the components of the secondary advance circuit are functioning correctly. For example the generator stator or pulser coil (as appropriate) may have been disturbed during the course of other work on the engine/gearbox unit, in which case it would be advisable to check that it has been refitted correctly, or one may be investigating the cause of a loss of power.
2 The ignition timing can be carried out using only a good quality 'strobe' timing lamp. It is recommended that one of the better quality xenon tube lamps is purchased, of the type which requires an external power source. The cheaper types may not produce enough light to be of any use, and may even produce a spurious reading.
3 Remove the inspection cap that is screwed into the top of the crankcase left-hand cover, then connect the strobe according to its manufacturer's instructions. It is useful to engage the aid of an assistant who can control the engine speed while the first person uses the timing lamp, and it is recommended that a piece of rag is placed over the inspection aperture, whenever the strobe is not being used to illuminate the timing mark, to reduce to a minimum the amount of oil that will be ejected as the engine rotates.
4 Start the engine and allow it to idle, then aim the lamp at the inspection aperture. The timing index mark (a straight line with the letter 'T' or 'O' adjacent) stamped on the rotor rim should be visible 'frozen' in place by the action of the strobe lamp and should be aligned exactly with the arrow cast on the crankcase cover (all GS 125 models) or exactly in the middle of the inspection aperture (DR125 S models). Increase engine speed slowly to no more than 4000 rpm: the timing index mark should move smoothly out of sight until the full advance position is reached. Stop the engine and disconnect the strobe, then refit the inspection plug and wash off all traces of oil from the engine. Do not forget to check the oil level and to top up if necessary.
5 If the timing marks were not in alignment, or if any doubts arise about the efficiency of the ignition advance the machine should be taken to a competent Suzuki dealer for an expert second opinion. If a discrepancy is encountered the fault will be in the pulser coil or the ignitor/CDI unit (as appropriate); these must be tested and renewed if necessary as described in the relevant Sections of this Chapter. It is worth checking, however, that the alternator rotor and stator are correctly fitted and securely fastened (see Sections 8 and 39 of Chapter 1) before any components are condemned unnecessarily.

10 Spark plug: checking and resetting the gap

1 The spark plug supplied as original equipment is the NGK DR8ES-L (or ND X24ESR-U), which will prove satisfactory in most operating conditions; alternatives are available to allow for varying altitudes, climatic conditions and the use to which the machine is put. If the spark plug is suspected of being faulty it can be tested only by the substitution of a brand new (not second-hand) plug of the correct make, type, and heat range: always carry a spare on the machine.
2 Note that the advice of a competent Suzuki dealer or similar expert should be sought before the plug heat range is altered from standard. The use of too cold, or hard, a grade of plug will result in fouling and the use of too hot, or soft, a grade of plug will result in engine damage due to the excess heat being generated. If the correct grade of plug is fitted, however, it will be possible to use the condition of the spark plug electrodes to diagnose a fault in the engine or to decide whether the engine is operating efficiently or not.

3 The final point on the selection of spark plug types and grades concerns the letter 'R' in the prefix mentioned above. This indicates that the plug has a resistor built in to suppress the radio and TV interference produced by the HT pulse, the manufacturers having decided that it is better to suppress the interference as close to its source as possible, i.e. in the spark plug itself. The ignition system has been designed and constructed with this type of spark plug in mind: its reliability and performance would be adversely affected if a standard, non-resistor spark plug were substituted for the correct type, therefore not only must the spark plug be of the correct heat range, it must also be a resistor type at all times.
4 It is advisable to carry a new spare spark plug on the machine, having first set the electrodes to the correct gap. Whilst spark plugs do not fail often, a new replacement is well worth having if a breakdown does occur. Ensure that the spare is of the correct heat range and type.
5 The correct electrode gap is 0.6 – 0.7 mm (0.024 – 0.028 in). The gap can be assessed using feeler gauges. If necessary, alter the gap by bending the outer electrode, preferably using a proper electrode tool. **Never** bend the centre electrode, otherwise the ceramic insulator will crack, and may cause damage to the engine if particles break away whilst the engine is running.
6 Before refitting a spark plug into the cylinder head coat the threads sparingly with a graphited grease to aid future removal. Use the correct size spanner when tightening the plug otherwise the spanner may slip and damage the ceramic insulator. The plug should be tightened by hand only at first and then secured with a quarter turn of the spanner so that it seats firmly on its sealing ring.
7 Never overtighten a spark plug otherwise there is risk of stripping the threads from the cylinder head, especially as it is cast in light alloy. A stripped thread can be repaired without having to scrap the cylinder head by using a 'Helicoil' thread insert. This is a low-cost service, operated by a number of dealers.

11 Spark plug (HT) lead and suppressor cap: examination

1 Erratic running faults and problems with the engine suddenly cutting out in wet weather can often be attributed to leakage from the high tension lead and spark plug cap. If this fault is present, it will often be possible to see tiny sparks around the lead and cap at night. One cause of this problem is the accumulation of mud and road grime around the lead, and the first thing to check is that the lead and cap are clean. It is often possible to

cure the problem by cleaning the components and sealing them with an aerosol ignition sealer, which will leave an insulating coating on both components.

2 Water dispersant sprays are also highly recommended where the system has become swamped with water. Both these products are easily obtainable at most garages and accessory shops. Occasionally, the suppressor cap or the lead itself may break down internally. If this is suspected, the components should be renewed.

3 Where the HT lead is permanently attached to the ignition coil, it is recommended that the renewal of the HT lead is entrusted to an auto-electrician who will have the expertise to solder on a new lead without damaging the coil windings.

4 When renewing the suppressor cap, be careful to purchase one that is suitable for use with resistor spark plugs. The traditional method of suppressing HT interference was to incorporate a resistor into the spark plug cap; now that resistor is sited in the spark plug itself, suppressing any interference at its source, the suppressor caps have much lower resistance values and the ignition system is designed with this in mind.

Chapter 4 Frame and forks

Refer to Chapter 7 for information relating to DR125 SF, SH, SJ 'Raider', GN125 and GZ125 'Marauder' models

Contents

Specifications

	GS125 models	DR125 S
Frame		
Type	Welded tubular steel	
Front forks		
Type	Telescopic, coil sprung, hydraulically damped	
Wheel travel	110 mm (4.3 in)	180 mm (7.1 in)
Fork spring minimum free length	467 mm (18.386 in)	540 mm (21.260 in), Z, D models 525 mm (20.669 in), E model
Fork oil level	205 mm (8.071 in)	174 mm (6.850 in)
Fork oil quantity – per leg	136 cc (4.79 fl oz)	177 cc (6.23 fl oz)
Recommended fork oil	SAE15 fork oil	SAE10 fork oil
Rear suspension	Hydraulically damped, coil sprung suspension units acting on swinging arm	Suzuki 'Full Floater' suspension system
Wheel travel	91 mm (3.6 in)	160 mm (6.3 in)
Swinging arm pivot bolt maximum runout	0.6 mm (0.02 in)	0.6 mm (0.02 in)
Suspension unit mounting bush maximum ID – top and bottom	N/App	15.2 mm (0.598 in)
Link arm top bush ID	N/App	15.03 – 15.10 mm (0.592 – 0.595 in)
Service limit	N/App	15.20 mm (0.598 in)
Link arm bottom bush ID	N/App	13.03 – 13.10 mm (0.513 – 0.516 in)
Service limit	N/App	13.20 mm (0.520 in)
Swinging arm pivot bush ID	20.02 – 20.06 mm (0.788 – 0.790 in)	20.02 – 20.06 mm (0.788 – 0.790 in)
Service limit	20.20 mm (0.795 in)	20.20 mm (0.795 in)

Torque wrench settings

Component	kgf m	lbf ft
Handlebar clamp bolts	1.2 – 2.0	8.5 – 14.5
Steering stem top bolt	3.5 – 5.5	25.5 – 40
Fork top bolts – GS125 models only	3.5 – 5.5	25.5 – 40
Top yoke pinch bolts – DR125S model only	2.0 – 3.0	14.5 – 21.5
Bottom yoke pinch bolts:		
GS125 models	2.5 – 3.5	18 – 25.5
DR125 S model	4.0 – 5.0	29 – 36

Damper rod Allen bolts:		
GS125 models	2.0 – 2.6	14.5 – 19
DR125 S model	1.5 – 2.5	11 – 18
Footrest mounting bolts – GS125 models only	3.6 – 5.2	26 – 37.5
Footrest mounting bolts – DR125 S model only:		
12 mm	3.6 – 5.2	26 – 37.5
10 mm	2.7 – 4.3	19.5 – 31
Pillion footrest mounting bolts – GS125 models only	2.7 – 3.6	19.5 – 26
Torque arm mounting nuts	1.0 – 1.5	7 – 11
Swinging arm pivot shaft nut	5.0 – 8.0	36 – 58
Suspension unit mounting nuts:		
GS125 models	2.0 – 3.0	14.5 – 21.5
DR125 S models	4.0 – 6.0	29 – 43.5
Suspension linkage fasteners – DR125 S model only:		
Rocker pivot shaft nut	4.5 – 7.0	32.5 – 50.5
Link arm top mounting bolt	4.0 – 6.0	29 – 43.5
Link arm bottom mounting bolts	1.8 – 2.8	13 – 20

1 General description

A welded tubular steel frame is employed, the frame being of the open diamond type which uses the engine as a stressed member. The front forks are of the conventional coil sprung, hydraulically damped type, and on all GS125 models the rear suspension is equally conventional, comprising two coil sprung, hydraulically damped suspension units acting on a tubular steel pivoted fork.

The DR125 S model is fitted with Suzuki's 'Full-Floater' rear suspension system. The steel box-section swinging arm incorporates a mounting point at its forward end for the single suspension unit bottom eye. The suspension unit top eye is attached to a rocker which pivots on needle roller bearings, the rear end of the rocker being connected via a link arm to a point on the swinging arm behind the suspension unit; thus the suspension unit is not attached directly to the frame and therefore is 'fully-floating'; this means that the suspension unit is compressed and extended from both ends at once as the swinging arm moves.

2 Front fork legs: removal and refitting

1 Place the machine securely on its centre stand or, for those models not equipped with a centre stand, a stout wooden box or paddock stand so that the front wheel is clear of the ground. The front wheel can then be removed according to the instructions given in the relevant Section of Chapter 5.

2 On GS125 ES models the caliper assembly must now be removed, if this was not done during wheel removal. Place a piece of wood between the pads, wedging it firmly to prevent the pads being ejected should the front brake lever be applied inadvertently. Remove the two caliper assembly/fork lower leg mounting bolts and withdraw the caliper assembly, taping it to a convenient point on the frame to keep it out of harm's way. Note that it is not necessary to disconnect the hydraulic hose.

3 Remove the four mudguard mounting bolts then carefully withdraw the mudguard and bridge piece, taking care not to damage the finish. Mudguard removal is not necessary on the DR125 S model.

4 On all GS125 models, the fork legs are retained in the top yoke by the fork top bolt. Fork leg removal therefore means that the fork top bolt is removed and that the bottom yoke pinch bolts are slackened. It will be necessary to release the handlebars to gain access to the top bolts – see Section 5. It should then be possible to slide the fork leg down and away from the yokes. On the DR125 S model, the fork legs are retained by pinch bolts in both top and bottom yokes. On these models it is necessary merely to slacken the pinch bolts and to slide the fork legs down and out of the yokes.

5 If the fork legs prove to be stuck in the yokes, apply penetrating fluid and attempt to rotate the legs by hand to free them. It may be necessary to completely remove the pinch bolts and to spring the clamps apart slightly with a large, flat-bladed screwdriver in order to achieve this. Great care must be taken not to distort or to break the clamp, as this will necessitate renewal of the complete yoke. If the leg is still reluctant to move, push a metal bar of suitable diameter through the spindle lug in the fork lower leg and tap firmly downwards on the protruding end of the bar to drive the fork leg from the yokes.

6 Once the legs have been removed, put them to one side to await stripping and examination. If they are not to be dismantled, ensure that they remain upright so that no fork oil is lost.

7 Reassembly is a straightforward reversal of the removal sequence, while noting the following points. On models fitted with a fork top bolt, the leg must be pushed up through the bottom yoke to the underside of the top yoke and retained there by firmly tightening the top bolt. On DR125 S models push the fork leg up through both yokes until the top of the chromed stanchion tube is flush with the upper surface of the top yoke. It may be necessary to remove the rubber plug from the top of the fork leg to ease this task. Once in place, the fork legs should be retained by tightening the top yoke pinch bolts as lightly as possible. For both types of forks, refitting the legs will be made easier if a small amount of grease or oil is smeared over the upper length of the stanchion. Tighten the gaiter clips securely.

8 When fitting the front mudguard and wheel back into the forks, tighten the spindle nut and other mounting bolts only lightly at first. Apply the front brake and push down on the handlebars several times so that the operation of the fork legs settles each component in its correct place. Using a torque wrench, tighten all nuts and bolts from the wheel spindle upwards to the fork top bolt or top yoke pinch bolt, to the torque settings given in the Specifications Section of this Chapter. This will ensure that the fork components can operate freely and easily, with no undue strain imposed from an overtightened bolt or an awkwardly-positioned part.

9 Be very careful to check fork operation, front brake adjustment and that all nuts and bolts are securely fastened before taking the machine out on the road.

2.3 Remove mudguard mounting bolts – GS125 models only

2.4a Remove handlebar clamp cover – GS125 ES models only

2.4b Unscrew handlebar clamp bolts – all GS125 models ...

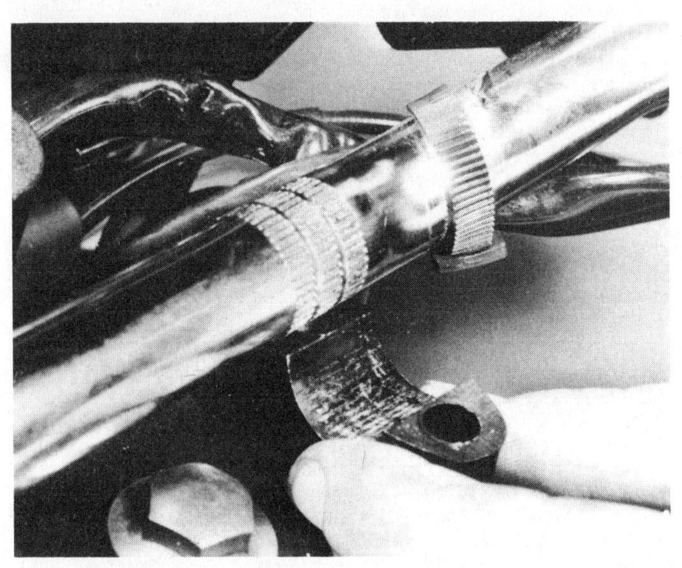

2.4c ... and displace clamp halves ...

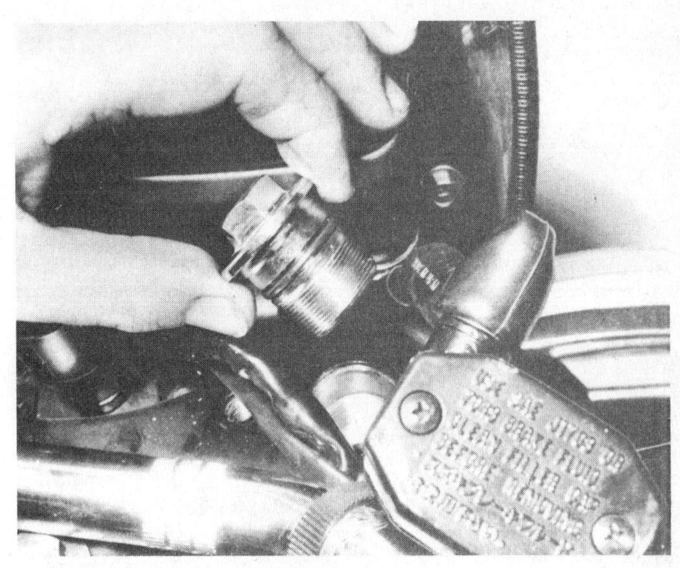

2.4d ... to gain access to fork top bolts

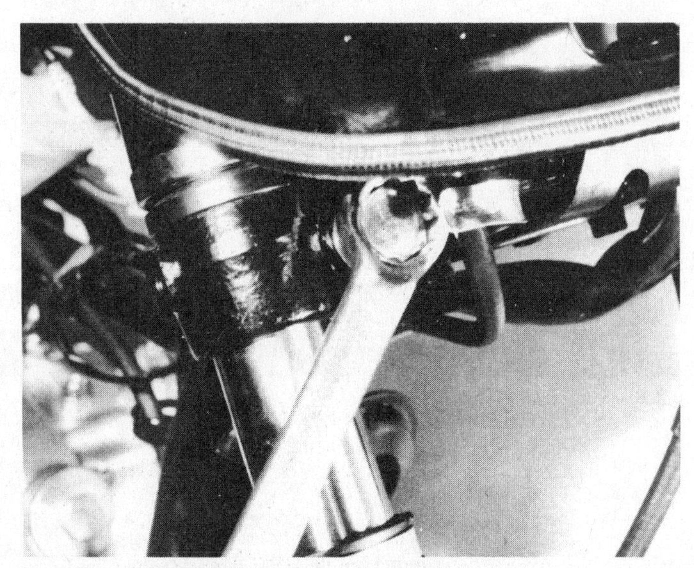

2.4e Slacken fork yoke pinch bolts to release stanchion

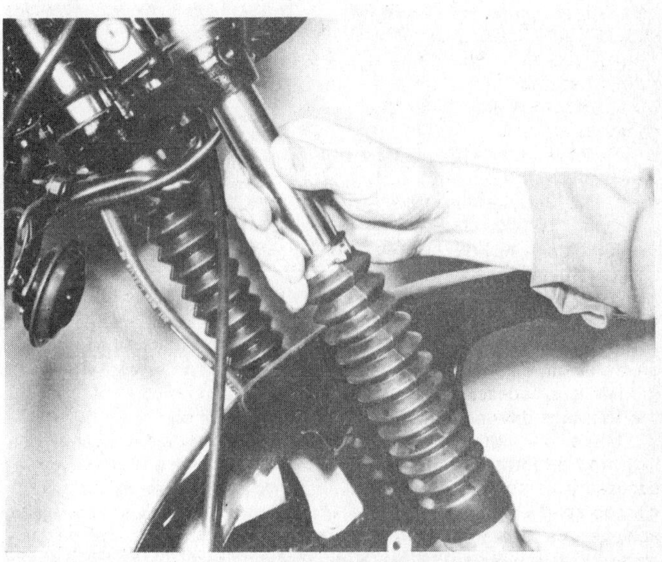

2.7 Smear grease over stanchion to aid refitting – tighten gaiter clip when fork leg is secured

3 Front fork legs: dismantling and reassembly

1 Dismantle and rebuild separately the fork legs so that there is no chance of exchanging components, thus promoting undue wear.
2 On all GS 125 models, use a suitable Allen key to unscrew the fork spring retaining nut from inside the stanchion. Invert the leg over a suitable container to tip out the fork spring, then allow the fork oil to drain fully.
3 On DR125 S models remove the fork spring by clamping the fork lower leg in a vice whose jaws have been suitably padded so as not to mark the alloy of the leg. It is recommended that the lower leg is clamped at the wheel spindle boss to avoid the possibility of distorting the leg itself. Remove the rubber plug from the top of the fork leg, exposing the metal top plug which is retained by a plain wire circlip. Using a metal rod of suitable diameter push the top plug down against spring pressure until the circlip can be displaced from its groove with a suitable sharp instrument such as a very small flat-bladed screwdriver. When the circlip has been removed, gradually relax the downward pressure on the top plug, allowing the fork spring to push the plug upwards out of the stanchion. Great care must be exercised to prevent the plug from flying out under spring pressure, thus causing injury or damage. Enlisting the aid of an assistant is recommended. After plug removal, withdraw the fork spring.
4 Invert the fork leg and pump the lower leg up and down several times to ensure that the maximum possible amount of oil is removed.
5 The next part of the dismantling procedure will require the use of a vice. Note that whenever a vice is used, its jaws must be padded, as previously mentioned, to avoid marking delicate alloy components or the sliding surface of the stanchion..
6 Clamp carefully, to avoid distortion, the fork lower leg in a vice and use an Allen key to unscrew the damper rod retaining bolt which is set in a recess in the base of the fork lower leg. In some cases the bolt will unscrew with ease, but it is more usual for the bolt to free itself from the lower leg and to then rotate with the damper rod assembly, so that nothing useful is achieved. In such a case, obtain a length of wooden dowel, grind a coarse taper on one end which will engage in the head of the damper rod, and insert the dowel down into the bore of the stanchion. The services of an assistant will now be required. Clamp a self-locking wrench to the protruding end of the dowel and with the assistant preventing the dowel from turning and simultaneously applying pressure via the dowel to the head of the damper rod, the damper rod will be locked in place so that the retaining bolt can be unscrewed. When working alone, use a longer length of dowel which can be clamped in the vice. By pushing down on the fork lower leg with one hand it should be possible to lock the damper rod firmly enough for the retaining bolt to be unscrewed.
7 On all models except the GS125 ESX, slacken the upper (and lower – DR models) gaiter clamp and remove the fork gaiter. On GS125 ESX models, prise the rubber dust seal off the top of the lower leg and slide it off the stanchion. On all models, carefully pull the stanchion out of the lower leg. Invert the stanchion to tip out the damper rod and spring, then invert the lower leg to tip out the damper rod seat.
8 The fork oil seal fitted to each leg should be renewed whenever the stanchion is removed and must be renewed if it is disturbed as the means used for removing the seal will almost certainly damage it. The seal is retained by a circlip which must be removed using a small, flat-bladed screwdriver to ease it away from its groove in the fork lower leg. Use a large, flat-bladed screwdriver to lever the seal from its housing. Take care not to scratch the internal surface of the seal housing with the edge of the screwdriver blade, and do not apply excessive pressure or there is a risk of the upper edge of the fork lower leg being cracked or distorted. If the seal appears difficult to move, heat the leg by pouring boiling water over its outer surface. This will cause the alloy leg to expand sufficiently to loosen the seal. Do not forget to displace the spacer fitted underneath the oil seal on DR 125 S models only.
9 Reassembly is a reversal of the dismantling procedure, but careful attention must be paid to certain points.
10 On DR125 S models only, insert the spacer into the top of the lower leg so that its raised edge faces upwards. Tap the spacer down until it rests in the machined shoulder. Clamp the lower leg securely in a vice by means of the wheel spindle boss. Coat the inner and outer diameters of the seal with the recommended fork oil and push the seal squarely into the bore of the fork lower leg by hand. Ensure that the seal is fitted squarely, then tap it fully into position using a hammer and a suitably sized drift such as a socket spanner, as shown in the accompanying photograph. The drift should bear only on the harder, outer diameter of the seal, never on the sealing lips themselves, and should have a smooth undamaged surface where it comes into contact with the seal. Tap the seal into the bore of the lower leg just enough to expose the circlip groove. Refit the retaining circlip securely in its groove.
11 Slide the damper rod rebound spring into place under the head of the damper rod and insert the damper rod assembly into the stanchion. Push the damper rod down the length of the stanchion until it projects fully from the stanchion lower end. Refit the damper rod seat over the damper rod end, using a smear of grease to stick it in place.
12 Smear the sliding surface of the stanchion with a light coating of fork oil and carefully insert the stanchion into the lower leg, taking great care not to damage the sealing lips of the oil seal. Push a fork spring or the length of dowel used on dismantling down into the stanchion and apply pressure on this to ensure that the damper rod or its seat is pressed firmly into the base of the lower leg. Check that the threads of the damper rod bolt are clean and dry, apply a few drops of thread locking compound and fit the damper rod bolt. Do not forget the sealing washer fitted under the head of the bolt. Tighten the bolt only partially at first, using an Allen key of suitable size. Maintain pressure on the head of the damper rod and push the stanchion firmly as far down into the lower leg as possible to centralise the damper rod in the stanchion. The damper rod bolt can then be tightened firmly to the recommended torque setting. Withdraw the spring or dowel from the stanchion.
13 On all models except the GS125 ESX, smear a light coating of grease over the stanchion sliding surface to provide additional protection against moisture and dirt, then fit the gatier and its retaining clip(s); leave tightening of the top clip until the fork is installed in the yokes and the gaiter can be positioned correctly. On GS125 ESX models, slide the dust seal down onto the lower leg and engage its inner lip in the lower leg groove.
14 Fill the leg with the correct amount of the specified grade of fork oil. Check carefully in the Specifications Section of this Chapter to find the correct quantity of fork oil required for each fork leg of your machine and acquire a suitably finely graduated measuring vessel to ensure that exactly that amount of oil is put in each leg. Note that a fork oil level is also given.
15 To measure the oil level in each leg, make up a dipstick from a piece of wire cut to the required length, as shown in the accompanying photograph. Check that the fork spring has been removed and push the stanchion fully into the fork lower leg. Pour the specified amount of oil into the fork leg and allow the level to settle. Add or remove oil as necessary. Once the correct oil level has been established in this way, the fork legs can be refilled to that level during routine maintenance when the forks are not completely dismantled. This will automatically allow for the presence of any residual fork oil which cannot be removed without fully dismantling the fork leg and will ensure smooth and consistent fork action.
16 When the fork oil has been poured in, and the level checked, pull the stanchion out of the leg as far as possible and insert the fork spring noting that the closer spaced coils must be at the top. On all GS125 models, refit and tighten securely the fork spring retaining nut. To refit the fork top plug on DR125 S models, enlist the services of an assistant to ease the operation, which is a straightforward reversal of the dismantling procedure. Take great care not to damage the sealing O-ring fitted to the top plug and ensure that the retaining circlip is securely engaged in its locating groove in the stanchion.
17 Refit the fork leg into the fork yokes following the instructions given in Section 2 of this Chapter.

4 Front fork legs: examination and renovation

1 Carefully clean and dry all the components of the fork leg. Lay them out on a clean work surface and inspect each one, looking for excessive wear, cracks, or other damage. All traces of oil, dirt, and swarf should be removed, and any damaged or worn components renewed.
2 Examine the sliding surface of the stanchion and the internal surface of the lower leg, looking for signs of scuffing which will indicate that excessive wear has taken place. Slide the stanchion into the lower leg so that it seats fully. Any wear present will be easily found by attempting to move the stanchion backwards and forwards, and from side to side, in the bore of the lower leg. It is inevitable that a certain degree of slackness will be found, especially when the test is repeated at different points as the stanchion is gradually withdrawn

from the lower leg, and it is largely a matter of experience accurately to assess the amount of wear necessary to justify renewal of either the stanchion or the lower leg. It is recommended that the two components be taken to a motorcycle dealer for an expert opinion to be given if there is any doubt about the degree of wear found. Note that while wear will only become a serious problem after a high mileage has been covered, it is essential that such wear is rectified by the renewal of the components concerned if the handling and stability of the machine are not to be impaired.

3 Check the outer surface of the stanchion for scratches or roughness; it is only too easy to damage the oil seal during the reassembly if these high spots are not eased down. The stanchions are unlikely to bend unless the machine is damaged in an accident. Any significant bend will be detected by eye, but if there is any doubt about straightness, roll down the stanchion tubes on a flat surface such as a sheet of plate glass. If the stanchions are bent, they must be renewed. Unless specialised repair equipment is available it is rarely practicable to effect a satisfactory repair to a damaged stanchion.

4 Check the stanchion sliding surface for pits caused by corrosion: if the gaiters (where fitted) are intact such damage will be rare. Such pits should be smoothed down with fine emery paper and filled, if necessary, with Araldite. Once the Araldite has set fully hard, use a fine file or emery paper to rub it down so that the original contour of the stanchion is restored.

5 After an extended period of service, the fork springs may take a permanent set. If the spring lengths are suspect, then they should be measured and the readings obtained compared with the service limits given in the Specifications Section of this Chapter. It is always advisable to fit new fork springs where the length of the original items has decreased beyond the service limit given. Always renew the springs as a set, never separately.

6 The piston ring fitted to the damper rod may wear if oil changes at the specified intervals are neglected. If damping has become weakened and does not improve as a result of an oil change, the piston ring should be renewed. Check also that the oilways in the damper rod have not become obstructed.

7 Closely examine the gaiter (where fitted) for splits or signs of deterioration. If found to be defective, it must be renewed as any ingress of dirt will rapidly accelerate wear of the oil seal and fork stanchion. It is advisable to renew any gasket washers fitted beneath bolt heads as a matter of course. The same applies to the O-rings fitted to the fork top bolts or top plugs, depending on which type of fork is being dismantled.

3.8a Remove circlip to gain access to fork oil seal

3.8b Be careful not to damage fork lower leg when removing oil seal

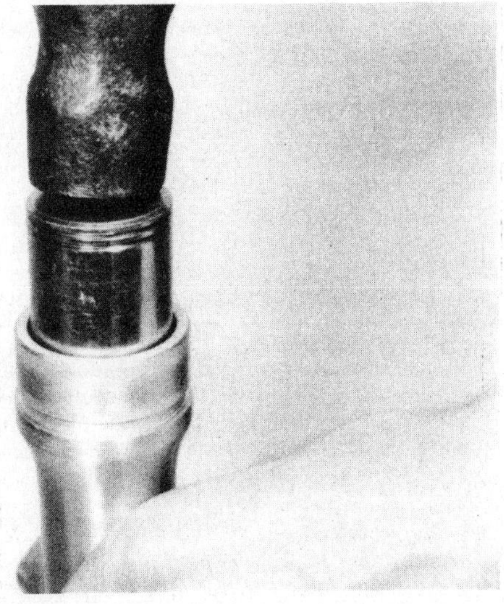

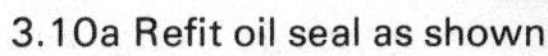

3.10a Refit oil seal as shown

3.10b Tap seal down far enough to expose circlip groove

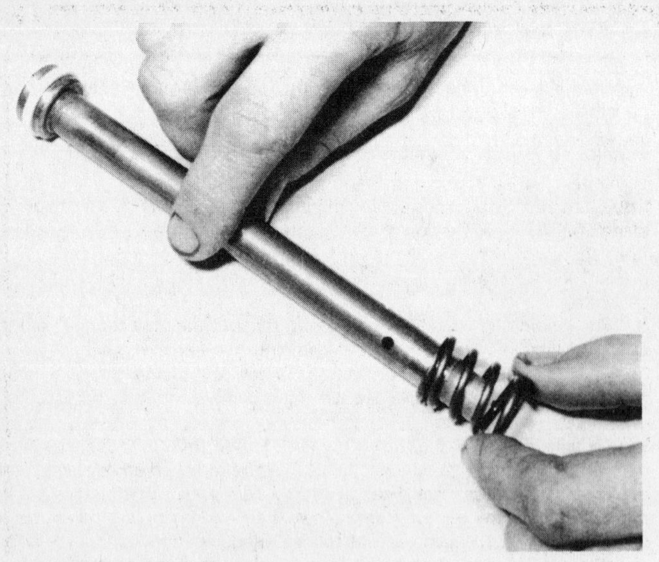

3.11a Rebound spring is refitted over damper rod ...

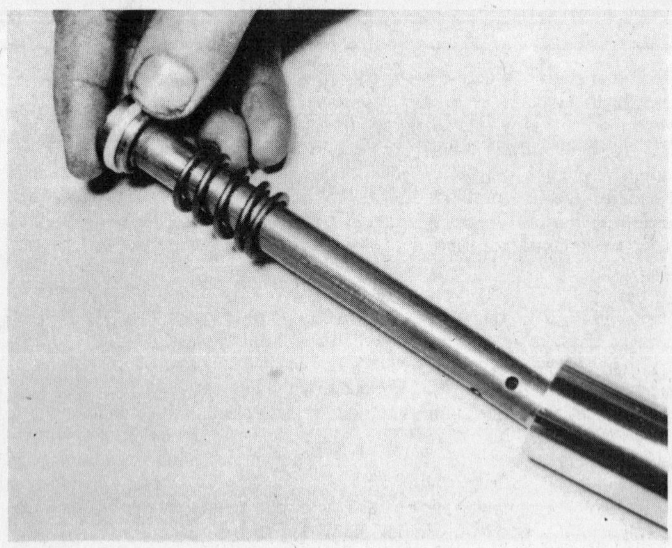

3.11b Insert damper assembly into stanchion

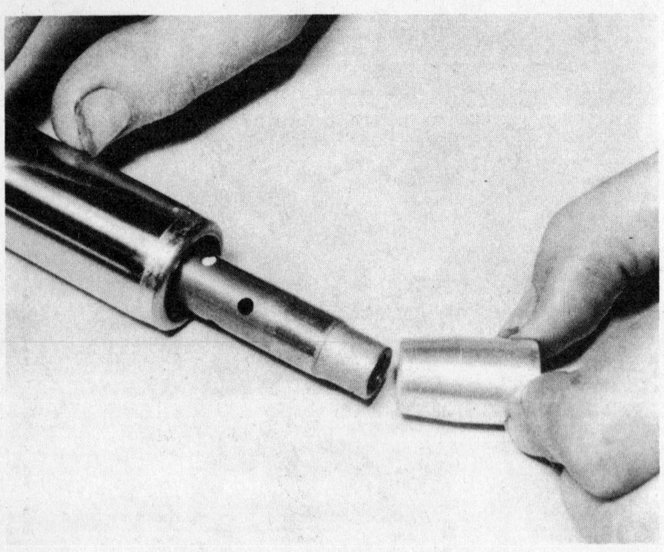

3.11c Refit seat to protruding end of damper rod ...

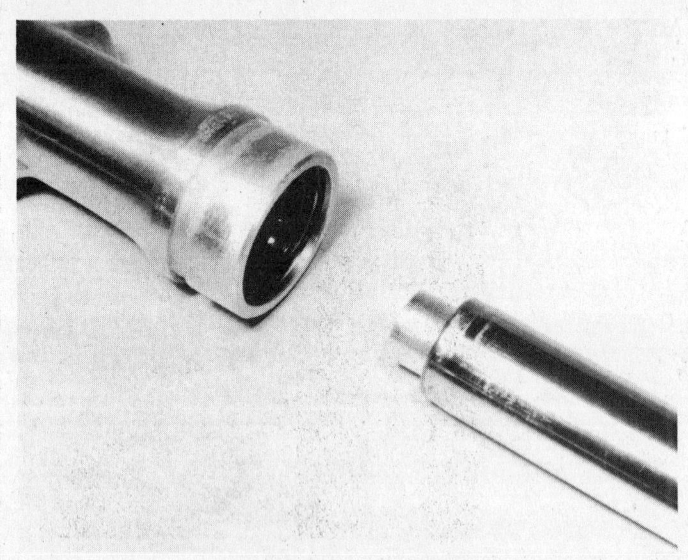

3.12a ... and insert stanchion into fork lower leg

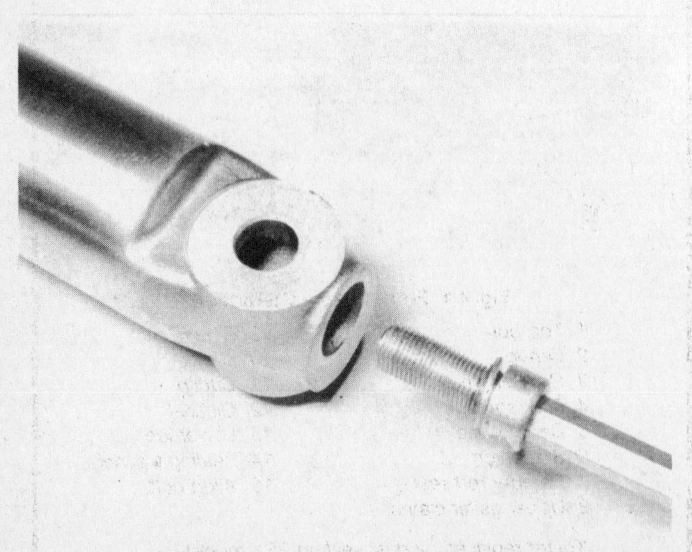

3.12b Do not omit sealing washer when refitting damper rod bolt

3.13 Smear stanchion with grease before refitting gaiter (except ESX model)

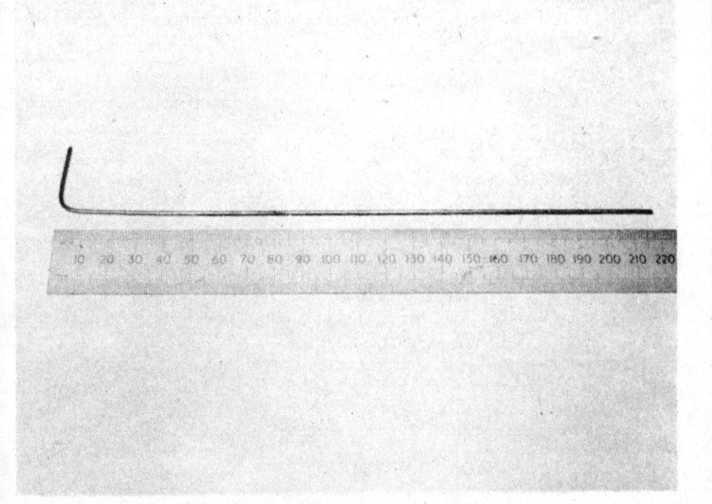

3.15a Dipstick can be made up from length of wire ...

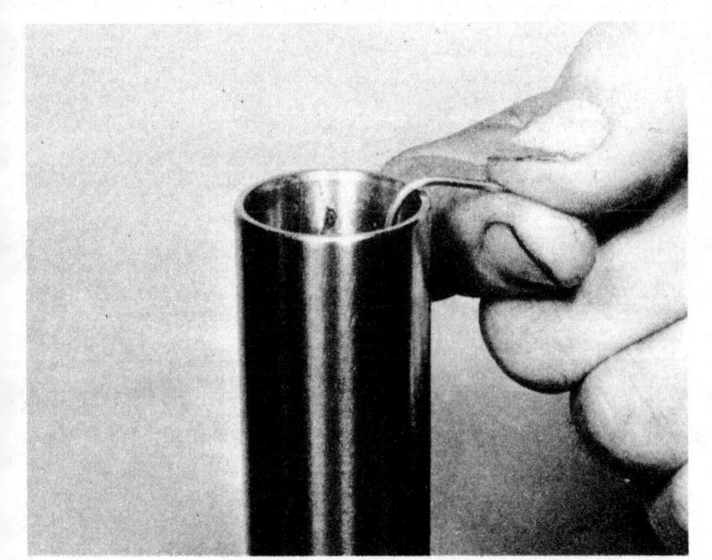

3.15b ... and used to establish correct fork oil level

3.16a Fork spring is refitted with closer spaced coils at top

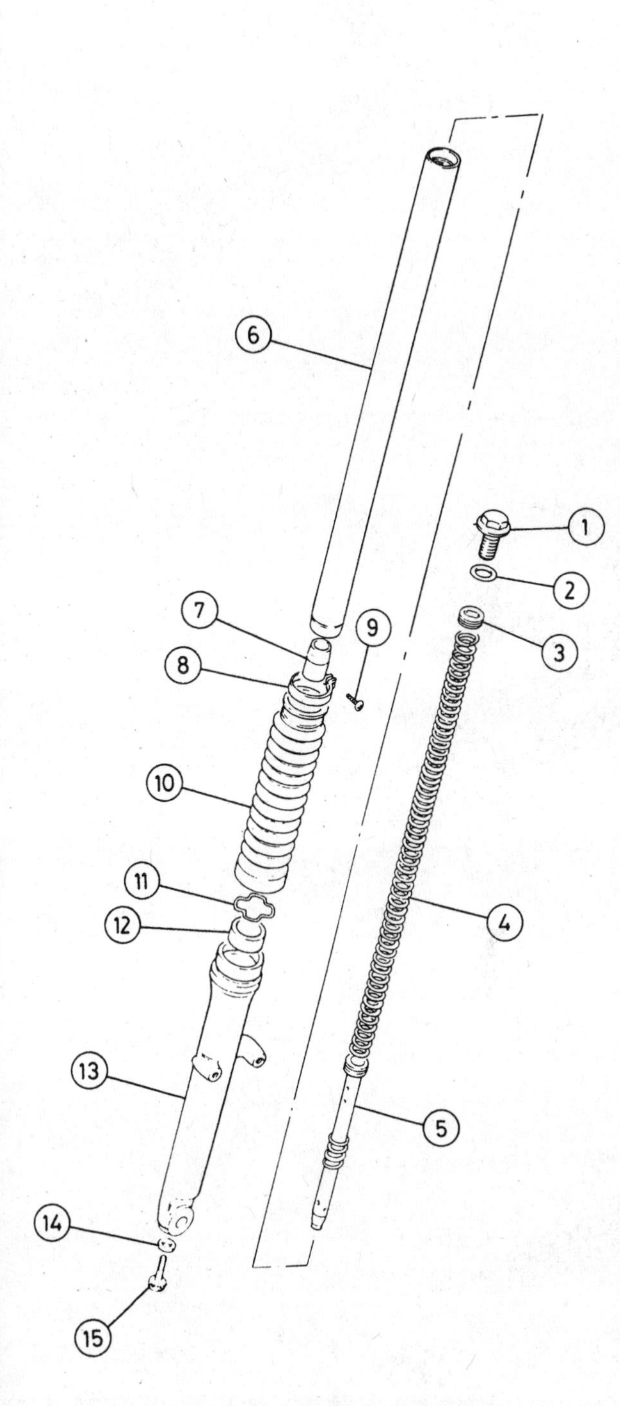

Fig. 4.1 Front forks – GS models

1 Top bolt
2 O-ring
3 Retaining nut
4 Fork spring
5 Damper rod
6 Stanchion
7 Damper rod seat
*8 Upper gaiter clamp**
*9 Screw**
*10 Gaiter**
11 Circlip
12 Oil seal
13 Lower leg
14 Sealing washer
15 Allen bolt

**Gaiter replaced by dust seal on ESX model*

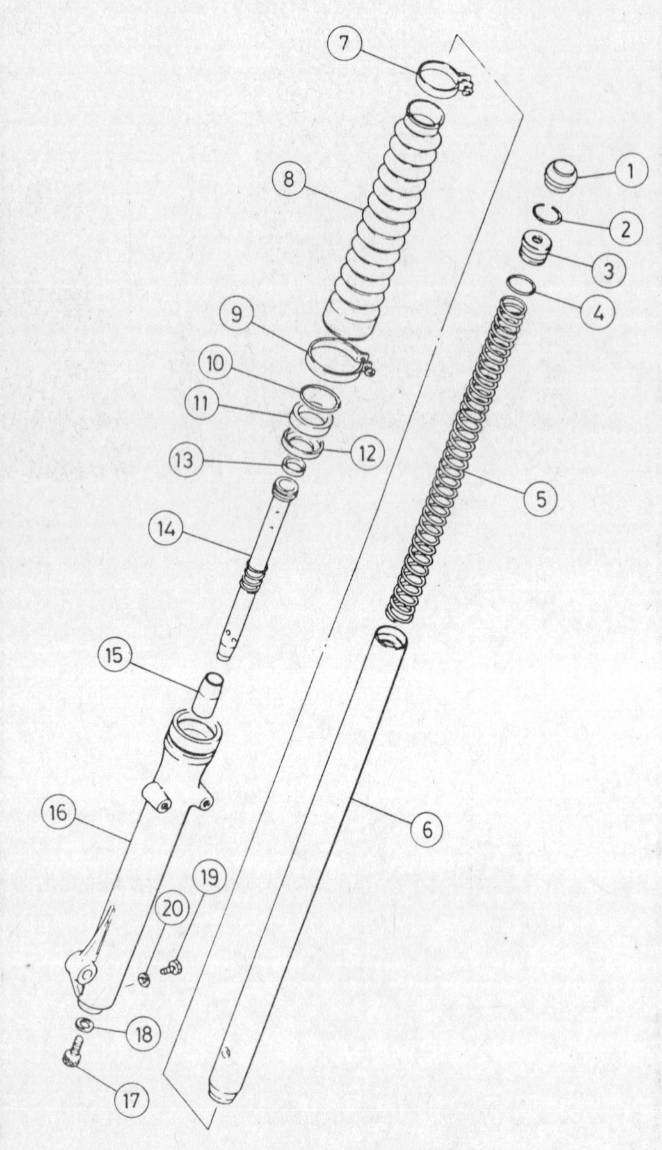

Fig. 4.2 Front forks – DR model

1 Rubber plug
2 Wire circlip
3 Top plug
4 O-ring
5 Fork spring
6 Stanchion
7 Upper gaiter clamp
8 Gaiter
9 Lower gaiter clamp
10 Circlip
11 Oil seal
12 Spacer
13 Damper rod piston ring
14 Damper rod
15 Damper rod seat
16 Lower leg
17 Allen bolt
18 Sealing washer
19 Drain bolt
20 Sealing washer

5 Steering head assembly: removal and refitting

1 Remove the front wheel, brake caliper (where necessary), front mudgard, and fork legs as described in Section 2 of this Chapter. The headlamp fairing fitted to GS125ES models is secured by three bolts: one on each side adjacent to the headlamp mounting brackets and one underneath the front of the fairing. Unscrew the bolts and withdraw the fairing.

2 To prevent short circuits, disconnect the battery before removing any electrical leads.

3 Remove the petrol tank as described in Chapter 2 or protect it with an old blanket, to prevent damage to the paintwork. The handlebars are secured to the top yoke on all GS125 models by two split clamps, each of which has two halves clamped by two long bolts passing through into the top yoke. Withdraw the handlebar clamp cover (GS125 ES only), remove the four bolts, withdraw the two upper clamp halves, and carefully move the handlebars backwards clear of the steering head area. On DR125 S models, unscrew the four handlebar clamp bolts, withdraw the clamps and carefully move the handlebars backwards clear of the steering head area.

4 The procedure from this stage onwards must depend on the work being undertaken, and much will depend on a commonsense approach and a measure of ingenuity on the part of the owner. If the machine is being dismantled to repair accident damage every component must be removed individually for repair or renewal, but if the steering head is being dismantled for greasing during the course of routine maintenance it is permissible to avoid as much of the preliminary dismantling as possible by keeping the components in major sub-assemblies and by removing only those items which will hinder the removal of the fork yokes.

5 On all GS125 models it will suffice to remove the headlamp rim and reflector unit so that the instrument cluster and ignition switch wiring can be disconnected, and to disconnect the speedometer and tachometer cables from the instruments so that the top yoke can be removed with the ignition switch and instrument cluster still attached. The headlamp brackets can then be lifted off the bottom yoke and the complete assembly allowed to hang down clear of the steering head area.

6 On DR125S models a similar course of action can be employed once the wiring is disconnected Slide the headlamp brackets out of the yokes, taking the headlamp shell and cowling with them. The instrument cluster can be detached if desired from the top yoke by removing the two retaining bolts.

7 Remove the large chromium plated bolt from its location through the centre of the top yoke. Using a soft-faced hammer, give the top yoke a gentle tap to free it from the steering head and lift it from position. Carry out a final check around the bottom yoke to ensure that the hydraulic brake hose (where fitted) the tachometer drive cable and any electrical leads have all been freed from their retaining clips. Note that if the bottom yoke is to be renewed or repaired, the high-mounted front mudguard fitted to DR125 S models must now be detached.

8 Support the weight of the bottom yoke and, using a C-spanner of the correct size, remove the steering head bearing adjusting ring. If a C-spanner is not available, a soft metal drift may be used in conjunction with a hammer to slacken the ring.

9 Remove the dust excluder and the cone of the upper bearing. The bottom yoke, complete with steering stem, can now be lowered from position. Ensure that any balls that fall from the bearings as the bearing races separate are caught and retained. It is quite likely that only the balls from the lower bearing will drop free, since those of the upper bearing will remain seated in the bearing cup. Full details of examining and renovating the steering head bearings are given in Section 6 of this Chapter.

10 Fitting of the steering head assembly is a direct reversal of that procedure used for removal, whilst taking into account the following points. It is advisable to position all eighteen balls of the lower bearing around the bearing cone before inserting the steering stem fully into the steering head. Retain these balls in position with grease of the recommended type and fill both bearing cups wth the same type of grease.

11 With the bottom yoke pressed fully home into the steering head, place the twenty-two balls into the upper bearing cup and fit the bearing cone followed by the dust excluder. Refit the adjusting ring and tighten it, finger-tight. The ring should now be tightened firmly to seat the bearings; the manufacturer recommends a torque setting of 4.0 – 5.0 kgf m (29 – 36 lbf ft). Turn the bottom yoke from lock to lock five or six times to settle the balls, then slacken the adjusting ring until all pressure is removed.

12 To provide the initial setting for steering head bearing adjustment, tighten the adjusting ring carefully until resistance is felt then loosen it by $\frac{1}{8}$ to $\frac{1}{4}$ of a turn. Remember to check that the adjustment is correct, as described in Routine Maintenance, when the steering head assembly has been reassembled and the forks and front wheel refitted.

13 Finally, whilst refitting and reconnecting all disturbed components, take care to ensure that all control cables, drive cables, electrical leads, etc are correctly routed and that proper reference is made to the list of torque wrench settings given in the Specifications Section of this Chapter and of Chapter 5. Check that the headlamp beam height has not been disturbed and ensure that all controls and instruments function correctly before taking the machine on the public highway.

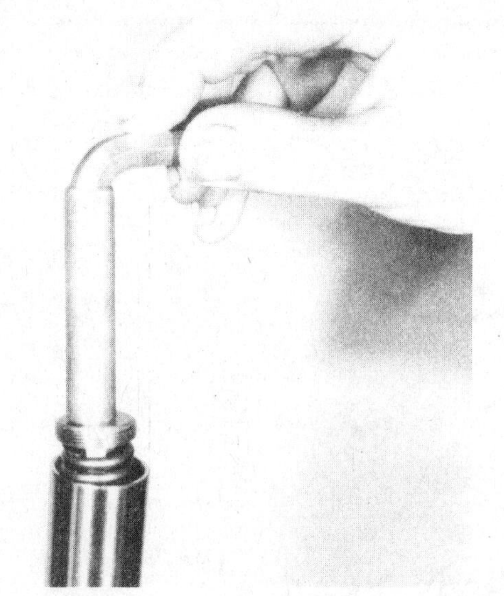
3.16b Slide tube over Allen key to aid refitting spring retaining nut – GS125 models only

5.1a GS125 ES headlamp fairing is retained by one bolt on each side at top ...

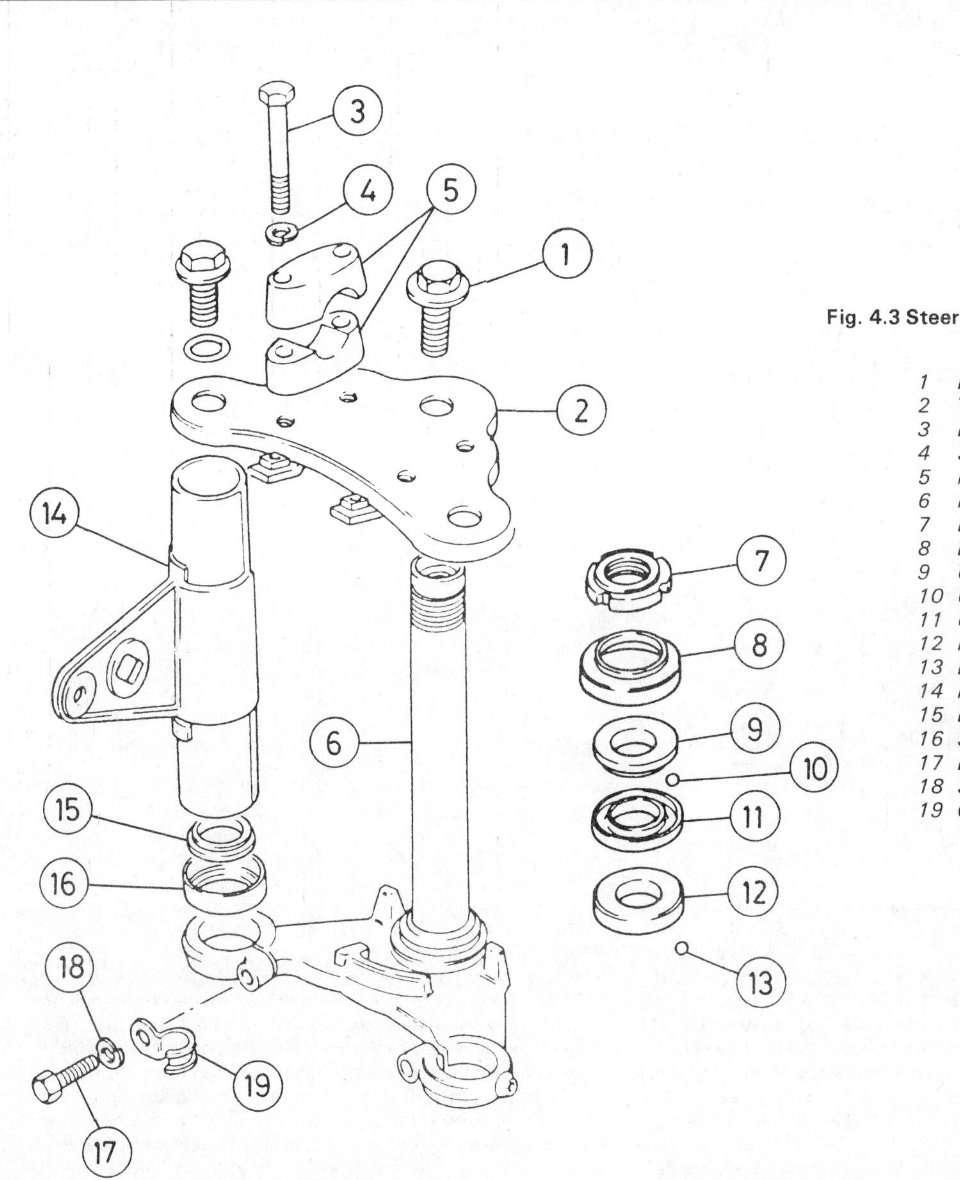

Fig. 4.3 Steering head assembly – GS models

1 Bolt
2 Top yoke
3 Bolt - 4 off
4 Spring washer - 4 off
5 Handlebar clamp - 4 off
6 Bottom yoke
7 Bearing adjustment ring
8 Dust excluder
9 Upper bearing cone
10 Upper bearing balls
11 Upper bearing cup
12 Lower bearing cup
13 Lower bearing balls
14 Headlamp bracket - 2 off
15 Damping rubber - 2 off
16 Spacer - 2 off
17 Bolt - 2 off
18 Spring washer - 2 off
19 Cable guide - 2 off

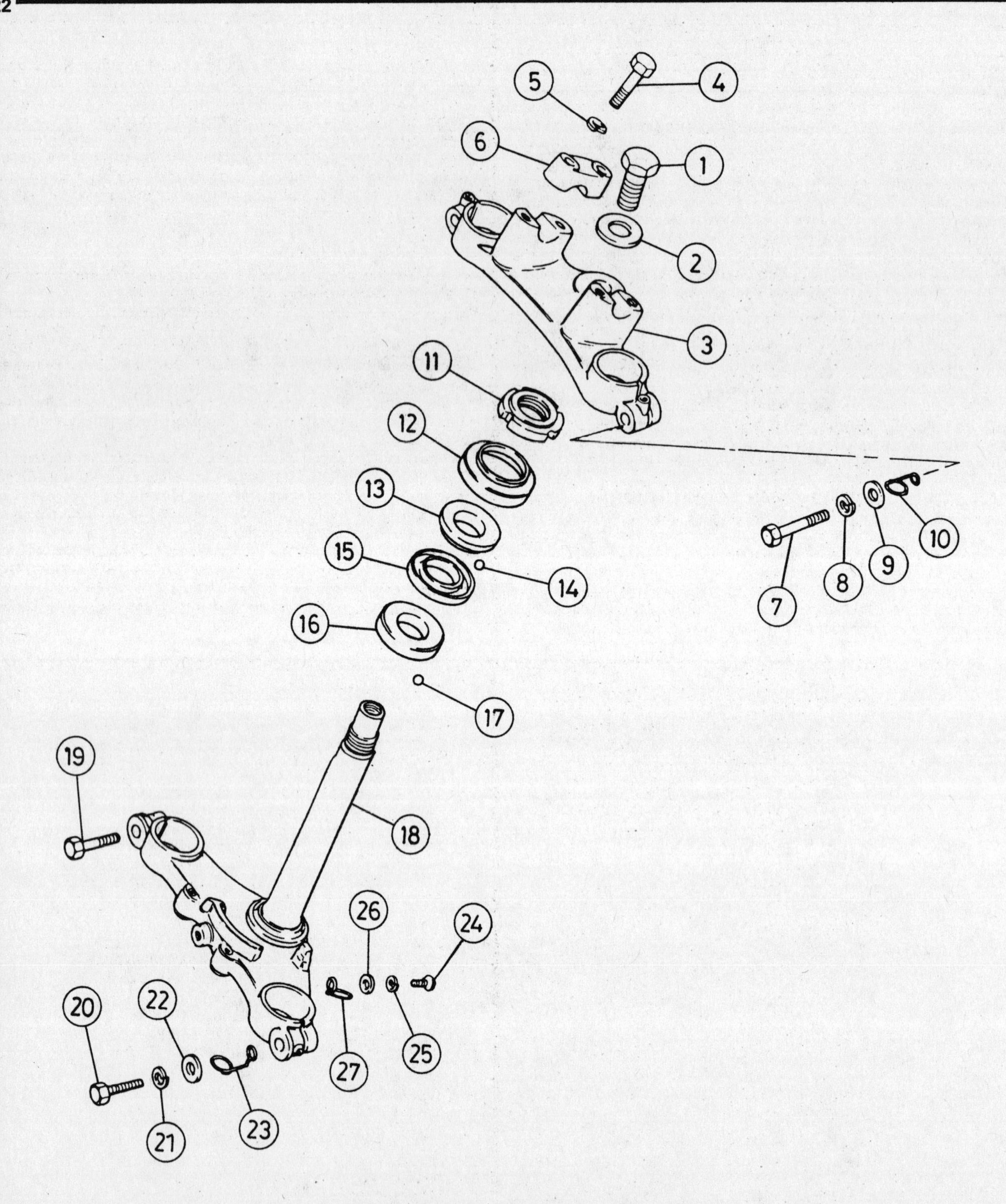

Fig. 4.4 Steering head assembly – DR model

1 Bolt
2 Washer
3 Top yoke
4 Bolt - 4 off
5 Spring washer - 4 off
6 Handlebar clamp - 2 off
7 Bolt - 2 off
8 Spring washer - 2 off
9 Washer - 2 off
10 Cable guide
11 Bearing adjusting ring
12 Dust excluder
13 Upper bearing cone
14 Upper bearing balls
15 Upper bearing cup
16 Lower bearing cup
17 Lower bearing balls
18 Bottom yoke
19 Bolt
20 Bolt
21 Spring washer
22 Washer
23 Cable guide
24 Screw
25 Spring washer
26 Washer
27 Cable guide

6 Steering head bearings: examination and renovation

1 Before commencing reassembly of the steering head component parts, take care to examine each of the steering head bearing races. The ball bearing tracks of their respective cup and cone bearings should be polished and free from any indentations or cracks. If wear or damage is evident, then the cups and cones must be renewed as complete sets.
2 Carefully clean and examine the balls contained in each bearing assembly. These should also be polished and show no signs of surface cracks or blemishes. If any one ball is found to be defective, then the complete set should be renewed. Remember that a complete set of these balls is relatively cheap and it is not worth the risk of refitting items that are in doubtful condition. Note that two different sizes of ball are used; the upper race being fitted with 6 mm ($\frac{1}{4}$ in) balls and the lower race with 8 mm ($\frac{5}{16}$ in) items. Be careful at all times not to confuse the two sizes and ensure that each race is fitted with balls of the correct size.
3 Twenty-two balls are fitted in the top bearing race and eighteen in the lower. This arrangement will leave a gap between any two balls but an extra ball must not be fitted, otherwise the balls will press against each other thereby accelerating wear and causing the steering action to be stiff.
4 The bearing outer races are a drive fit in the steering head and may be removed by passing a long drift through the inner bore of the steering head and drifting out the defective item from the opposite end. The drift must be moved progressively around the race to ensure that it leaves the steering head evenly and squarely.
5 The lower of the two inner races fits over the steering stem and may be removed by carefully drifting it up the length of the stem with a flat-ended chisel or a similar tool. Again, take care to ensure that the race is kept square to the stem.
6 Fitting of the new bearing races is a straightforward procedure whilst taking note of the following points. Ensure that the race locations within the steering head are clean and free of rust; the same applies to the steering stem. Lightly grease the stem and head locations to aid fitting of the races and drift each race into position whilst keeping it square to its location. Fitting of the outer races into the steering head will be made easier if the opposite end of the head to which the race is being fitted has a wooden block placed against it to absorb some of the shock as the drift strikes the race.

7 Frame: examination and renovation

1 The frame is unlikely to require attention unless accident damage has occurred. In some cases, renewal of the frame is the only satisfactory remedy if the frame is badly out of alignment. Only a few frame specialists have the jigs and mandrels necessary for resetting the frame to the required standard of accuracy, and even then there is no easy means of assessing to what extent the frame may have been over-stressed.
2 After the machine has covered a considerable mileage, it is advisable to examine the frame closely for signs of cracking or splitting at the welded joints. Rust corrosion can also cause weakness at these joints. Minor damage can be repaired by welding or brazing, depending on the extent and nature of the damage.
3 Remember that a frame which is out of alignment will cause handling problems and may even promote 'speed wobbles'. If misalignment is suspected, as a result of an accident, it will be necessary to strip the machine completely so that the frame can be checked, and if necessary, renewed.

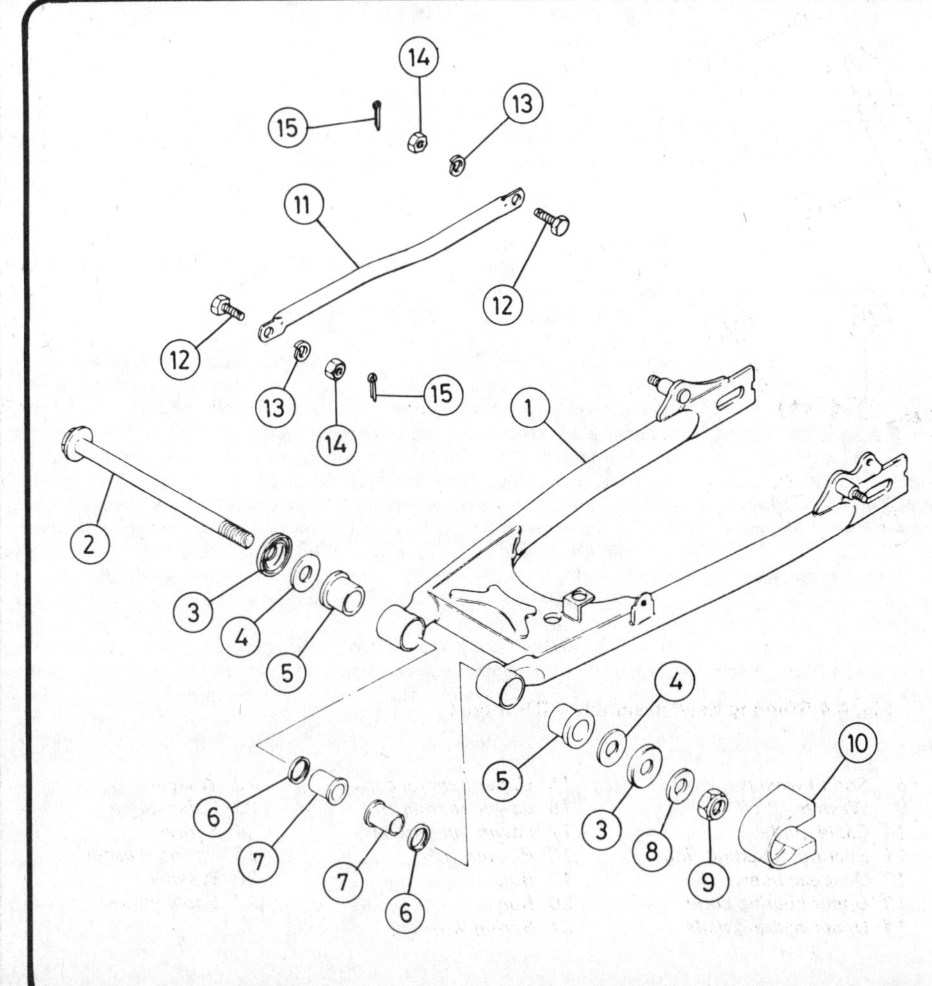

Fig. 4.5 Swinging arm – GS models

1 *Swinging arm*
2 *Pivot bolt*
3 *Sealing cap - 2 off*
4 *Washer - 2 off*
5 *Outer bush - 2 off*
6 *Oil seal - 2 off*
7 *Inner bush - 2 off*
8 *Washer*
9 *Nut*
10 *Nylon chain buffer*
11 *Brake torque arm*
12 *Bolt - 2 off*
13 *Spring washer - 2 off*
14 *Nut - 2 off*
15 *Split pin - 2 off*

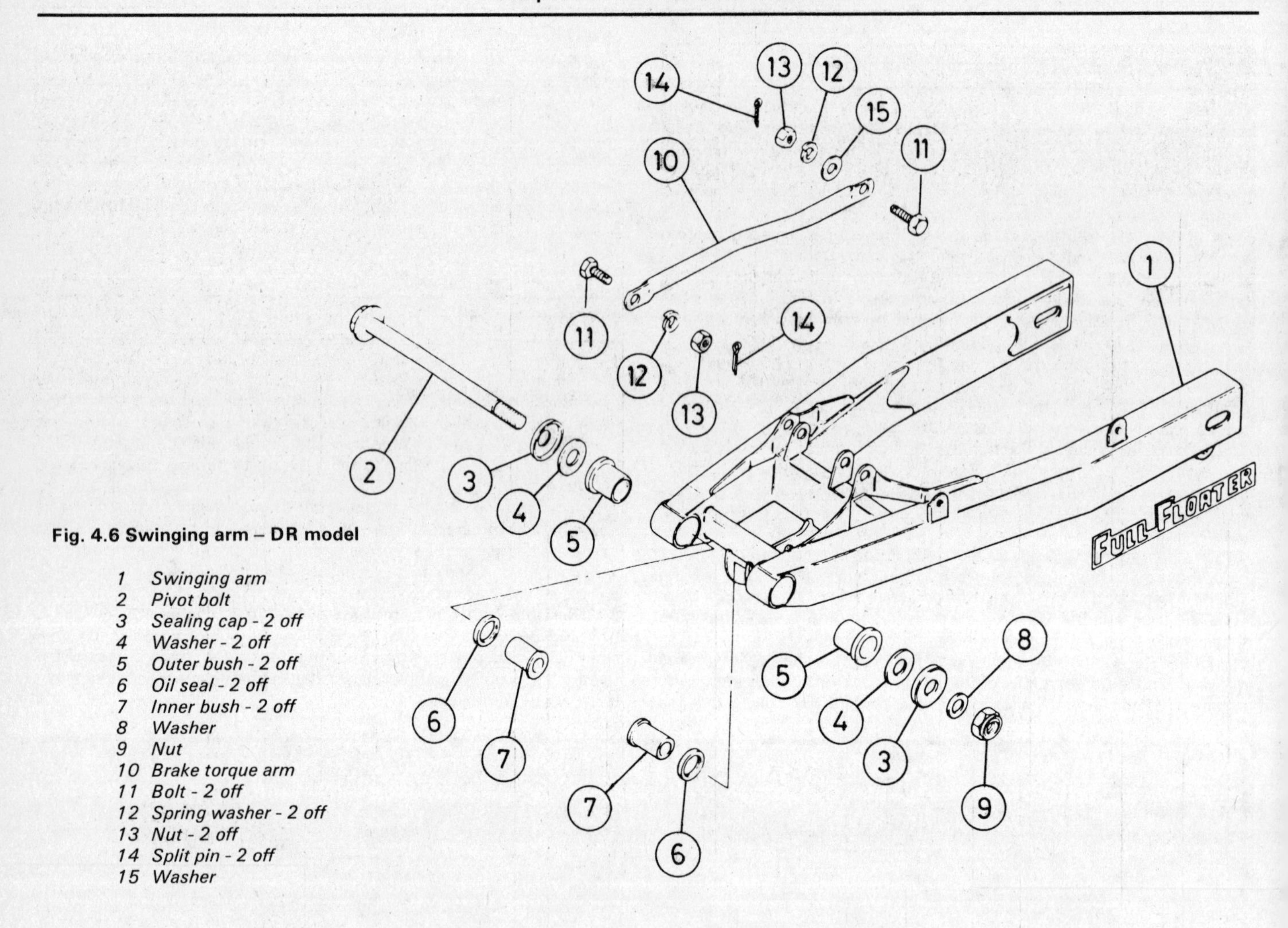

Fig. 4.6 Swinging arm – DR model

1 Swinging arm
2 Pivot bolt
3 Sealing cap - 2 off
4 Washer - 2 off
5 Outer bush - 2 off
6 Oil seal - 2 off
7 Inner bush - 2 off
8 Washer
9 Nut
10 Brake torque arm
11 Bolt - 2 off
12 Spring washer - 2 off
13 Nut - 2 off
14 Split pin - 2 off
15 Washer

8 Swinging arm: removal and refitting

1 Remove the rear wheel following the instructions given in the relevant Section of Chapter 5.

All GS125 models

2 Remove, if required, the brake torque arm and the chainguard from the swinging arm, then slacken all four suspension unit mounting nuts. Remove the two suspension unit bottom mounting nuts and their washers, then pull the units sideways off their swinging arm mounting lugs. Note carefully the position and number of the plain washers at each mounting.

3 Remove the swinging arm pivot bolt securing nut and its washer, then withdraw the pivot bolt. If the bolt proves stubborn, apply a good quantity of penetrating fluid, allow time for it to work, then displace the bolt using a hammer and a metal drift. Withdraw the swinging arm.

4 Reassembly is the reverse of the dismantling procedure. Check that the pivot bolt and the passages through which it fits are completely clean and free from corrosion, then apply liberal quantities of grease to the surface of the bolt and to the passages, not forgetting the crankcase. Check that all bushes and sealing caps are in place in the swinging arm pivot lugs and that all are well greased. Offer up the swinging arm and insert the pivot bolt from right to left, then refit the retaining nut and its plain washer. Tighten the nut to a torque setting of 5.0 – 8.0 kgf m (36 – 58 lbf ft). Refit the suspension units, ensuring that the plain washers are correctly refitted, and tighten all four suspension unit mounting nuts to a torque setting of 2.0 – 3.0 kgf m (14.5 – 21.5 lbf ft). Complete the remainder of the reassembly work.

5 Before taking the machine out on the road, check that all the nuts and bolts are securely fastened, and that the rear suspension, rear brake, and chain tension are adjusted correctly and working properly.

DR125 S models

6 Although it is possible to separate the suspension linkage from the swinging arm by removing the suspension unit bottom mounting bolt and the two link arm bottom mounting bolts so that either the swinging arm or the linkage can be withdrawn individually, it is recommended that the complete suspension assembly is removed as described below. This is a simple operation requiring only the removal of the swinging arm and linkage rocker pivot bolts and will permit the complete assembly to be closely examined.

7 Remove the seat and both sidepanels, then unclip the rubber mudflap from around the frame horizontal bracing tube. Remove the retaining nuts from the swinging arm pivot bolt and the linkage rocker pivot bolt, then withdraw first the rocker pivot bolt, then the swinging arm pivot bolt and manoeuvre the complete assembly away from the machine. If either pivot bolt proves stubborn, apply a good quantity of penetrating fluid, allow time for it to work, then displace the bolt using a hammer and a metal drift.

8 Reassembly is the reverse of the dismantling procedure. Check that both pivot bolts and the passages through which they fit are completely clear and free from corrosion, then apply liberal quantities of grease to both bolts and to their matching passages. Check that all bushes and sealing caps are in place in the swinging arm and linkage rocker pivot lugs and that all are well greased.

9 Offer up the suspension assembly and retain it by pushing through from right to left the swinging arm pivot bolt. Manoeuvre the linkage rocker into position and refit its pivot bolt, then refit both pivot bolt retaining nuts and their washers. Tighten both nuts to the torque setting given in the Specifications Section of this Chapter.

10 Complete the remainder of the reassembly work. Before taking the machine out on the road, check that all nuts and bolts are securely fastened, using the torque settings specified, and that the rear suspension, rear brake and chain tension are adjusted correctly and working properly.

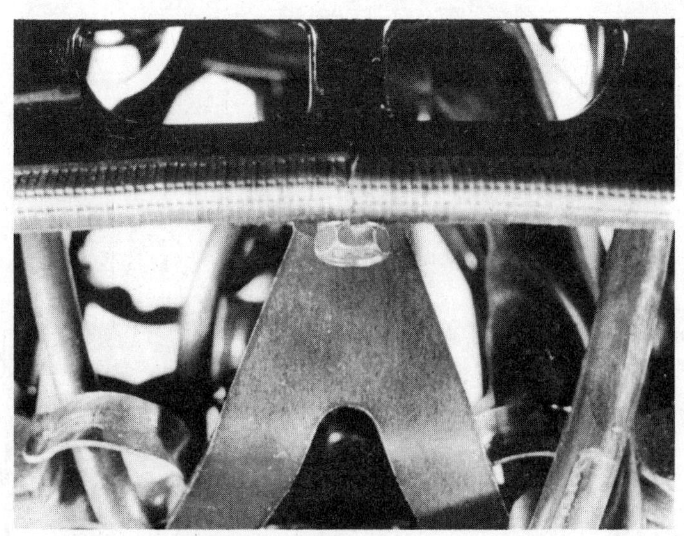
5.1b ... and by one bolt at the front

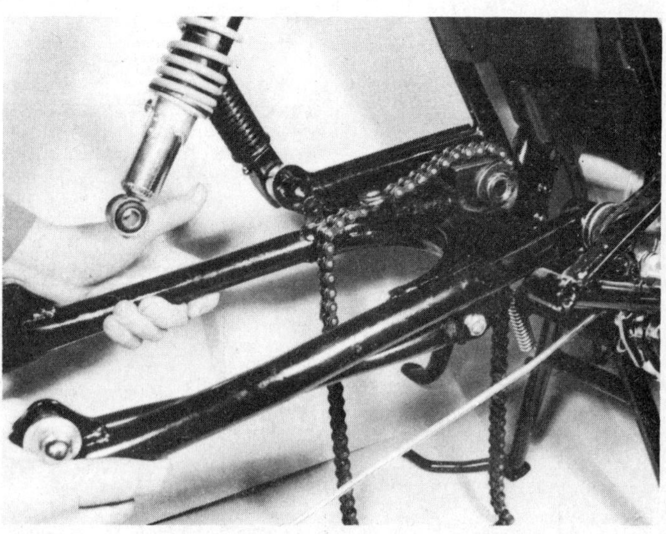
8.4a Do not forget to loop chain over swinging arm before arm is refitted

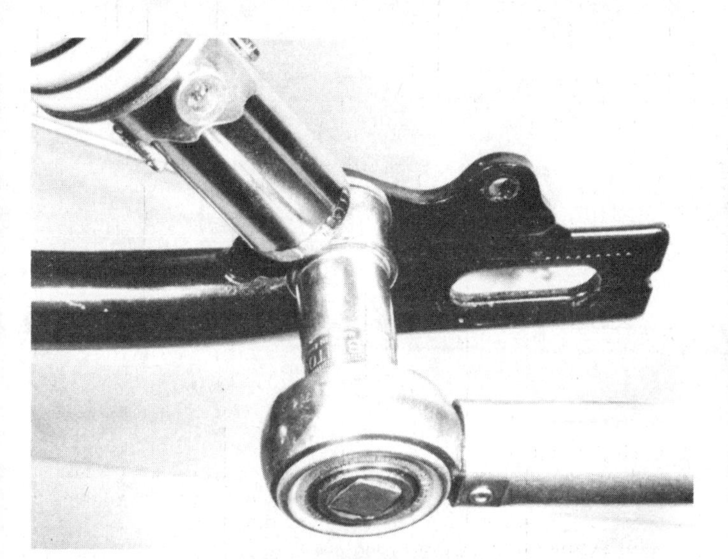
8.4b Ensure plain washers are correctly located before tightening suspension unit mounting nuts

8.7a Withdraw linkage rocker pivot bolt

9 Swinging arm and suspension linkage: examination and renovation

1 On DR125 S models, dismantle the rear suspension by removing the suspension unit mounting bolts and the link arm mounting bolts. If required, the mudflap can be unbolted from the link arm and the rubber covers slid off the link arm lower ends.
2 Dismantle as far as possible all components, removing the sealing caps and pushing out the hardened metal inner sleeves and dust seals (where fitted) from each bearing. Thoroughly clean all components, removing all traces of dirt, corrosion and old grease.
3 Inspect closely all components looking for obvious signs of wear such as heavy scoring, or for damage such as cracks or distortion due to accidental impact. Any obviously damaged or worn component must be renewed. If the painted finish has deteriorated it is worth taking the opportunity to repaint the affected area, ensuring that the surface is correctly prepared beforehand.
4 Check the swinging arm pivot bolt for wear. If the shank of the bolt is seen to be stepped or badly scored, then it must be renewed. Remove all traces of corrosion and hardened grease from the bolt before checking it for straightness by rolling it on a flat surface, such as a sheet of plate glass, whilst attempting to insert a feeler gauge of 0.6 mm (0.02 in) thickness beneath it. Alternatively, place the bolt on two V-blocks and measure the amount of runout on its shank with a dial gauge. If the amount of runout measured exceeds 0.6 mm (0.02 in), the bolt must be renewed. Note that a bent pivot bolt will prevent the swinging arm from moving smoothly about its axis.
5 Detach the nylon buffer from the left-hand pivot lug and inspect it for signs of excessive wear. If it is considered that the buffer no longer serves as an adequate means of protecting the metal of the pivot lug against interference from the final drive chain, then it must be renewed.
6 Measure the inside diameter of the swinging arm pivot bushes, renewing them if they are worn to more than 20.2 mm (0.795 in). If measuring equipment is not available check the fit of the hardened metal inner sleeve or spacer in the bushes; little or no free play should be discernible. The bushes at the top and bottom of the link arm can be assessed in the same way and should be renewed also if found to be worn to more than the limit given in the Specifications Section of this Chapter. The two needle roller bearings in the rocker pivot should be examined closely and renewed if there are any signs of wear or damage, or if free play is discernible when the inner sleeve is inserted.
7 Displace the swinging arm pivot bushes by inserting a suitable drift into each pivot lug from the end opposite to the bush shoulder and tapping with a hammer; the suspension linkage bushes or bearings can be removed in a similar manner, or a drawbolt arrange-

ment can be applied as shown in the accompanying illustration. Note that removal will almost certanly damage the bush or bearing; new components should be refitted whenever they are disturbed.

8 On refitting bushes or bearings, clean thoroughly the bearing housing, removing any burrs or traces of corrosion, then coat both surfaces with grease. Assemble a drawbolt arrangement as shown in the accompanying illustration and carefully draw the bush or bearing into position, being careful to keep it absolutely square to its housing at all times. Note that all outer bushes, whether of synthetic material or metal, are fitted with their raised collars to the outside, and the rocker pivot needle roller bearings are fitted with the punch marks in one end also facing to the outside.

9 All dust seals, sealing caps and bearing covers should be examined closely and renewed if at all worn or damaged; ideally the dust seals should be renewed whenever they are disturbed.

10 Note that shims are fitted beneath each swinging arm pivot sealing cap to eliminate endfloat along the axis of the pivot bolt. When the swinging arm is installed in the frame and the pivot bolt retaining nut is tightened to the specified torque setting, there should be no discernible free play when the swinging arm is pulled and pushed from side to side, but the arm should be free to move; add (or subtract) shims as necessary to achieve this.

11 On reassembly use a high-quality molybdenum disulphide based grease to pack each bearing housing. Smear grease over the inside and outside of each inner sleeve as it is refitted, and to the sealing lips of each dust seal or sealing cap. Be very careful to pack fully the rocker pivot needle roller bearings.

12 Referring to the accompanying illustrations to assist the correct refitting of components, reassemble the swinging arm and (where applicable) the suspension linkage. When working on a DR125 S model, ensure that all linkage nuts and bolts are tightened securely to the torque setting given in the Specifications Section of this Chapter.

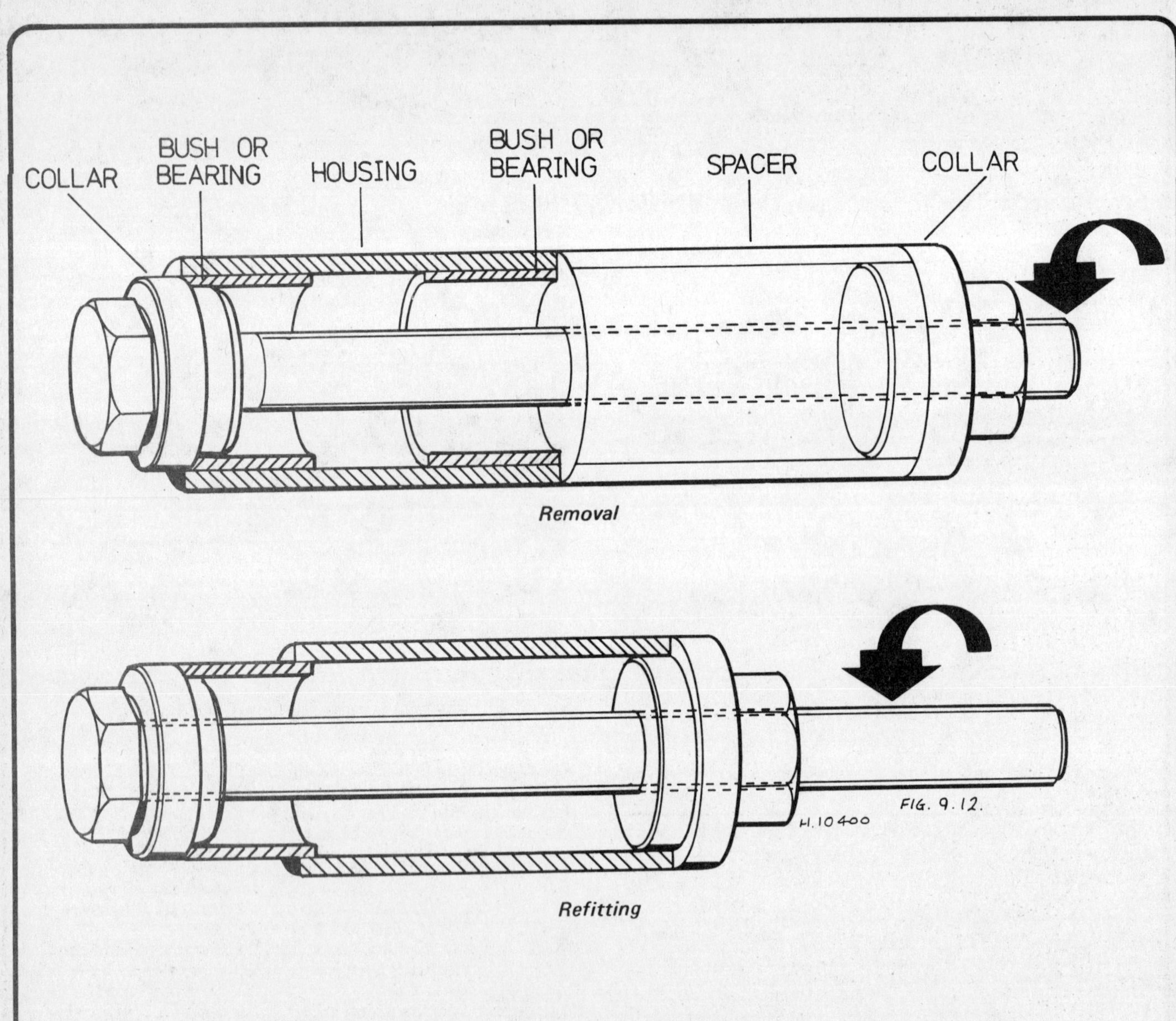

Fig. 4.7 Drawbolt tool for removing and refitting suspension linkage and swinging arm bearings

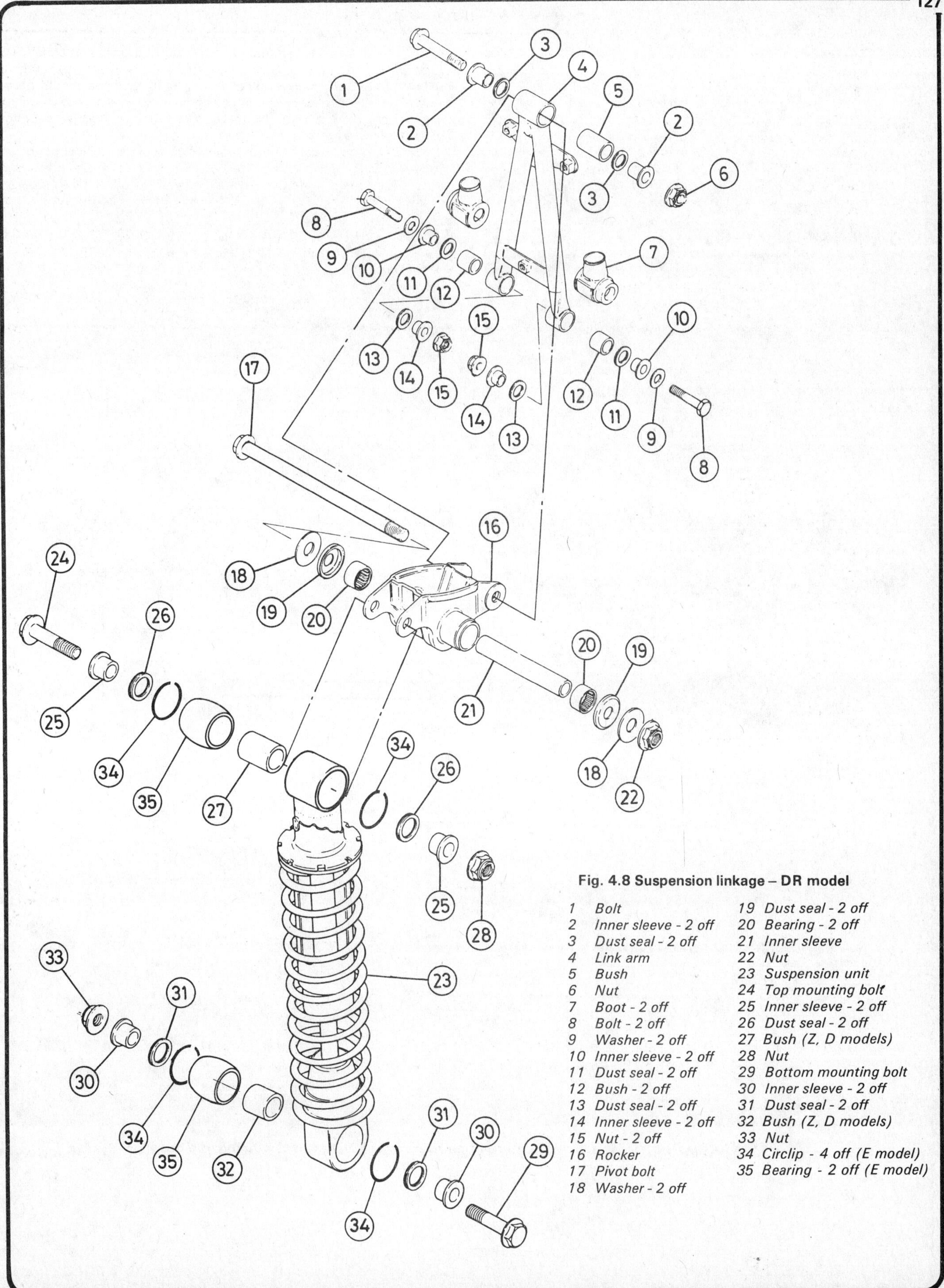

Fig. 4.8 Suspension linkage – DR model

1 Bolt
2 Inner sleeve - 2 off
3 Dust seal - 2 off
4 Link arm
5 Bush
6 Nut
7 Boot - 2 off
8 Bolt - 2 off
9 Washer - 2 off
10 Inner sleeve - 2 off
11 Dust seal - 2 off
12 Bush - 2 off
13 Dust seal - 2 off
14 Inner sleeve - 2 off
15 Nut - 2 off
16 Rocker
17 Pivot bolt
18 Washer - 2 off
19 Dust seal - 2 off
20 Bearing - 2 off
21 Inner sleeve
22 Nut
23 Suspension unit
24 Top mounting bolt
25 Inner sleeve - 2 off
26 Dust seal - 2 off
27 Bush (Z, D models)
28 Nut
29 Bottom mounting bolt
30 Inner sleeve - 2 off
31 Dust seal - 2 off
32 Bush (Z, D models)
33 Nut
34 Circlip - 4 off (E model)
35 Bearing - 2 off (E model)

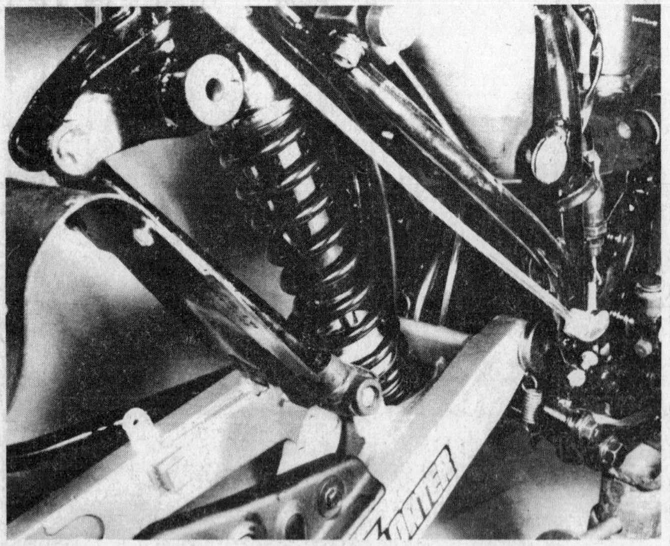

8.7b Rear suspension is removed as a complete assembly

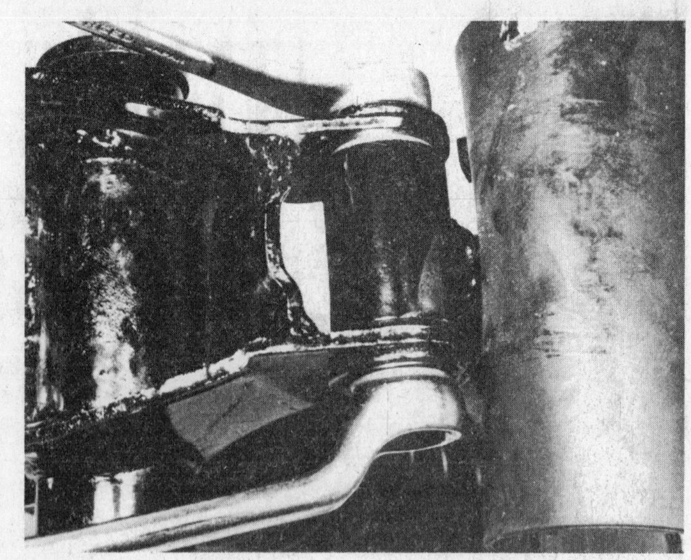

9.1a DR125 S model – unscrew pivot bolts to allow ...

9.1b ... separation of suspension linkage components

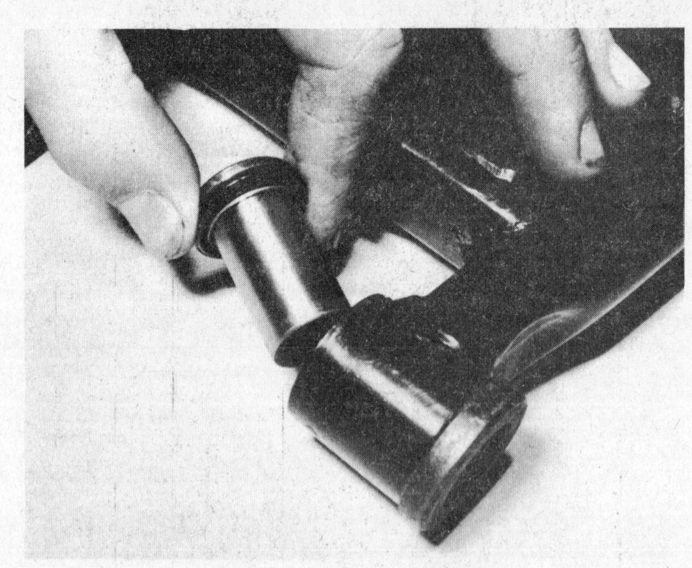

9.2a Pivot bearing inner sleeves can be pushed out ...

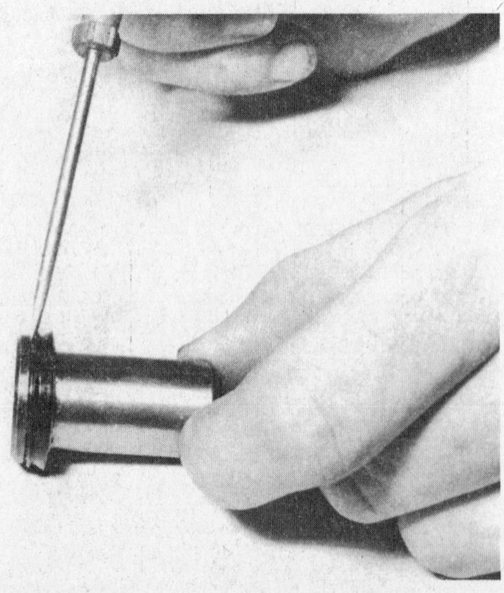

9.2b ... and seals displaced for renewal

9.6 Check swinging arm (and link arm) pivot bushes for wear – renew if worn

9.9a Use tube of suitable size to press dust seals on to inner sleeves ...

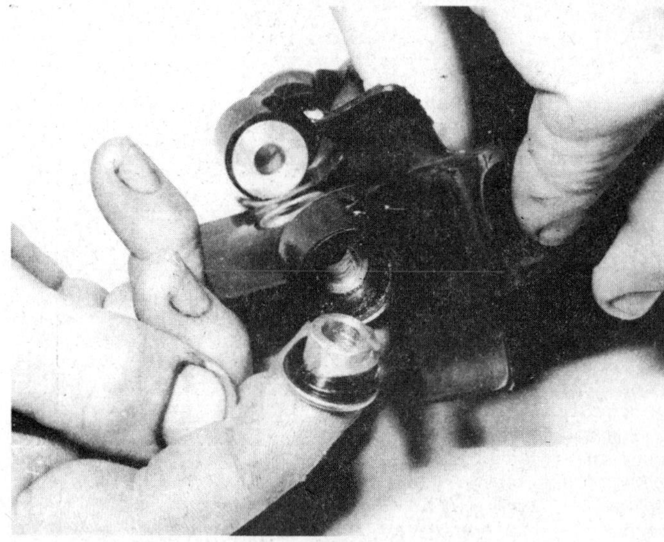

9.9b ... and apply thick smear of grease before refitting

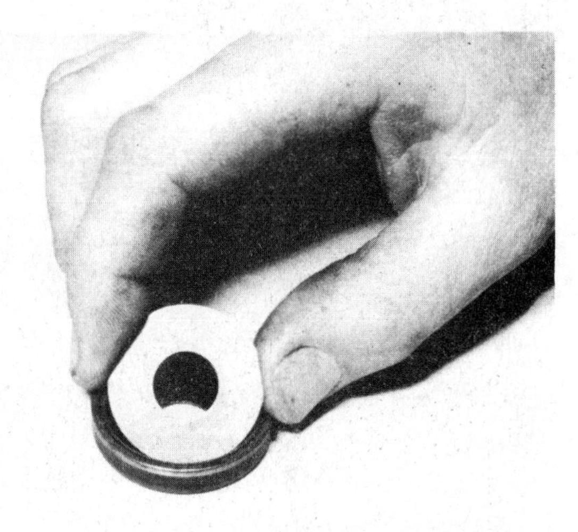

9.10 Shims are fitted in sealing caps to eliminate swinging arm pivot endfloat

9.11a Pack bearings with recommended grease ...

9.11b ... insert inner sleeve ...

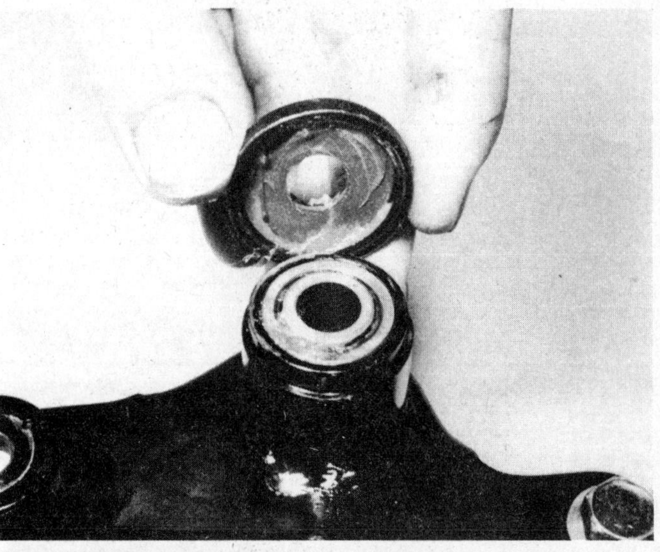

9.11c ... and grease sealing caps before refitting

10 Rear suspension unit(s): removal, examination and refitting

1 On all GS125 models the suspension units are withdrawn by removing the four retaining cap nuts and plain washers, then pulling each unit sideways off its mounting lugs, noting that the pillion grab handle must be displaced first to permit the removal of the left-hand unit. Refitting is a straightforward reversal of the above, but care must be taken to refit correctly the plain washers. Tighten all four retaining nuts to a torque setting of 2.0 – 3.0 kgf m (14.5 – 21.5 lbf ft).
2 On DR125 S models remove the seat and both sidepanels, then remove both suspension unit top and bottom mounting bolts. Unclip the rubber mudflap from the frame horizontal bracing tube, then remove the linkage rocker pivot bolt. Taking care not to displace the sealing caps from the rocker pivot bearings, tilt backwards the suspension linkage until the unit can be manoeuvred clear and lifted out of the machine. Refitting is a straightforward reversal of the above; tighten the rocker pivot bolt retaining nut to a torque setting of 4.5 – 7.0 kgf m (32.5 – 50.5 lbf ft) and the suspension unit mounting bolt retaining nuts to a torque setting of 4.0 – 6.0 kgf m (29 – 43.5 lbf ft).
3 The suspension units fitted to the GS125 models can be regarded as sealed; if any wear or damage is found they must be renewed as a matched pair of complete units. Look for signs of oil leakage around the damper rod, worn mounting bushes, and any other signs of damage.
4 The DR125 S suspension unit is examined in the same way and should be renewed if found to be damaged or worn; again it is a sealed unit for which no replacement parts are available for repair or reconditioning. The only exception to this is that the mounting bushes or bearings can be renewed if worn.
5 Displace the two metal collars and the dust seals from each mounting eye, thoroughly clean all components then refit the collars and feel for traces of free play. Check also that the mounting bolts are a reasonably close fit in the collars. If any wear is discovered the affected component must be renewed. On DR125 SE models, first displace the two circlips retaining each bearing. The bush or bearing can be displaced using either a hammer and a suitably-sized pair of sockets as shown in the accompanying photograph, or by using a version of the drawbolt arrangement described in the previous Section. Refit the bushes or bearings using a drawbolt, being careful to position each exactly in the centre of the mounting eye. Pack each bearing with high quality molybdenum disulphide based grease, and on DR125 SE models refit the retaining circlips. Do not forget to renew the dust seals if these are damaged or worn.
6 The spring preload setting of each suspension unit is adjustable through five positions to compensate for varying loads. On all GS125 models the setting can be easily altered without moving the units from the machine by using a tommy bar in the hole directly below the springs. Turning clockwise will increase the spring tension and stiffen the rear suspension, turning anti-clockwise will lessen the spring tension and therefore soften the ride. As a general guide the third softest setting is recommended for road use only, when no pillion passenger is carried. The hardest setting should be used when a heavy load is carried, and during high-speed riding. The intermediate positions may be used as conditions dictate.
7 The DR125 S rear spring preload is altered in the same way, but a C-spanner or pin spanner must be used instead of a tommy bar. In view of the very restricted access, it is recommended that the suspension unit is removed from the machine as described above, for the preload setting to be altered. The manufacturer's recommended method, which can be used as an alternative, is to remove the seat, both sidepanels, the air filter casing assembly, and the carburettor, thus gaining access to the spring adjusting ring while the unit is in place.

11 Footrests, stands and controls: examination and renovation

1 At regular intervals all footrests and stands and the brake pedal and gearchange lever or linkage should be checked and lubricated. Check that all mounting nuts and bolts are securely fastened using the recommended torque wrench settings where these are given. Check that any securing split-pins are correctly fitted.
2 Check that the bearing surfaces at all pivot points are well greased and unworn; renewing any component that is excessively worn. If lubrication is required, dismantle the assembly to ensure that grease can be packed fully into the bearing surface. Return springs, where fitted, must be in good condition with no traces of fatigue and must be securely mounted.
3 If accident damage is to be repaired, check that the damaged component is not cracked or broken. Such damage may be repaired by welding, if the pieces are taken to an expert, but since this will destroy the finish, renewal is usually the most satisfactory course of action. If a component is merely bent it can be straightened after the affected area has been heated to a full cherry red using a blowlamp or welding torch. Again the finish will be destroyed, but painted surfaces can be repainted easily, while chromed or plated surfaces can be only replated, if the cost is justified.

12 Speedometer and tachometer heads: removal, examination and refitting

1 These instruments must be carefully handled at all times and must never be dropped or held upside down. Dirt, oil, grease and water all have an equally adverse effect on them, and so a clean working area must be provided if they are to be removed.
2 The instrument heads are very delicate and should not be dismantled at home. In the event of a fault developing, the instrument should be entrusted to a specialist repairer or a new unit fitted. If a replacement unit is required it is well worth trying to obtain a good secondhand item from a motorcycle breaker in view of the high cost of a new instrument.
3 Remember that a speedometer in correct working order is a statutory requirement in the UK. Apart from this legal necessity, reference to the odometer readings is the most satisfactory means of keeping pace with the maintenance schedules.
4 The instrument clusters are removed as a complete assembly as described in Section 5 of this Chapter. Remove the fairing or headlamp cowling (where fitted) to provide sufficient clearance, and the headlamp unit so that the relevant electrical leads can be disconnected. Disconnect the drive cables at their upper ends by unscrewing the knurled retaining rings. Slacken and remove the two bolts securing the instrument panel to the top yoke, and withdraw the panel assembly. On GS125 ES and DR125 S models, separate the two halves of the panel assembly by removing the cap nuts and washers, and screws (where fitted) which secure the panel top half and by lifting the panel top half away. Each instrument is secured to the panel bottom half by two plated screws (DR125S models only). Slacken and remove these and lift the instrument away. Refitting is the reverse of the above.

13 Speedometer and tachometer drive cables: examination and renovation

1 It is advisable to detach the speedometer and tachometer drive cables from time to time in order to check whether they are adequately lubricated and whether the outer cables are compressed or damaged at any other point along their run. A jerky or sluggish movement at the instrument head can often be attributed to a cable fault.
2 To grease the cable, uncouple both ends and withdraw the inner cable. After removing any old grease, clean the inner cable with a petrol soaked rag and examine the cable for broken strands or other damage. Do not check the cable for broken strands by passing it through the fingers or palm of the hand, this may well cause a painful injury if a broken strand snags the skin. It is best to wrap a piece of rag around the cable and pull the cable through it, any broken strands will snag the rag.
3 Regrease the cable with high melting point grease, taking care not to grease the last six inches closest to the instrument head. If this precaution is not observed, grease will work into the instrument and immobilise the sensitive movement.
4 The cables on all models are secured at both ends by large knurled rings which must be tightened or slackened using a suitable pair of pliers. Do not overtighten the knurled rings, or they will crack, necessitating renewal of the complete cable.
5 When refitting drive cables, always ensure that they have smooth, easy runs to minimise wear, and check that the cables are secured where necessary by any clamps or ties provided for the purpose of keeping the cables away from any hot or moving parts.

9.12 Rubber covers must be refitted to protect bearings from road dirt

10.1 Grab rail must be withdrawn to permit suspension unit removal

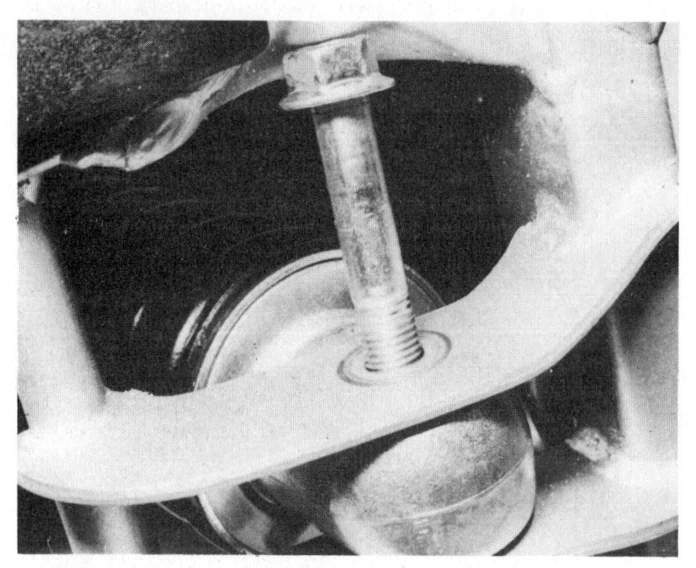

10.2 Mounting bolt is refitted from left – note cutout in swinging arm

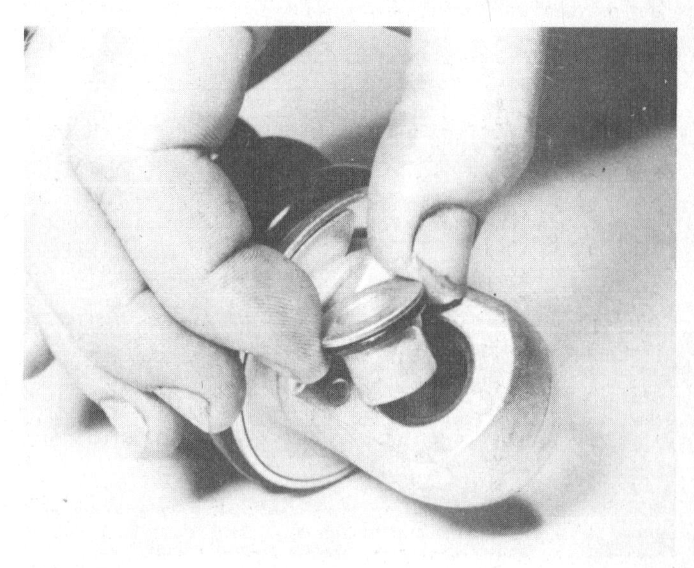

10.5a Withdraw inner sleeve and seal by hand

10.5b Bush or bearing can be drifted out as shown

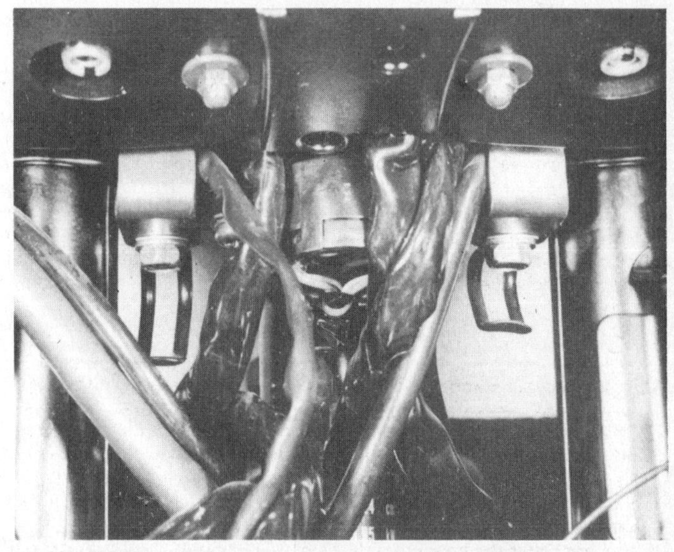

12.4 Instrument panel is secured to top yoke by two bolts

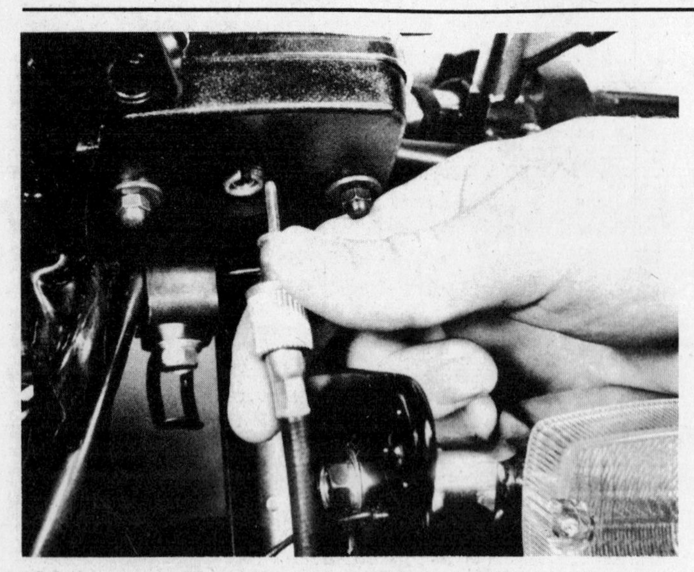

13.2 Instrument drive cables are retained by a large knurled ring at each end

14 Speedometer and tachometer drives: location and examination

Disc brake models

1 Models fitted with a front brake of the disc type have a speedometer gearbox fitted over the front wheel spindle on the right-hand side of the wheel hub. Drive is transmitted through slots cast in the hub centre which engages with the drive plate of the gearbox. Provided that the gearbox is detached and repacked with grease from time to time, very little wear should be experienced. In the event of failure, the complete unit should be renewed, as it is only available as an assembly and cannot therefore be reconditioned. Use a suitable pair of circlip pliers to remove the drive plate retaining circlip, whereupon the drive plate and washer can be lifted away. Pack a small amount of good quality high melting-point grease into the gearbox and refit the drive plate, thrust washer and circlip.

Drum brake models

2 Models fitted with a front brake of the drum type have the speedometer drive located in the front brake backplate; it is therefore necessary to withdraw the front wheel from the machine in order to gain access to the drive components.

3 To dismantle the drive assembly, remove the retaining circlip from the centre of the backplate, then lift out the thrust washer, driveplate, drive gear and second thrust washer. Turn the backplate over and remove the small grub screw from the boss which houses the worm drive components. Pull the outer bush from its housing. If this should prove difficult to accomplish, attach the outer body of an old speedometer cable by means of its knurled ring, and pull the outer bush carefully away. The speedometer drive pinion can then be removed, taking care not to lose the two thrust washers which fit over each end.

4 Carefully clean and inspect all the component parts. Any wear or damage will be immediately obvious and should be rectified by the renewal of the part concerned. The most likely areas of wear are on the tangs of the driveplate or on the teeth of either the drive gear or the drive pinion. If wear is encountered on the teeth of either of the latter components it is advisable to renew both together. Note also the condition of the large oil seal set in the backplate. If this seal shows any sign of damage or deterioration it must be renewed to prevent grease from the drive assembly working through to the brake linings.

5 Reassemble the speedometer drive assembly in the reverse order of dismantling. Lightly lubricate all components with a good quality high melting-point grease.

All models

6 The tachometer drive assembly is located in a separate housing on top of the cylinder head cover, and can be removed individually as described in Section 6 of Chapter 1. Refitting is described in Section 41 of Chapter 1; note that the tachometer assembly must be fitted after the cylinder head cover has been secured to avoid damage to the camshaft.

7 Wear or damage is unlikely and can be cured only by the renewal of the affected component. Always renew the sealing band around the driven gear housing spigot, and be careful to check the condition of the oil seal fitted in the housing. This must be renewed if damaged or worn to prevent oil from working up the cable and into the instrument.

Chapter 5 Wheels, brakes and tyres

Refer to Chapter 7 for information relating to DR125 SF, SH, SJ 'Raider', GN125 and GZ125 'Marauder' models

Contents

Specifications

Wheels	**GS125 ES**	**GS125**	**DR125 S**
Type	Cast alloy	Steel, wire spoked	Steel, wire spoked
Size:			
Front	1.60 x 18	1.60 x 18	1.60 x 21
Rear	1.60 x 18	1.40 x 18	1.85 x 18
Rim runout limit:			
Radial and axial	2.0 mm (0.08 in)	2.0 mm (0.08 in)	2.0 mm (0.08 in)
Spindle runout limit	0.25 mm (0.01 in)	0.25 mm (0.01 in)	0.25 mm (0.01 in)

Brakes			
Type:			
Front	Hydraulic disc	Drum	Drum
Rear	Drum	Drum	Drum
Front and rear drum brakes:			
Drum ID	130 mm (5.12 in)		
Service limit	130.7 mm (5.15 in)		
Friction material minimum thickness	1.5 mm (0.06 in)		
Brake disc:			
Thickness	3.8 – 4.2 mm (0.15 – 0.17 in)		
Service limit	3.0 mm (0.12 in)		
Maximum runout	0.3 mm (0.012 in)		
Master cylinder:			
Bore	12.700 – 12.743 mm (0.4999 – 0.5017 in)		
Piston diameter	12.657 – 12.684 mm (0.4983 – 0.4994 in)		
Brake caliper:			
Bore	33.960 – 34.036 mm (1.3370 – 1.3400 in)		
Piston diameter	33.884 – 33.934 mm (1.3340 – 1.3360 in)		

Tyres	**GS125 ES, GS125**	**DR125 S**
Size:		
Front	2.75 x 18-4PR	2.75 x 21-4PR
Rear	3.00 x 18-4PR	4.10 x 18-4PR
Pressures:		
Front – solo	24 psi	22 psi
Rear – solo	28 psi	24 psi
Front – pillion	24 psi	22 psi
Rear – pillion	32 psi	28 psi
Manufacturer's recommended minimum tread depth:		
Front	1.6 mm (0.06 in)	4.0 mm (0.16 in)
Rear	2.0 mm (0.08 in)	4.0 mm (0.16 in)

Torque wrench settings

Component	kgf m	lbf ft
Front wheel spindle nut	3.6 – 5.2	26 – 37.5
Rear wheel spindle nut	5.0 – 8.0	36 – 58
Drum brake cam lever pinch bolts	0.5 – 0.8	3.5 – 6
Torque arm retaining nuts	1.0 – 1.5	7 – 11
Front hub flange bolts – DR125 S only	0.6 – 0.9	4.5 – 6.5
Brake disc mounting bolts	1.5 – 2.5	11 – 18
Brake caliper mounting bolts	1.5 – 2.5	11 – 18
Caliper axle bolt – GS125 ESD, ESF, ESK and ESM only	1.5 – 2.2	11 – 16
Caliper bleed nipple	0.6 – 0.9	4.5 – 6.5
Master cylinder clamp bolts	0.5 – 0.8	3.5 – 6
Brake hose union bolts	2.5 – 3.5	18 – 25
Sprocket mounting nuts:		
GS125 models	1.8 – 2.8	13 – 20
DR125 S models	2.5 – 4.0	18 – 29

1 General description

DR125 S and GS125 models are equipped with wheels of a traditional design, in which a full-width alloy hub is laced to a chromed steel rim by means of steel spokes. The GS125 ES model is fitted with one-piece cast alloy wheels of Suzuki's own design. In all cases, conventional tubed tyres are used. Wheel diameters and tyre sizes vary between the different models; these sizes and dimensions are shown in the Specifications Section of this Chapter.

The DR125 S and GS125 models are fitted with a cable operated, single leading shoe front brake which operates inside a full-width hub in the front wheel. The GS125 ES model is fitted with a front disc brake of the hydraulically-operated type. All models are fitted with a rod-operated, single leading shoe rear brake which operates inside a full-width hub.

2 Front wheel: examination and renovation – wire spoked wheels

1 Spoked wheels can go out of true over periods of prolonged use and like any wheel, as the result of an impact. The condition of the hub, spokes and rim should therefore be checked at regular intervals.

2 Make the machine as stable as possible, if necessary using blocks beneath the crankcase as extra support. Spin the wheel and ensure that there is no brake drag. If necessary, remove the disc pads (disc brake models) or slacken the brake adjuster (drum brake models) until the wheel turns freely. In the case of rear wheels it is advisable, though not essential, to remove the final drive chain.

3 Slowly rotate the wheel and examine the rim for signs of serious corrosion or impact damage. Slight deformities, as might be caused by running the wheel along a curb, can often be corrected by adjusting spoke tension. More serious damage may require a new rim to be fitted, and this is best left to an expert. Whilst this is not an impossible undertaking at home, there is an art to wheel building, and a professional wheel builder will have the facilities and parts required to carry out the work quickly and economically. Badly rusted steel rims should be renewed in the interests of safety as well as appearance. Where light alloy rims are fitted corrosion is less likely to be a serious problem, though neglect can lead to quite substantial pitting of the alloy.

4 If it has been decided that a new rim is required, in most cases it should be possible to have a light alloy rim fitted in place of an original plated steel item. This will have a marginal effect in terms of weight reduction, but will prove far more corrosion resistant.

5 Assuming the wheel to be undamaged it will be necessary to check it for runout. This is best done by arranging a temporary wire pointer so that it runs close to the rim. The wheel can now be turned and any distortion noted. Check for lateral distortion and for radial distortion, noting that the latter is less likely to be encountered if the wheel was set up correctly from new and has not been subject to impact damage.

6 The rim should be no more than 2.0 mm (0.08 in) out of true in either plane. If a significant amount of distortion is encountered check that the spokes are of approximately equal tension. Adjustment is effected by turning the square-headed spoke nipples with the appropriate spoke key. This tool is obtainable from most good motorcycle shops or tool retailers.

7 With the spokes evenly tensioned, any remaining distortion can be pulled out by tightening the spokes on one side of the hub and slackening the corresponding spokes from the opposite hub flange. This will allow the rim to be pulled across whilst maintaining spoke tension.

8 If more than slight adjustment is required it should be noted that the tyre and inner tube should be removed first to give access to the spoke ends. Those which protrude through the nipple after adjustment should be filed flat to avoid the risk of puncturing the tube. It is essential that the rim band is in good condition as an added precaution against chafing. In an emergency, use a strip of duct tape as an alternative; unprotected tubes will soon chafe on the nipples.

9 Should a spoke break, a replacement item can be fitted and retensioned in the normal way. Wheel removal is usually necessary for this operation, although complete removal of the tyre can be avoided if care is taken. A broken spoke should be attended to promptly because the load normally taken by that spoke is transferred to adjacent spokes which may fail in turn.

3 Front wheel: examination and renovation – cast alloy wheels

1 Carefully check the complete wheel for cracks and chipping, particularly at the spoke roots and the edge of the rim. As a general rule a damaged wheel must be renewed as cracks will cause stress points which may lead to sudden failure under heavy load. Small nicks may be radiused carefully with a fine file and emery paper (No 600 - No 1000) to relieve the stress. If there is any doubt as to the condition of a wheel, advice should be sought from a reputable dealer or specialist repairer.

2 Each wheel is covered with a coating of lacquer, to prevent corrosion. If damage occurs to the wheel and the lacquer finish is penetrated, the bared aluminium alloy will soon start to corrode. A whitish grey oxide will form over the damaged area, which in itself is a protective coating. This deposit however, should be removed carefully as soon as possible and a new protective coating of lacquer applied.

3 Check the lateral run out at the rim by spinning the wheel and placing a fixed pointer close to the rim edge. If the maximum run out is greater than 2.0 mm (0.08 in) the manufacturer recommends that the wheel be renewed. This is however, a counsel of perfection; a run out somewhat greater than this can probably be accommodated without noticeable effect on steering. No means is available for straightening a warped wheel without resorting to the expense of having the wheel skimmed on all faces. If warpage was caused by impact during an accident, the safest measure is to renew the wheel complete. Worn wheel bearings may cause rim run out. These should be renewed.

4 Front wheel: removal and refitting

Disc brake models

1 Place the machine securely on its centre stand on firm, level, ground and raise the front wheel clear of the ground by placing wooden blocks or a similar support underneath the front of the

engine/gearbox unit. Using a suitable pair of pliers, unscrew the knurled ring securing the speedometer cable to its drive gearbox, and withdraw the speedometer cable. Remove the split pin from the wheel spindle nut, then remove the spindle nut and its plain washer (plain flanged nut on GS125 ESX model). Using a tommy bar placed through the hole in the spindle end, pull out the spindle and carefully lower the wheel to the ground. If corrosion has formed to any great extent it will be necessary gently to tap the spindle out, using a hammer and a long metal drift.

2 It is not necessary to remove the brake caliper during the course of wheel removal, but great care must be taken not to disturb the caliper as the wheel rim and tyre pass it. If the caliper is moved across on its spindles, the clearance between the pads will be reduced and refitting of the wheel will be made much more difficult. To avoid the possibility of the pads being ejected through the accidental application of the front brake lever while the wheel is removed, always wedge a suitably sized piece of wood between the pads to ensure that they cannot move.

3 Refitting is the reverse of the removal procedure. Ensure that the wheel spindle is completely clean and free from dirt and corrosion, then smear high-melting point grease along its length. Remove the wooden wedge from between the pads and check that the shouldered spacer is correctly located on the disc, or left-hand side of the wheel hub. Similarly, ensure that the speedometer drive gearbox is correctly fitted, with its drive tangs engaged in the slots in the right-hand side of the hub. Offer the wheel assembly up to the forks, sliding the disc carefully between the brake pads and ensuring that both the gearbox and spacer remain in place. Check that the plain washer is fitted under the head of the wheel spindle, and push the spindle through from left to right. Refit the plain washer (not GS125 ESX) and spindle nut, tightening the nut by hand only.

4 Rotate the speedometer drive gearbox until the embossed arrow and 'UP' mark are facing vertically upwards; this will ensure that the speedometer cable is correctly routed. Connect the speedometer drive cable, rotating the wheel to ensure that the drive blade in the gearbox meshes correctly with the cable end. Tighten the knurled retaining ring securely. Using a torque wrench, tighten the wheel spindle retaining nut to a torque setting of 3.6 – 5.2 kgf m (26 – 37.5 lbf ft). Secure the nut with a new split-pin (not GS125 ESX, which uses a plain nut).

5 Before taking the machine out on the road, operate the front brake lever several times until the pads are moved back into contact with the disc and full brake pressure is restored. Check that the front wheel rotates freely, and that the speedometer and front brake are operating correctly.

Drum brake models

6 Support the machine securely on a strong wooden box placed underneath the engine/gearbox unit so that the front wheel is clear of the ground. The front brake cable on GS125 models is disconnected by unscrewing the adjusting nut from the lower, threaded, end of the cable. This will permit the cable to be pulled clear of the brake operating arm and from the cable stop in the brake backplate. On DR125 S models, slacken the cable retaining clamp at the top of the fork lower leg, slide the rubber sleeve away from the adjuster threads and slacken fully the adjuster lower locknut. Withdraw the split pin from the top of the brake operating arm and push out the clevis pin to release the cable. Withdraw the cable from the brake backplate.

7 Disconnect the speedometer cable by slackening its lower knurled retaining ring with a suitable pair of pliers. Withdraw the split pin from the wheel spindle retaining nut, then remove the nut and the plain washer beneath it. Using a tommy bar placed through the hole in the spindle head, pull out the spindle and carefully lower the wheel to the ground.

8 If corrosion has formed on the wheel spindle, sticking it in place, it will be necessary to use a hammer and a long metal drift gently to tap the spindle out. Take care not to damage the threads on the spindle end.

9 On refitting the front wheel, ensure that the brake backplate is replaced so that the two tangs of the speedometer drive assembly are correctly located in the slots provided in the hub, and that the spacer is pushed firmly into the right-hand side of the hub. Ensure that the wheel spindle is clean and completely free from corrosion, then smear high-melting point grease along its length. Replace the plain washer under the spindle head. Offer the wheel up to the forks ensuring that the raised lug on the fork left-hand lower leg engages correctly between the two raised ribs cast in the brake backplate. Push the spindle through from left to right, replace the plain washer and spindle retaining nut, then tighten the nut by hand only.

10 Connect the speedometer drive cable, rotating the front wheel to ensure that the cable end meshes correctly with the drive pinion. Tighten securely the knurled retaining ring. Connect the front brake cable by reversing the cable removal procedures, then spin the front wheel and apply the front brake lever sharply to centralise the shoes on the drum. Maintain pressure on the front brake lever, clamping the shoes against the drum, and tighten the wheel spindle retaining nut to a torque setting of 3.6 – 5.2 kgf m (26 – 37.5 lbf ft). Secure the nut with a new split pin, spreading the split-pin ends to ensure that it cannot work loose. Spin the front wheel and gradually tighten the front brake cable by turning the adjusting nut until the shoes can be heard rubbing lightly on the brake drum as it rotates. Turn the adjusting nut back until the noise disappears and then tighten the locknut (DR125 S models only). Spin the front wheel and apply the front brake sharply once or twice to settle both shoes and cable. Recheck the adjustment, altering it if necessary. This procedure should give the recommended adjustment setting which is that a distance of 20 – 30 mm (0.8 – 1.2 in) should be left between the handlebar lever ball end and the throttle twistgrip when the brake lever is firmly applied. Tighten the cable clamp (DR125 S models only).

11 Before taking the machine out on the road, check that the front wheel rotates easily and that the speedometer and the front brake are working correctly.

5 Front wheel bearings: removal, examination and refitting

1 Remove the front wheel as described in the previous Section. On disc brake wheels, withdraw the left-hand spacer and speedometer gearbox, while on drum brake wheels withdraw the brake backplate and right-hand spacer. On DR125 S models only, unscrew the three bolts and displace the hub right-hand flange plate.

2 Position the wheel on a work surface with its hub well supported by wooden blocks so that enough clearance is left beneath the wheel to drive the bearing out. Ensure the blocks are placed as close to the bearing as possible, to lessen the risk of distortion occurring to the hub casting whilst the bearings are being removed or fitted.

3 Place the end of a long-handled drift against the upper face of the lower bearing and tap the bearing downwards out of the wheel hub. The spacer located between the two bearings may be moved sideways slightly in order to allow the drift to be positioned against the face of the bearing. Move the drift around the face of the bearing whilst drifting it out of position, so that the bearing leaves the hub squarely.

4 With the one bearing removed, the wheel may be lifted and the spacer withdrawn from the hub. Invert the wheel and remove the second bearing, using a similar procedure to that used for the first. On GS125 models, the dust seal which fits against the right-hand bearing will be driven out as the bearing is removed. This seal should be closely inspected for any indication of damage, hardening or perishing and renewed if necessary. It is advisable to renew this seal as a matter of course if the bearings are found to be defective.

5 Remove all the old grease from the hub and bearings, giving the latter a final wash in petrol. Check the bearings for signs of play or roughness when they are turned. If there is any doubt about the condition of a bearing, it should be renewed.

6 If the original bearings are to be refitted, then they should be repacked with the recommended grease before being fitted into the hub. New bearings must also be packed with the recommended grease. Ensure that the bearing recesses in the hub are clean and both bearings and recess mating surfaces lightly greased to aid fitting. Check the condition of the hub recesses for evidence of abnormal wear which may have been caused by the outer race of a bearing spinning. If evidence of this happening is found, and the bearing is a loose fit in the hub, then it is best to seek advice from a Suzuki Service Agent or a competent motorcycle engineer. Alternatively a proprietary product such as Loctite Bearing Fit may be used to retain the bearing outer race; this will mean, however, that the bearing housing must be carefully cleaned and degreased before the locking compound can be used.

7 With the wheel hub and bearing thus prepared, proceed to fit the bearings and central spacer as follows. With the hub again well supported by the wooden blocks, drift the first of the two bearings into position. To do this, use a soft-faced hammer in conjunction with a socket or length of metal tube which has an overall diameter which is slightly less than that of the outer race of the bearing, but which does

not bear at any point on the bearing sealed surface or inner race. Tap the bearing into place against the locating shoulder machined in the hub, remembering that the sealed surface of the bearing must always face outwards. With the first bearing in place, invert the wheel, insert the central spacer and pack the hub centre no more than $\frac{2}{3}$ full with high-melting point grease. Proceed to fit the second bearing, using the same procedure as given for the first. Take great care to ensure that each of the bearings enters its housing correctly, that is, square to the housing, otherwise the housing surface may be broached. On GS125 models only, fit the seal against the right-hand bearing using the same method as for fitting bearings. On DR125 S models, refit the flange plate to the hub right-hand side, apply thread locking compound to the threads of the bolts and refit the three bolts, tightening them to a torque setting of 0.6 – 0.9 kgf m (4.5 – 6.5 lbf ft). Refit the wheel to the machine as described in Section 4.

6 Front brake disc: examination, removal and refitting

1 The brake disc can be checked for wear and for warpage whilst the front wheel is still in the machine. Using a micrometer, measure the thickness of the disc at the point of greatest wear. If the measurement is much less than the recommended service limit of 3.0 mm (0.12 in), then the disc should be renewed. Check the warpage (runout) of the disc by setting up a suitable pointer close to the outer periphery of the disc and spinning the front wheel slowly. If the total warpage is more than 0.3 mm (0.012 in), the disc should be renewed. A warped disc, apart from reducing the braking efficiency, is likely to cause juddering during braking and will also cause the brake to bind when it is not in use.

2 The brake disc should also be checked for bad scoring on its contact area with the brake pads. If any of the above mentioned faults are found, then the disc should be removed from the wheel for renewal or for repair by skimming. Such repairs should be entrusted only to a reputable engineering firm. A local motorcycle dealer may be able to assist in having the work carried out.

3 To detach the disc, first remove the wheel by following the procedure given in Section 4 of this Chapter. The disc is retained in position by four bolts, which are locked in pairs by tab washers. Bend down the ears of these washers and remove the bolts. The disc can now be eased off the hub boss.

4 Fit the disc by reversing the removal procedure. Do not reuse the same tabs at the tab washer ends; renew the washers if necessary. Avoid placing any strain on the disc by tightening the four retaining bolts evenly and in a diagonal sequence to a torque setting of 1.5 - 2.5 kgf m (11 - 18 lbf ft).

4.1a Remove wheel spindle retaining nut ...

4.1b ... and withdraw wheel spindle

4.3a Shouldered spacer fits in hub left-hand side ...

4.3b .. and speedometer drive gearbox tangs engage in slots in hub right-hand side

4.4a 'Up' mark on gearbox must face upwards to align gearbox correctly

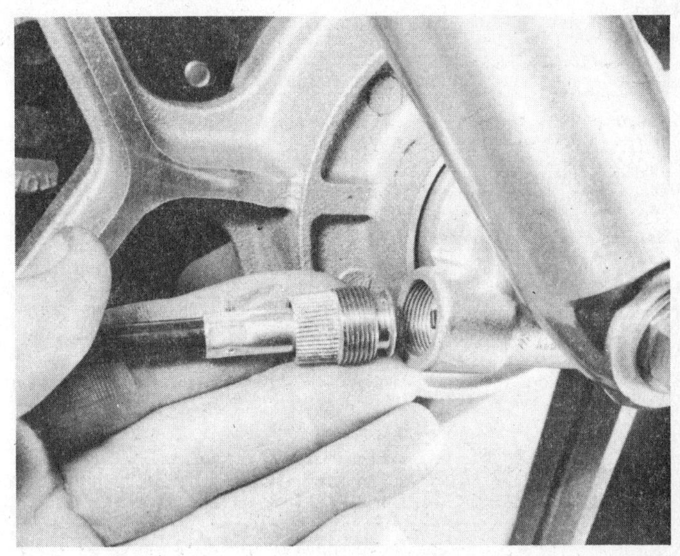

4.4b Rotate wheel to assist engagement of speedometer cable

4.4c Secure spindle nut by fitting new split-pin as shown

4.6a DR125 S model only – slacken cable clamp ...

4.6b .. remove split-pin and press out clevis pin to release brake cable

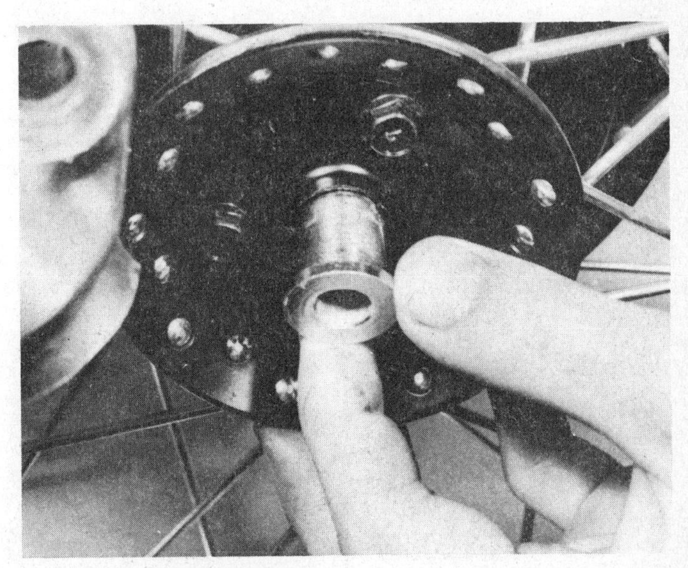

4.9a Hub right-hand spacer is refitted as shown

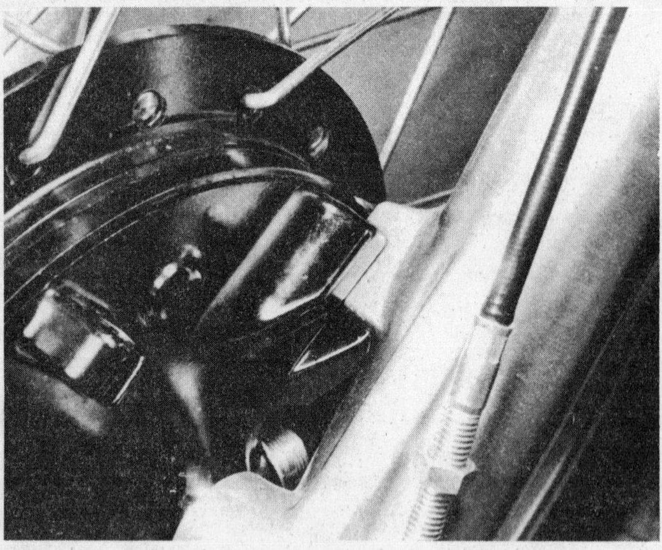

4.9b Lug on fork lower leg must engage with slot in backplate

5.4a Pass drift through hub to tap out opposite bearing ...

5.4b ... and withdraw centre spacer

5.7a Bearings are refitted always with sealed surface facing outwards ...

5.7b ... and are drifted into place as shown

6.4 Disc retaining bolts are tightened to correct torque setting and secured with lock washers

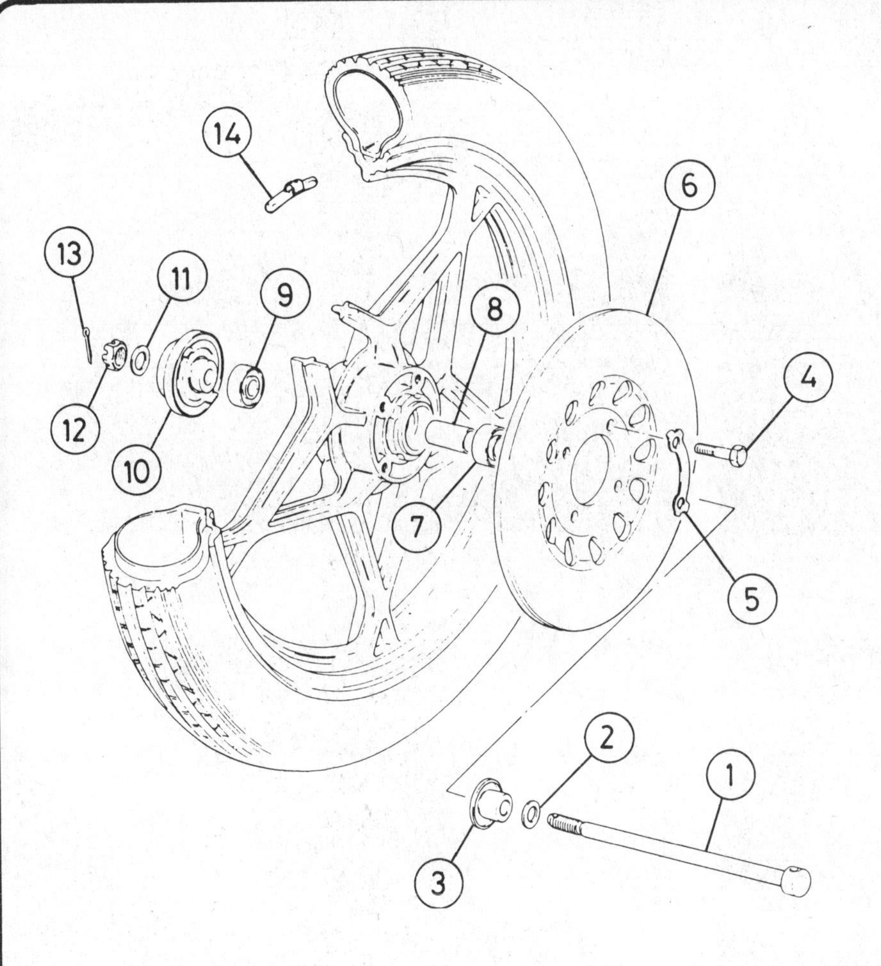

Fig. 5.1 Front wheel – GS125 ES model

1 *Wheel spindle*
2 *Washer*
3 *Spacer/dust cover*
4 *Bolt – 4 off*
5 *Tab washer – 2 off*
6 *Brake disc*
7 *Left-hand bearing*
8 *Spacer*
9 *Right-hand bearing*
10 *Speedometer drive gearbox*
11 *Washer**
12 *Nut**
13 *Split pin**
14 *Balance weight AR*

**Plain flanged nut on GS125 ESX*

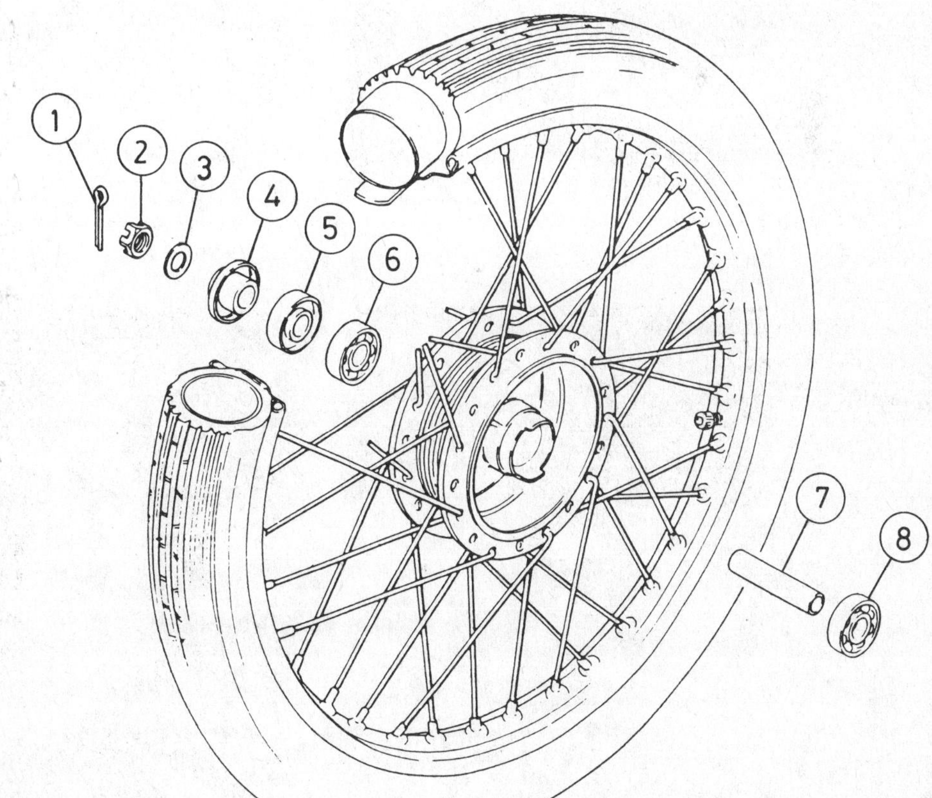

Fig. 5.2 Front wheel – GS125

1 *Split pin*
2 *Nut*
3 *Washer*
4 *Spacer/dust cover*
5 *Oil seal*
6 *Right-hand bearing*
7 *Spacer*
8 *Left-hand bearing*

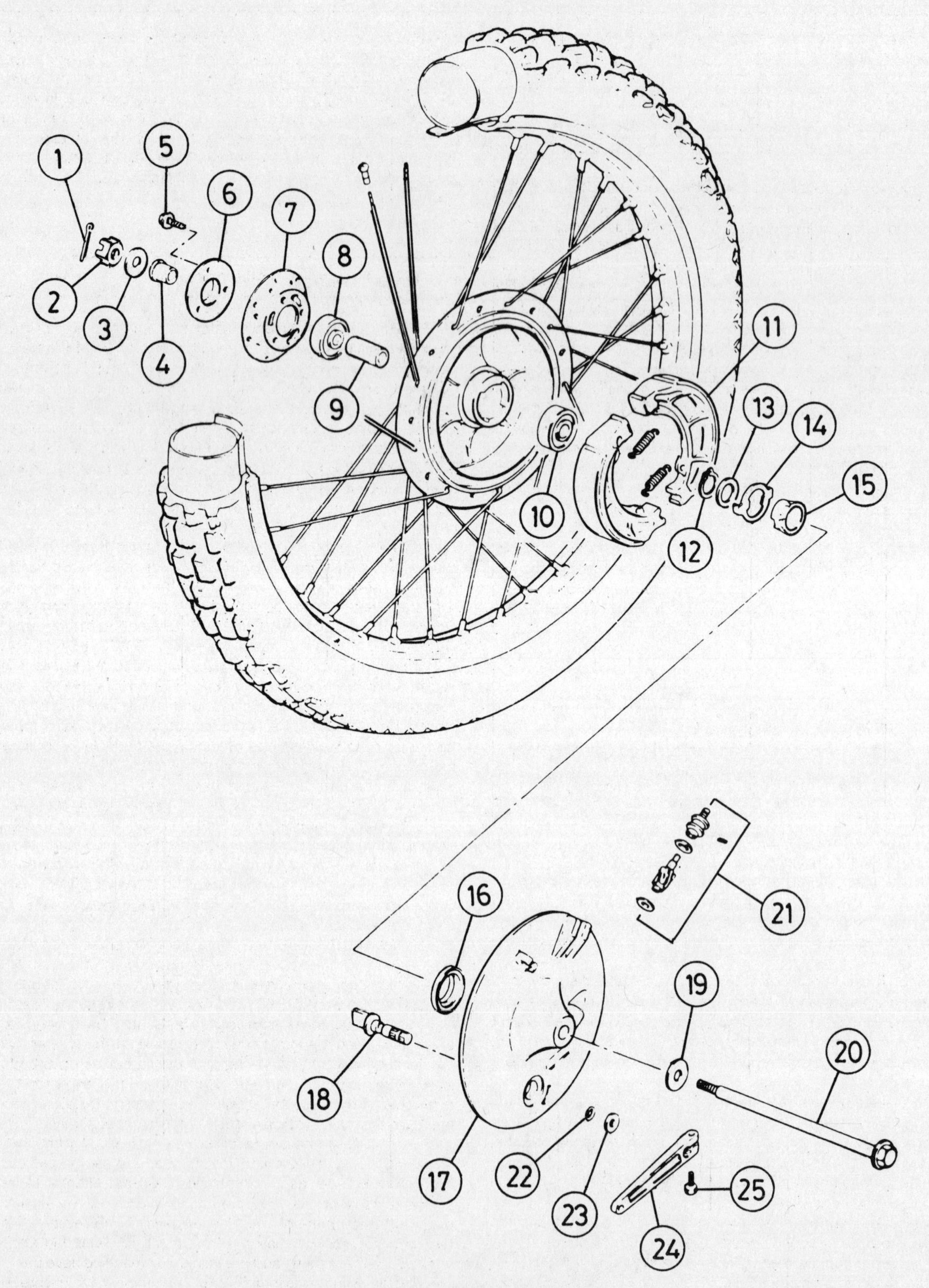

Fig. 5.3 Front wheel – DR model

1 *Split pin*
2 *Nut*
3 *Washer*
4 *Spacer*
5 *Bolt - 3 off*
6 *Bearing retaining plate*
7 *Hub flange plate*
8 *Right-hand bearing*
9 *Spacer*
10 *Left-hand bearing*
11 *Brake shoe - 2 off*
12 *Circlip*
13 *Thrust washer*
14 *Speedometer drive plate*
15 *Speedometer drive gear*
16 *Oil seal*
17 *Brake backplate*
18 *Operating cam*
19 *Washer*
20 *Wheel spindle*
21 *Speedometer driven gear assembly*
22 *O-ring*
23 *Washer*
24 *Operating lever*
25 *Bolt*

7 Hydraulic disc brake: general

1 The disc brake fitted to the GS125 ES model is, as previously mentioned, of the hydraulically-operated type. A number of precautions must be taken when dealing with such a system, to avoid the disastrous consequences of brake failure.

2 The hydraulic system must be kept free from air bubbles. Any air in the system will be compressed when the brake lever is operated instead of transmitting braking effort to the disc. It follows that efficiency will be impaired, and given sufficient air, can render the brake inoperative. If any part of the hydraulic system is disturbed, the system must always be bled to remove any air. See Section 8 for details. It is vital that all hoses, pipes and unions are examined regularly and renewed if damage, deterioration or leakage is suspected.

3 Hydraulic fluid is specially formulated for given applications and must always be of the correct type. Any fluid conforming to SAE J1703 or DOT 3 or 4 may be used; other types may not be suitable. On no account should any other type of oil or fluid be used. Old or contaminated fluid must be discarded. It is dangerous to use old fluid which may have been degraded to the point where it will boil in the caliper creating air bubbles in the system. Note that brake fluid will attack and discolour paintwork and plastics. Care must be taken to avoid contact and any accidental splashes washed off immediately.

4 Cleanliness is more important with hydraulic systems than in any other single area of the motorcycle. Dirt will rapidly destroy seals, allowing fluid to leak out or air to be drawn in. Water, even in the form of moist air, will be absorbed by the fluid which is hygroscopic. Fluid degraded by water has a lowered boiling point and can boil in use. The master cylinder and any cans of fluid must be kept securely closed to prevent this.

8 Bleeding the hydraulic brake system

1 If the brake action becomes spongy, or if any part of the hydraulic system is dismantled (such as when a hose is replaced) it is necessary to bleed the system in order to remove all traces of air. The procedure for bleeding the hydraulic system is best carried out by two people.

2 Check the fluid level in the reservoir and top up with new fluid of the specified type if required. Keep the reservoir at least half full during the bleeding procedure; if the level is allowed to fall too far air will enter the system requiring that the procedure be started again from scratch. Refit the reservoir cover to prevent the ingress of dust or the ejection of a spout of fluid.

3 Remove the dust cap from the caliper bleed nipple and clean the area with a rag. Place a clean glass jar below the caliper and connect a pipe from the bleed nipple to the jar. A clear plastic tube should be used so that air bubbles can be more easily seen. Place some clean hydraulic fluid in the glass jar so that the pipe is immersed below the fluid surface throughout the operation.

4 If parts of the system have been renewed, and thus the system must be filled, open the bleed nipple about one turn and pump the brake lever until fluid starts to issue from the clear tube. Tighten the bleed nipple and then continue the normal bleeding operation as described in the following paragraphs. Keep a close check on the reservoir level whilst the system is being filled.

5 Operate the brake lever as far as it will go and hold it in this position against the fluid pressure. If spongy brake operation has occurred it may be necessary to pump rapidly the brake lever a number of times until pressure is achieved. With pressure applied, loosen the bleed nipple about half a turn. Tighten the nipple as soon as the lever has reached its full travel and then release the lever. Repeat this operation until no more air bubbles are expelled with the fluid into the glass jar. When this condition is reached the air bleeding operation should be complete, resulting in a firm feel to the brake operation. If sponginess is still evident continue the bleeding operation; it may be that an air bubble trapped at the top of the system has yet to work down through the caliper.

6 When all traces of air have been removed from the system, top up the reservoir and refit the diaphragm cover. Check the entire system for leaks, and check also that the brake system in general is functioning efficiently before using the machine on the road.

7 Brake fluid drained from the system will almost certainly be contaminated, either by foreign matter or more commonly by the absorption of water from the air. All hydraulic fluids are to some degree hygroscopic, that is, they are capable of drawing water from the atmosphere, and thereby degrading their specifications. In view of this, and the relative cheapness of the fluid, old fluid should always be discarded.

8 Great care should be taken not to spill hydraulic fluid on any painted cycle parts; it is a very effective paint stripper. Also, the plastic glasses in the instrument heads, and most other plastic parts, will be damaged by contact with this fluid.

9 Master cylinder: examination and renovation

1 The master cylinder and hydraulic fluid reservoir form a combined assembly which is clamped to the right-hand side of the handlebars. The moving parts consist of a piston and cup assembly which is operated directly by the handlebar brake lever. The piston and cup assembly is usually reliable and longlasting, but if worn, will reduce brake efficiency considerably due to lost hydraulic pressure. Such wear is normally indicated by a fluid leak around the brake lever end of the master cylinder, but can also be revealed by the spongy feel of the brake lever caused by the entry of air into the system past the defective seals. Such wear must be rectified instantly by carrying out the following procedure.

2 Slacken and remove the rear view mirror from the master cylinder assembly. On GS125 ESZ, ESD and ESF models slacken and remove the headlamp rim retaining screw and withdraw the headlamp unit. Trace the lead from the front brake light switch down into the headlamp shell and disconnect the two wires. Withdraw the lead, releasing any wiring ties which retain it to the handlebars. On GS125 ESK, ESM, ESR and ESX models, it should be possible to disconnect the switch wiring at the switch terminals. The hydraulic fluid must now be drained. Place a clean container beneath the caliper unit and run a clear plastic tube from the caliper bleed nipple to the container. Unscrew the bleed nipple by one full turn and drain the system by operating the brake lever repeatedly until no more fluid can be seen issuing from the nipple.

3 Position a pad of clean rag beneath the point where the brake hose joins the master cylinder. This simple precaution is essential to prevent brake fluid from dripping onto, and therefore damaging, any plastic and painted components located beneath the hose union once the union bolt is removed. Detach the rubber cover from the head of the union bolt and remove the bolt. Once any excess fluid has drained from the union connection, wrap the end of the hose in rag or polythene and then attach it to a point on the handlebars.

4 Remove the brake lever by unscrewing its shouldered pivot screw and locknut. The fluid reservoir cover is retained by two long screws which should now be unscrewed so that the cover and diaphragm can be withdrawn. On GS125 ESZ, ESD and ESF models, the reservoir body is also retained by these screws but is located by a deep spigot on its underside which fits into a matching recess in the master cylinder body. The joint is sealed by an O-ring and should not be disturbed unless fluid leakage or accident damage necessitates the separation of the two components. If separation should be necessary, it should now be carried out while the master cylinder body is still clamped firmly to the handlebars and can be easily held. Grasp the reservoir body with one hand while steadying the handlebars with the other, and rotate the body to lift it out of the master cylinder as if unscrewing it. Do not attempt to lever the two components apart as they are delicate and will be easily marked or distorted to the point at which renewal will be necessary. Whenever this joint is disturbed, the sealing O-ring must be discarded and a new item fitted. Once the draining and preliminary dismantling have been carried out, the master cylinder assembly can be removed by unscrewing the two clamp bolts.

5 Use the flat of a small screwdriver carefully to prise out the rubber dust seal boot. This will expose a retaining circlip which must be removed using a pair of circlip pliers which have long, straight jaws. With the circlip removed, the piston and cup assembly can be pulled out. Be very careful to note the exact order in which these components are fitted.

6 Note that if a vice is used to hold the master cylinder at any time during dismantling and reassembly, its jaws must be padded with soft alloy or wooden covers and the master cylinder must be wrapped in soft cloth to obviate any risk of the assembly being marked or distorted.

7 Place all the master cylinder component parts in a clean container

and wash each part thoroughly in new brake fluid. Lay the parts out on a sheet of clean paper and examine each one as follows.

8 Inspect the unit body for signs of stress failure around both the brake lever pivot lugs and the handlebar mounting points. Carry out a similar inspection around the hose union boss. Examine the cylinder bore for signs of scoring or pitting. If any of these faults are found, then the unit body must be renewed.

9 Inspect the surface of the piston for signs of scoring or pitting and renew it if necessary. It is advisable to discard all the components of the piston assembly as a matter of course as the replacement cost is relatively small and does not warrant re-use of components vital to safety. The same advice applies to the O-ring fitted between the reservoir and the unit body. Inspect the threads of the brake hose union bolt for any signs of failure and renew the bolt if in the slightest doubt. Renew each of the gasket washers located one either side of the hose union.

10 Check before reassembly that any traces of contamination remaining within the reservoir body have been removed. Inspect the diaphragm to see that it is not perished or split. It must be noted at this point that any reassembly work must be undertaken in ultra-clean condition. Particles of dirt entering the component will only serve to score the working points of the cylinder and thereby cause early failure of the system.

11 When reassembling and fitting the master cylinder, follow the removal and dismantling procedures in the reverse order whilst paying particular attention to the following points. Make sure that the piston components are fitted the correct way round and in the correct order. Immerse all of these components in new brake fluid prior to reassembly and refer to the figure accompanying this text when in doubt as to their fitted positions. When refitting the master cylinder assembly to the handlebar, position the assembly so that the reservoir will be exactly horizontal when the machine is in use. Tighten the clamp top bolt first, and then the bottom bolt, to a torque setting of 0.5 – 0.8 kgf m (3.5 – 6.0 lbf ft). Connect the brake hose to the master cylinder, ensuring that a new sealing washer is placed on each side of the hose union, and tightening the hose union bolt to a torque setting of 2.5 – 3.5 kgf m (18 – 25 lbf ft). Finally, replace the rubber union cover.

12 Bleed the brake system after refilling the reservoir with new hydraulic fluid, then check for leakage of fluid whilst applying the brake lever. Push the machine forward and bring it to a halt by applying the brake. Do this several times to ensure that the brake is operating correctly before taking the machine for a test run. During the run, use the brakes as often as possible and on completion, recheck for signs of fluid loss.

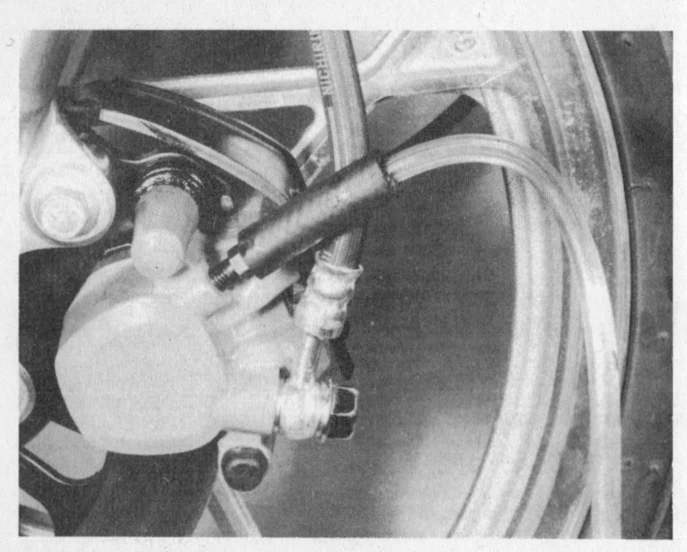

8.3 Connect clear plastic pipe to bleed nipple

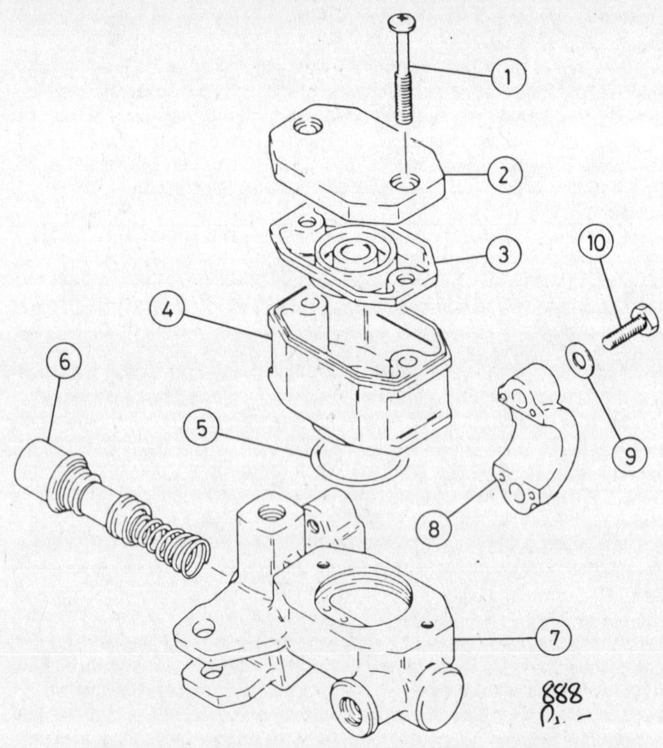

Fig. 5.4 Front brake master cylinder – GS125 ESZ, ESD and ESF (one-piece reservoir and body on later models)

1 *Screw - 2 off*
2 *Cover*
3 *Diaphragm*
4 *Reservoir*
5 *O-ring*
6 *Piston and cup assembly*
7 *Master cylinder body*
8 *Handlebar clamp*
9 *Washer - 2 off*
10 *Bolt - 2 off*

8.6 Tighten securely reservoir cover retaining screws – check for fluid leaks

10 Brake hose: examination

1 An external flexible brake hose is used as a means of transmitting hydraulic pressure to the caliper unit once the front brake lever is applied.

2 When the brake assembly is being overhauled, or at any time during a routine maintenance or cleaning procedure, check the condition of the hose for signs of leakage, damage or deterioration or scuffing against any cycle components. Any such damage will mean that the hose must be renewed immediately. The union connections at either end of the hose must also be in good condition, with no stripped

threads or damaged sealing washers. Do not tighten these union bolts over the recommended torque setting of 2.5 – 3.5 kgf m (18 – 25 lbf ft) as they are easily sheared if overtightened.
3 Suzuki recommend that the brake hose be renewed every four years in the interests of safety, irrespective of its apparent condition.

11 Brake caliper: examination and renovation

1 The GS125 ESZ, ESR and ESX models use a caliper made by Tokico, whereas that fitted to the GS125 ESD, ESF, ESK and ESM models is made by Aisin; the two types are identical in specification and very similar in appearance but differ slightly in dismantling procedures. Identify the type to be dismantled.
2 Remove the caliper cover which is retained by two screws, then remove the caliper and withdraw the pads as described in Routine Maintenance.
3 On the Tokico caliper, the mounting bracket will slide out of the caliper body. Remove the rubber seals protecting each axle pin. On the Aisin caliper, displace the protruding cap and unscrew the axle bolt, then slide the mounting bracket out of the caliper and remove the three rubber seals protecting the axle pin and bolt. Press out the bush and collar which fit around the axle bolt.
4 The piston must now be ejected from the caliper bore. The simplest way of achieving this is to apply repeatedly the front brake lever, using normal hydraulic pressure to force the piston out. If this method is used, wrap a piece of clean rag around the caliper to prevent the piston from falling clear and being damaged. Take great care to catch the escaping hydraulic fluid in a clean container and to prevent any of the fluid from splashing on to plastic or painted metal components. When the piston has been displaced, drain all surplus hydraulic fluid into the container by pumping gently the front brake lever, and then disconnect the hydraulic hose by unscrewing the caliper union bolt. An alternative to the above method will require a source of compressed air. Attach a length of clear plastic tubing to the bleed nipple, placing the tube lower end in a suitable container. Open the bleed nipple by one full turn and pump gently on the front brake lever to drain as much fluid as possible from the hydraulic system. When no more fluid can be seen issuing from the bleed nipple, tighten it down again, withdraw the plastic tube and disconnect the hydraulic hose at the caliper union. Wrap a large piece of cloth loosely around the caliper and apply a jet of compressed air to the union orifice. Be careful not to use too high an air pressure or the piston may be damaged.
5 When the piston has been removed, it should be placed in a clean container to ensure that it cannot be damaged and the piston seals should be picked out of the caliper bore and discarded. Wrap the exposed end of the hydraulic hose in clean rag or polythene to prevent the entry of dirt. Note that if the piston cannot be ejected using either of the above methods, the caliper assembly must be detached from the machine and taken to a motorcycle dealer for expert help. It is possible, however, that the piston surface and caliper bore are so badly damaged by corrosion that the only satisfactory solution will be the renewal of the caliper and piston assembly as a single unit.
6 Clean the caliper components thoroughly, only in hydraulic brake fluid. **Never** use petrol or cleaning solvent for cleaning hydraulic brake parts otherwise the rubber components will be damaged. Discard all the rubber components as a matter of course. The replacement cost is relatively small and does not warrant re-use of components vital to safety. Check the piston and caliper cylinder bore for scoring, rusting or pitting. If any of these defects are evident it is unlikely that a good fluid seal can be maintained and for this reason the components should be renewed.
7 Inspect the shank of each axle pin or bolt for any signs of damage or corrosion and clean or renew each one as necessary. Check that the bush and collar (Aisin caliper) are unworn, renewing them if necessary. Remove the bleed nipple and check that it has not become blocked. Check the condition of the nipple sealing cap and renew it, if necessary.
8 Assemble the caliper unit by reversing the dismantling sequence. Note that assembly must be undertaken under ultra-clean conditions. Particles of dirt will score the bearing surfaces of moving parts and cause early failure.
9 When assembling the unit, pay attention to the following points. When fitting the new piston seal, take care to ensure that it is not twisted on its retaining groove. Apply a generous amount of new brake fluid to the surface of the caliper bore and to the periphery of the piston before pushing the piston slowly into position whilst taking care not to damage the piston seal. Refit the outer piston seal, again taking care that it is correctly seated. Lubricate lightly the sliding surface of each mounting bracket axle pin with brake caliper grease, also the axle bolt, collar and bush (Aisin caliper only). Press in the bush and collar and refit the rubber seals over the axle pin bosses in the caliper. Insert the mounting bracket onto the caliper body. On the Aisin caliper only, refit the axle bolt, tightening it to a torque setting of 1.5 – 2.2 kgf m (11 – 16 lbf ft), then press into place the protruding cap. Check that the mounting bracket slides easily in and out, then refit the pad spring.
10 Refit the brake pads and refit the caliper assembly to the fork leg by following the procedure given in Routine Maintenance. Before reconnecting the brake hose union to the caliper, check the condition of the two gasket washers located one either side of the union. Renew these washers if necessary and then fit and tighten the union bolt to a torque loading of 2.5 – 3.5 kgf m (18 – 25 lbf ft).
11 Refill the master cylinder reservoir with new hydraulic brake fluid and bleed the system by following the procedure given in Section 8 of this Chapter. On completion of bleeding, carry out a check for leakage of fluid whilst applying the brake lever. Push the machine forward and bring it to a halt by applying the brake. Do this several times to ensure that the brake is operating correctly before taking the machine out on the road.

11.9 Use only special brake caliper grease to lubricate axle pins

11.10 Note correct position of pad spring – Tokico caliper shown

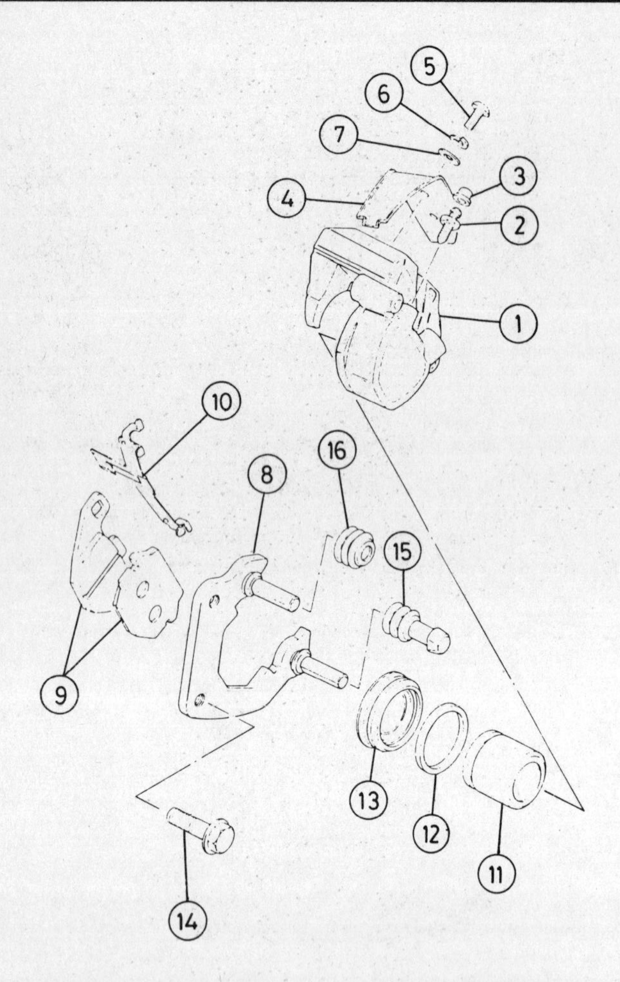

Fig. 5.5 Front brake caliper – GS125 ESZ, ESR and ESX models

1	*Caliper*	9	*Brake pads*
2	*Bleed nipple*	10	*Anti-rattle spring*
3	*Cap*	11	*Piston*
4	*Caliper cover*	12	*Piston seal*
5	*Screw - 2 off*	13	*Piston boot*
6	*Spring washer - 2 off*	14	*Bolt*
7	*Washer - 2 off*	15	*Rubber seal*
8	*Mounting bracket*	16	*Rubber seal*

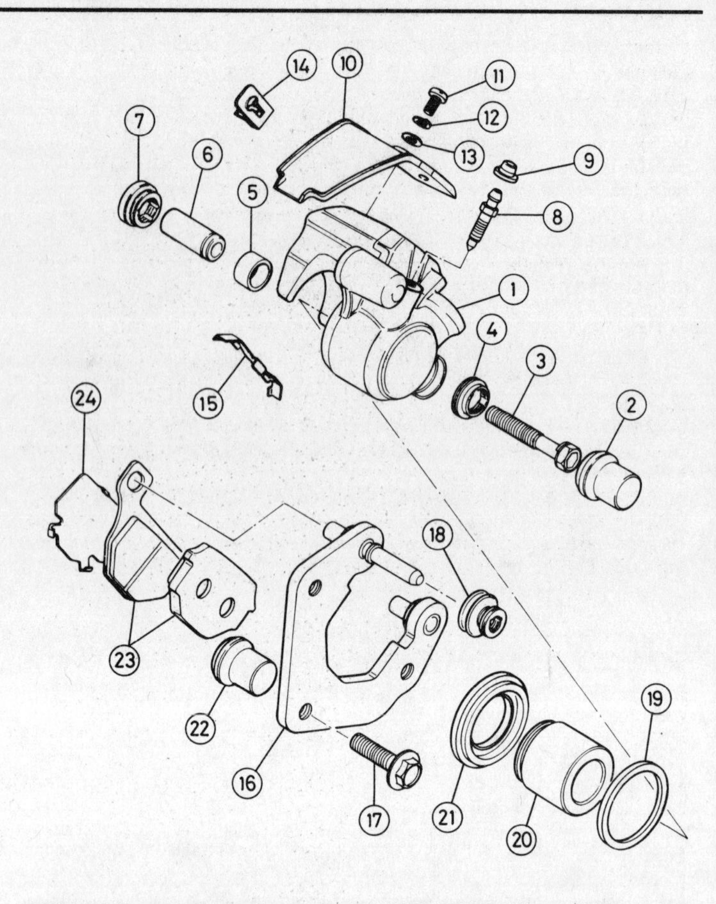

Fig. 5.6 Front brake caliper – GS125 ESD, ESF, ESK and ESM models

1	*Caliper*	13	*Washer - 2 off*
2	*Protruding cap*	14	*Brake pad spring*
3	*Bolt*	15	*Anti-rattle spring*
4	*Rubber seal*	16	*Mounting bracket*
5	*Bush*	17	*Bolt - 2 off*
6	*Collar*	18	*Rubber seal*
7	*Rubber seal*	19	*Piston seal*
8	*Bleed nipple*	20	*Piston*
9	*Cap*	21	*Piston boot*
10	*Caliper cover*	22	*Protruding cap*
11	*Screw - 2 off*	23	*Brake pads*
12	*Spring washer - 2 off*	24	*Shim*

12 Front drum brake: examination, renovation and adjustment

1 The brake assembly can be withdrawn from its hub after removal of the wheel from the machine.

2 Examine the condition of the brake linings. If they are worn beyond the specified limit then the brake shoes should be renewed. The linings are bonded on and cannot be supplied separately.

3 If oil or grease from the wheel bearings has badly contaminated the linings, the brake shoes should be renewed. There is no satisfactory way of degreasing the lining material. Any surface dirt on the linings can be removed with a stiff-bristled brush. High spots on the linings should be carefully eased down with emery cloth.

4 Examine the drum surface for signs of scoring, wear beyond the service limit or oil contamination. All of these conditions will impair braking efficiency. Remove all traces of dust, preferably using a brass wire brush, taking care not to inhale any of it, as it is of an asbestos nature, and consequently harmful. Remove oil or grease deposits, using a petrol soaked rag.

5 If deep scoring is evident, due to the linings having worn through to the shoe at some time, the drum must be skimmed on a lathe, or renewed. Whilst there are firms who will undertake to skim the drum without dismantling the wheel, it should be borne in mind that excessive skimming will change the radius of the drum in relation to the brake shoes, thereby reducing the friction area until extensive bedding in has taken place. Also full adjustment of the shoes may not be possible. If in doubt about this point, the advice of one of the specialist engineering forms who undertake this work should be sought.

6 Note that it is a false economy to try to cut corners with brake components; the whole safety of both machine and rider being dependent on their good condition.

7 Removal of the brake shoes is accomplished by folding the shoes together so that they form a 'V'. With the spring tension relaxed, both shoes and springs may be removed from the brake backplates as an assembly. Detach the springs from the shoes and carefully inspect them for any signs of fatigue or failure. If in doubt, compare them with a new set of springs.

8 Before fitting the brake shoes, check that the brake operating cam is working smoothly and is not binding in its pivot. The cam can be removed by withdrawing the retaining bolt on the operating arm and pulling the arm off the shaft. Before removing the arm, it is advisable to mark its position in relation to the shaft, so that it can be relocated correctly.

9 Remove any deposits of hardened grease or corrosion from the bearing surface of the brake cam shoe by rubbing it lightly with a strip of fine emery paper or by applying solvent with a piece of rag. Lightly grease the length of the shaft and the face of the operating cam prior

to reassembly. Clean and grease the pivot stub which is set in the backplate.

10 Check the condition of the O-ring which prevents the escape of grease from the end of the camshaft. If it is in any way damaged or perished, then it must be renewed before the shaft is relocated in the backplate. Relocate the camshaft and align and fit the operating arm with the O-ring and plain washer. The bolt and nut retaining the arm in position on the shaft should be torque loaded to 0.5 – 0.8 kgf m (3.5 – 6.0 lbf ft).

11 Before refitting existing shoes, roughen the lining surface sufficiently to break the glaze which will have formed in use. Glasspaper or emery cloth is ideal for this purpose but take care not to inhale any of the asbestos dust that may come from the lining surface.

12 Fitting the brake shoes and springs to the brake backplate is a reversal of the removal procedure. Some patience will be needed to align the assembly with the pivot and operating cam whilst still retaining the spring in position; once they are correctly aligned though, they can be pushed back into position by pressing downwards in order to snap them into position. Do not use excessive force, as there is risk of distorting the brake shoes permanently. Adjustment of the front brake is fully described in the relevant paragraphs of Section 4 of this Chapter. Take great care to check that the brake functions efficiently before using the machine on the road.

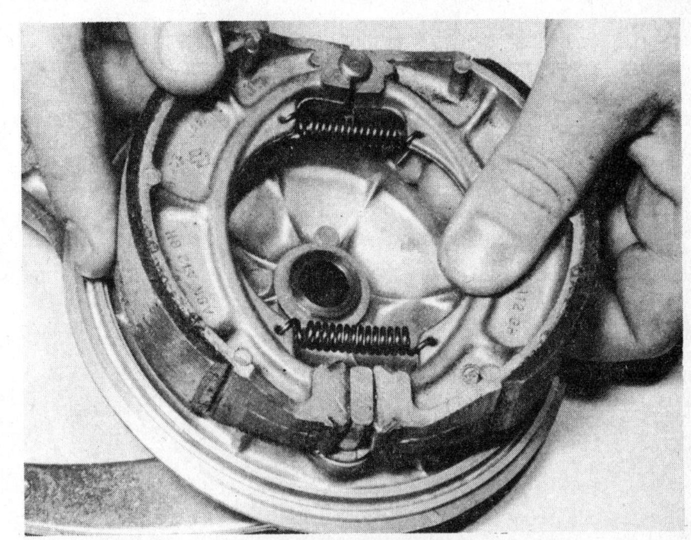

12.7 Fold brake shoes together to remove or refit

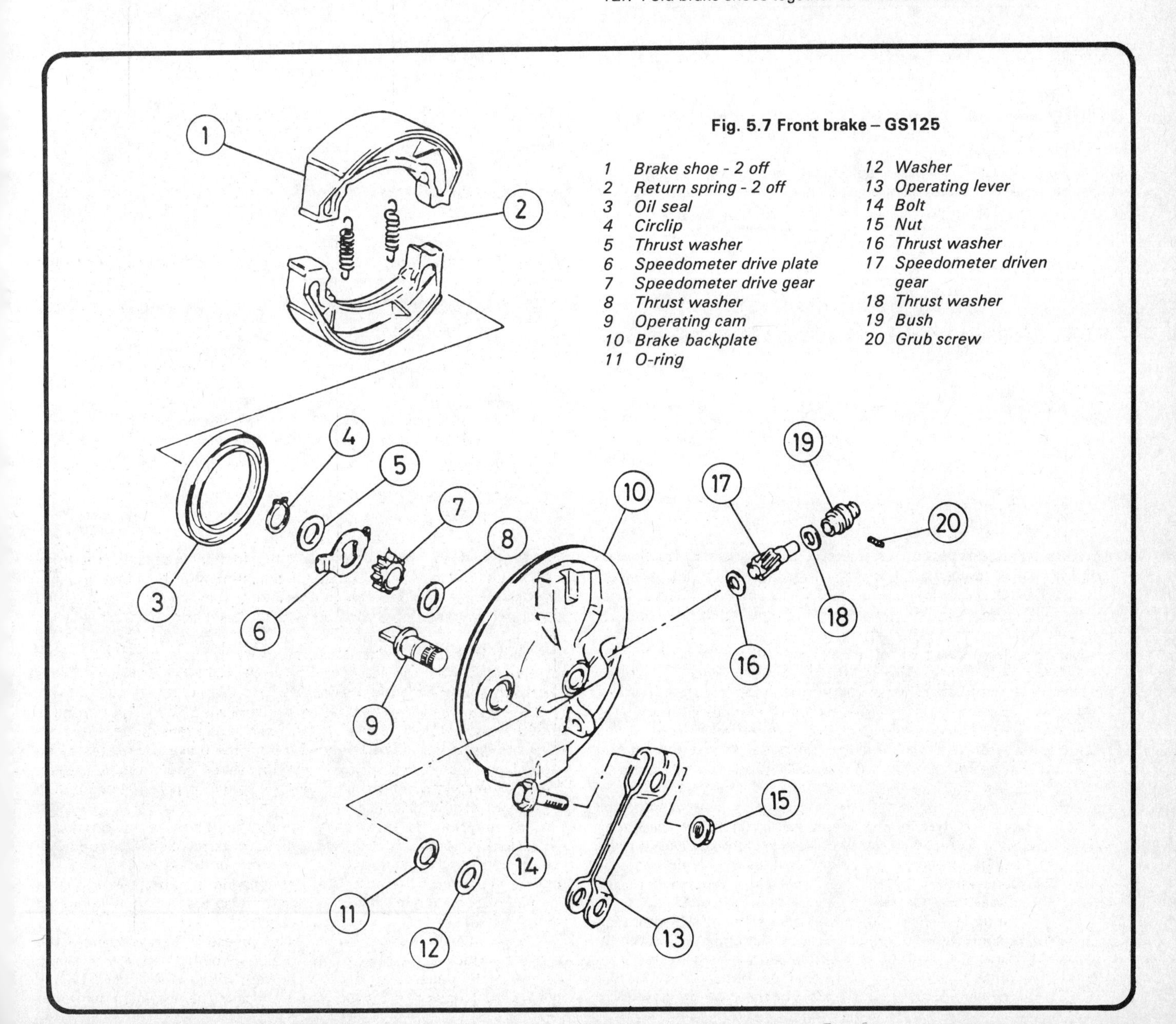

Fig. 5.7 Front brake – GS125

1 *Brake shoe - 2 off*
2 *Return spring - 2 off*
3 *Oil seal*
4 *Circlip*
5 *Thrust washer*
6 *Speedometer drive plate*
7 *Speedometer drive gear*
8 *Thrust washer*
9 *Operating cam*
10 *Brake backplate*
11 *O-ring*
12 *Washer*
13 *Operating lever*
14 *Bolt*
15 *Nut*
16 *Thrust washer*
17 *Speedometer driven gear*
18 *Thrust washer*
19 *Bush*
20 *Grub screw*

13 Rear wheel: examination and renovation

The rear wheel is identical in design and construction to that fitted to the front. Refer, therefore, to Section 2 or to Section 3 of this Chapter, whichever is relevant to the type of wheel fitted to the machine being worked on, for details of examination and renovation techniques.

14 Rear brake: examination, renovation and adjustment

The drum brake fitted to the rear wheel of all models is essentially similar in design and construction to the unit fitted to the front of the DR125 S and GS125 models. Refer, therefore, to Section 12 of this Chapter for details of examination and renovation techniques. One basic difference between the two units is that the rear brake is rod operated and the slightly different technique of adjustment that is therefore required is described in Routine Maintenance.

15 Rear wheel: removal and refitting

1 Raise the wheel clear of the ground by placing the machine on its centre stand or on a strong wooden box or similar support placed under the engine/gearbox unit.

2 Remove the R-clip or split-pin which secures the torque arm retaining nut, then remove the nut and spring washer. Pull the torque arm off its mounting bolt. Detach the brake operating rod by unscrewing the adjusting nut from the threaded end of the rod and then depressing the brake pedal so that the rod is pulled clear of the brake cam operating arm; the adjusting nut, trunnion, brake return spring and the plain washer should then be refitted on to the operating rod to prevent their loss.

3 Withdraw the split pin (not GS125 ESX) and remove the spindle nut and plain washer (two plain washers and guide plate on DR125 S models). Withdraw the spindle, using a hammer and a metal drift if necessary, and catch the spacers as they drop clear. Disengage the chain from the sprocket, loop the chain over the swinging arm and withdraw the wheel from the machine.

4 Remove the brake backplate from the hub right-hand side and the sprocket carrier (GS125 models only) from the hub left-hand side.

5 On reassembly, check that the spindle, chain adjusters and spacers are clean and free from corrosion, then smear grease over the length of the spindle. Insert the sprocket carrier into the hub left-hand side ensuring that its cast vanes engage fully with the slots in the rubber block, and insert the hub left-hand spacer into the sprocket carrier oil seal (all GS125 models only). Refit the brake backplate.

6 Offer up the wheel and engage the chain on the rear sprocket, then push the spindle through from left to right, ensuring that all components are correctly refitted. On DR125 S models, note that the longer length of the guide plates faces to the front. Refit the guide plate, one or two plain washers (as applicable) and the spindle nut. Tighten the spindle nut by hand only and refit the brake torque arm and brake rod.

7 Check and adjust the chain tension and rear brake as described in Routine Maintenance. Tighten the spindle nut to a torque setting of 5.0 – 8.0 kgf m (36 – 58 lbf ft) and fit a new split pin (not GS125 ESX, which uses a plain nut) to secure it. Do not forget to fit the split pin or R-clip to the torque arm retaining nut.

16 Rear wheel bearings: removal, examination and refitting

As the rear wheel hub assembly is essentially the same as that fitted to the front wheel, examination and renovation techniques are the same as those described in Section 5 of this Chapter. The rear wheel first must be detached from the machine as described in Section 15 of this Chapter, and the brake backplate and cush-drive assembly must be removed from the hub in order to gain access to the bearings. On DR125 S models there is no need to remove the sprocket before removing the bearings.

15.5a Vanes on sprocket carrier must engage fully with slots in cush drive rubber

15.5b Insert spacer into sprocket carrier oil seal

15.6a Push wheel spindle through from left to right

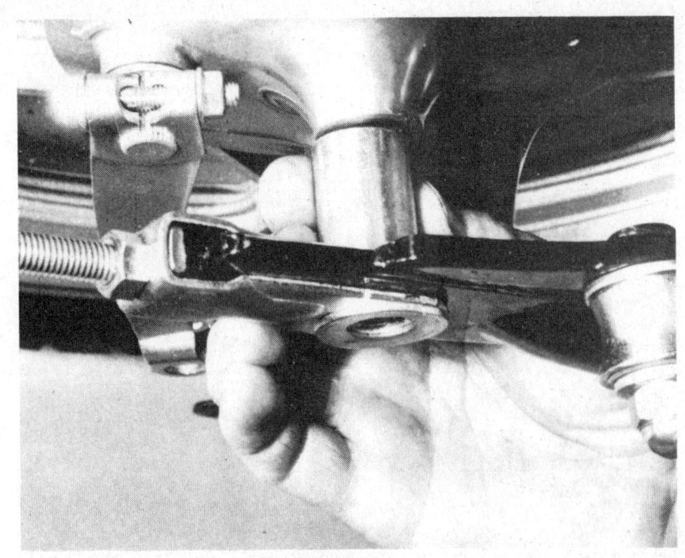

15.6b GS125 models – align hub right-hand spacer and chain adjuster before spindle is pushed fully home

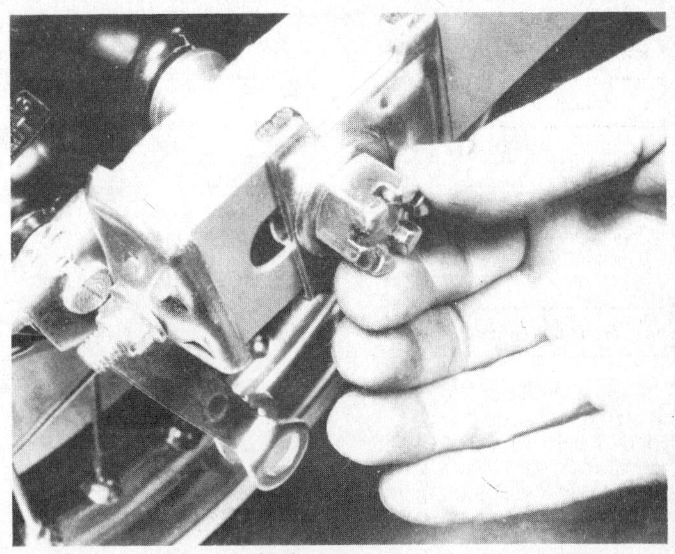

15.6c DR125 S – spindle guide plate longer length faces to front

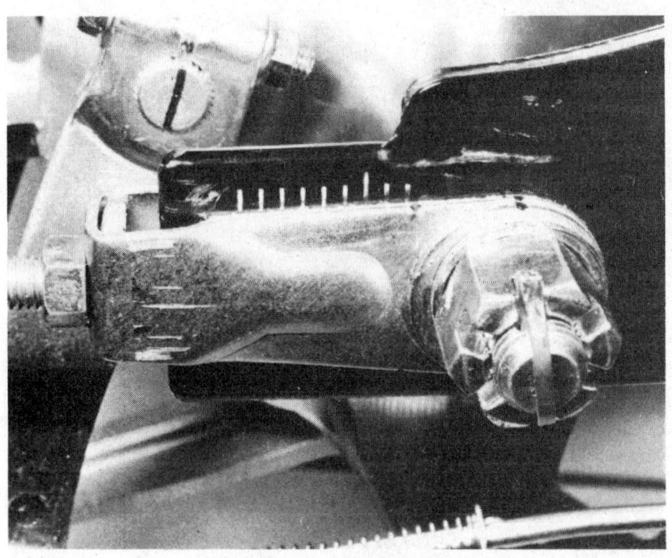

15.7a Fit new split-pins as shown to secure spindle nut ...

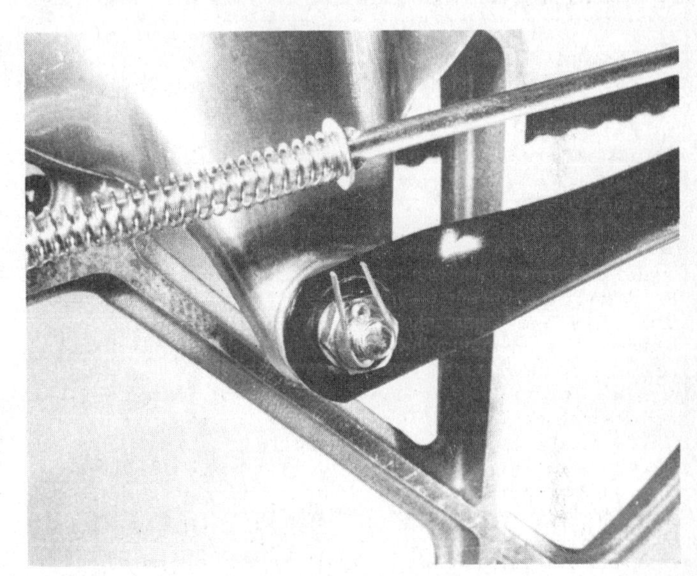

15.7b ... and brake torque arm nut

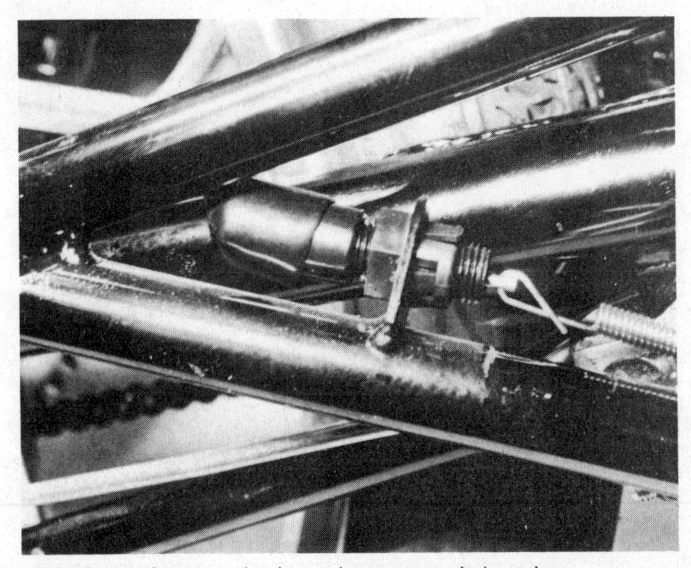

15.7c Do not forget to check stop lamp rear switch setting

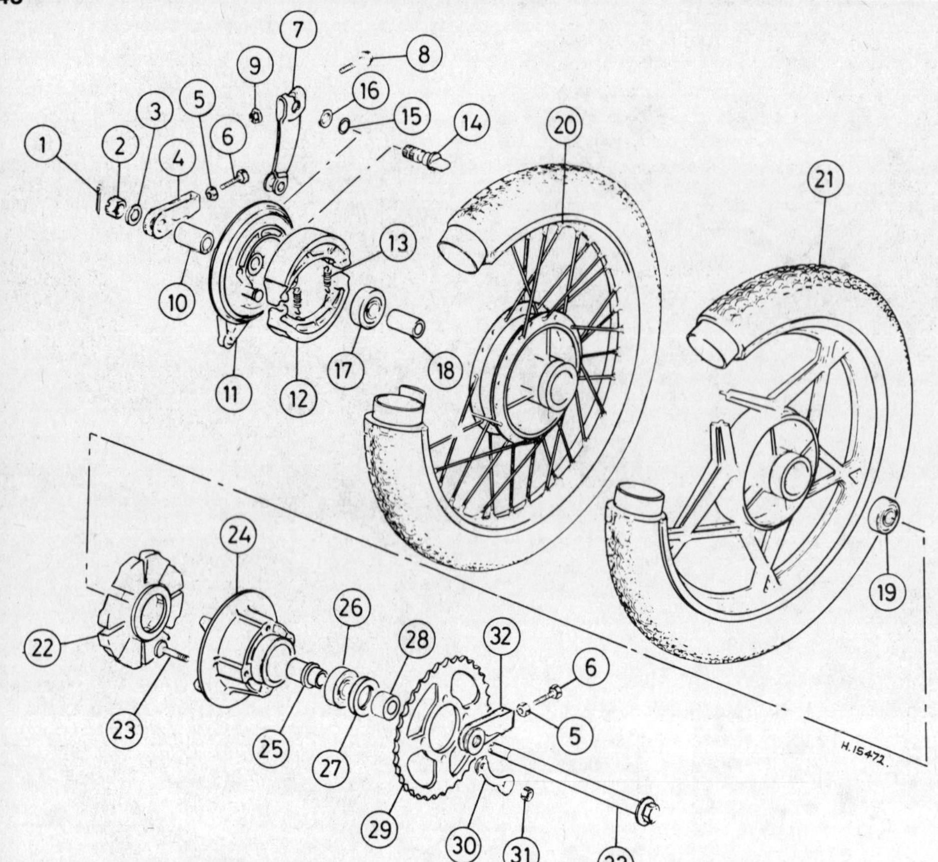

Fig. 5.8 Rear wheel – GS models

1 Split pin
2 Nut
3 Washer
4 Right-hand chain adjuster
5 Locknut - 2 off
6 Adjuster bolt - 2 off
7 Operating lever
8 Bolt
9 Nut
10 Right-hand spacer
11 Brake backplate
12 Brake shoe - 2 off
13 Return spring - 2 off
14 Operating cam
15 O-ring
16 Washer
17 Right-hand bearing
18 Spacer
19 Left-hand bearing
20 Rear wheel - spoked type
21 Rear wheel - cast alloy type
22 Cush drive rubbers
23 Bolt - 4 off
24 Sprocket carrier
25 Shouldered spacer
26 Sprocket carrier bearing
27 Oil seal
28 Left-hand spacer
29 Sprocket
30 Tab washer - 2 off
31 Nut - 4 off
32 Left-hand chain adjuster
33 Wheel spindle

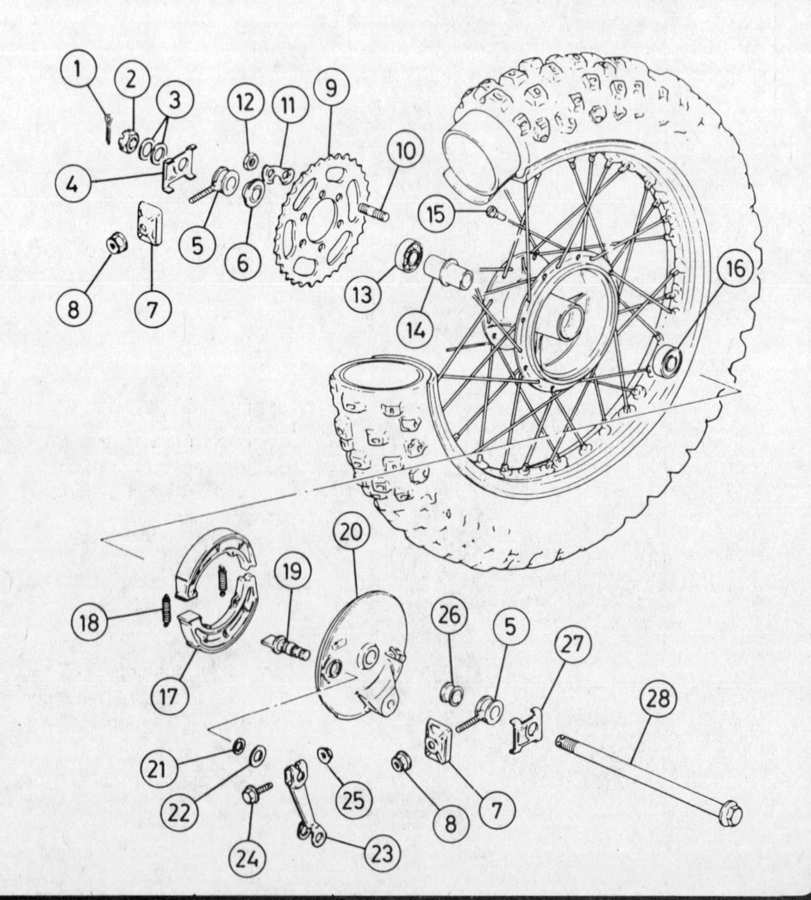

Fig. 5.9 Rear wheel – DR models

1 Split pin
2 Nut
3 Washers
4 Spindle guide plate
5 Chain adjuster
6 Flanged spacer
7 End plate - 2 off
8 Adjuster nut - 2 off
9 Sprocket
10 Stud - 6 off
11 Tab washer - 3 off
12 Nut - 6 off
13 Left-hand bearing
14 Spacer
15 Spoke nipple
16 Right-hand bearing
17 Brake shoe - 2 off
18 Return spring - 2 off
19 Operating cam
20 Brake backplate
21 O-ring
22 Washer
23 Operating lever
24 Bolt
25 Nut
26 Spacer
27 Spindle guide plate
28 Wheel spindle

Tyre changing sequence - tubed tyres

Deflate tyre. After pushing tyre beads away from rim flanges push tyre bead into well of rim at point opposite valve. Insert tyre lever adjacent to valve and work bead over edge of rim.

Use two levers to work bead over edge of rim. Note use of rim protectors

Remove inner tube from tyre

D

When first bead is clear, remove tyre as shown

When fitting, partially inflate inner tube and insert in tyre

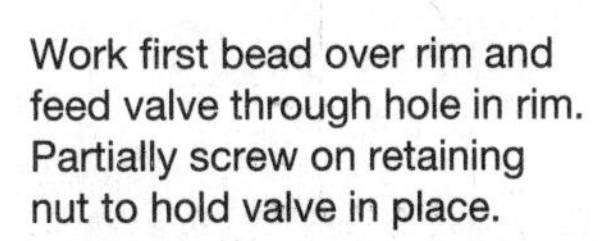

F

Work first bead over rim and feed valve through hole in rim. Partially screw on retaining nut to hold valve in place.

Check that inner tube is positioned correctly and work second bead over rim using tyre levers. Start at a point opposite valve.

H

Work final area of bead over rim whilst pushing valve inwards to ensure that inner tube is not trapped

17 Rear sprocket and cush drive: examination and renovation

1 Two types of sprocket mounting are employed; if any work is to be carried out on the sprocket or its mountings, the rear wheel assembly first must be removed from the machine as described in Section 15 of this Chapter so that the sprocket (and cush-drive) can be detached from the hub.

DR125 S model

2 The sprocket is mounted directly on the wheel hub, being retained by four nuts threaded on to studs; the nuts are secured in pairs by lock washers. Due to the lack of a transmission shock absorber the mounting studs are subjected to high shock loads and must be examined frequently.
3 Sprocket removal is accomplished by bending back the locking tabs of the tab washers and by unscrewing the four retaining nuts. Remove the tab washers and pull the sprocket away from the hub.
4 The sprocket need only be renewed if the teeth are hooked or badly worn. It is considered bad practice to renew one sprocket on its own, both drive sprockets should be renewed as a pair, preferably with a new final drive chain. If this recommendation is not observed, rapid wear resulting from the running of old and new parts together will necessitate even earlier replacement on the next occasion.
5 Carefully clean the hub mounting boss and examine the hub, looking for cracks and any signs of wear. Check that the studs are in good condition; the sprocket must be secure on its mountings. To check this, refit temporarily the sprocket and check for excessive movement in its normal direction of rotation. Examine also the nuts and the lock washers, renewing any component which appears to be in any way worn or damaged.
6 If any of the studs are to be renewed, they should be removed by locking two nuts together on the exposed thread of the stud and by using an open-ended spanner to turn the lower nut as if unscrewing it. This will remove the stud in all normal circumstances. If, however, the stud is sheared off, it must be withdrawn by drilling and using a stud extractor; this procedure is difficult and should be entrusted to an expert to ensure that the hub is not damaged. Refit each stud by locking two nuts together on its threads. Apply a few drops of thread locking compound to the threads at the inner end of the stud and screw it into the hub using a spanner to tighten the stud by means of the outer of the two nuts until the stud is screwed into the hub as far as possible and securely tightened.
7 Refitting is the reverse of the dismantling sequence. Apply thread locking compound to the threads of each stud and tighten the retaining nuts to a torque setting of 2.5 – 4.0 kgf m (18 – 19 lbf ft). Bend up an unused portion of the lock washer against one of the flats of each nut.

GS125 models

8 The sprocket carrier bearing is removed by tapping out the shouldered spacer from behind, thus drifting the bearing from position. The bearing oil seal will be displaced with the bearing, or it can be removed by using a flat-bladed screwdriver to lever it from its housing. Renew the oil seal as a matter of course if it is disturbed. Examine the carrier bearing as described in Section 5 of this Chapter and renew it if necessary.
9 The sprocket is removed from the carrier by flattening the locking tabs of the two tab washers, then unscrewing the four retaining nuts. Examine carefully the bolts, nuts, and tab washers, renewing any component that is found to be damaged or worn. The sprocket need only be renewed if the teeth are hooked, chipped, or badly worn. Comparison with a new component will give a good idea of the degree of wear, but remember that it is bad practice to renew just one sprocket or to fit a new chain to worn sprockets and for this reason the front and rear sprockets and the chain should always be renewed as a complete set.
10 The only other component in the cush-drive assembly which will deteriorate appreciably is the rubber block. Check carefully that the rubber is not splitting or breaking up and check that the sprocket cannot move an excessive amount in relation to the wheel. If any fault is found in the rubber block, it must be renewed.
11 Reassembly is a straightforward reversal of the dismantling procedure. Lubricate the cush-drive rubber with a solution of soapy water or with an aerosol spray of water-dispersant fluid such as WD40. This will make the surface of the rubber slippery, thus ensuring that the separate pads of the block will slide easily into place. Pack the sprocket carrier bearing with high melting-point grease and drift it into place, followed by the carrier oil seal. On refitting the sprocket, push the four mounting bolts through from the rear of the carrier, apply thread locking compound to the threads of each one, then refit the sprocket, the two tab washers, and the four nuts. Tighten the nuts to a torque setting of 1.8 - 2.8 kgf m (13 - 20 lbf ft), then secure each nut by bending up a previously unused part of the tab washer against one flat of the nut.

18 Tyres: removal, repair and refitting

1 At some time or other the need will arise to remove and replace the tyres, either as a result of a puncture or because replacements are necessary to offset wear. To the inexperienced, tyre changing represents a formidable task, yet if a few simple rules are observed and the technique learned, the whole operation is surprisingly simple.
2 To remove the tyre from either wheel, first detach the wheel from the machine. Deflate the tyre by removing the valve core, and when the tyre is fully deflated, push the bead away from the wheel rim on both sides so that the bead enters the centre well of the rim. Remove the locking ring and push the tyre valve into the tyre itself.
3 Insert a tyre lever close to the valve and lever the edge of the tyre over the outside of the rim. Very little force should be necessary; if resistance is encountered it is probably due to the fact that the tyre beads have not entered the well of the rim, all the way round. If aluminium rims are fitted, damage to the soft alloy by tyre levers can be prevented by the use of plastic rim protectors.
4 Once the tyre has been edged over the wheel rim, it is easy to work round the wheel rim, so that the tyre is completely free from one side. At this stage the inner tube can be removed.
5 Now working from the other side of the wheel, ease the other edge of the tyre over the outside of the wheel rim that is furthest away. Continue to work around the rim until the tyre is completely free from the rim.
6 If a puncture has necessitated the removal of the tyre, reinflate the inner tube and immerse it in a bowl of water to trace the source of the leak. Mark the position of the leak, and deflate the tube. Dry the tube, and clean the area around the puncture with a petrol soaked rag. When the surface has dried, apply rubber solution and allow this to dry before removing the backing from the patch, and applying the patch to the surface.
7 It is best to use a patch of self vulcanizing type, which will form a permanent repair. Note that it may be necessary to remove a protective covering from the top surface of the patch after it has sealed into position. Inner tubes made from a special synthetic rubber may require a special type of patch and adhesive, if a satisfactory bond is to be achieved.
8 Before replacing the tyre, check the inside to make sure that the article that caused the puncture is not still trapped inside the tyre. Check the outside of the tyre, particularly the tread area to make sure nothing is trapped that may cause a further puncture.
9 If the inner tube has been patched on a number of past occasions, or if there is a tear or large hole, it is preferable to discard it and fit a replacement. Sudden deflation may cause an accident, particularly if it occurs with the rear wheel.
10 To replace the tyre, inflate the inner tube for it just to assume a circular shape but only to that amount, and then push the tube into the tyre so that it is enclosed completely. Lay the tyre on the wheel at an angle, and insert the valve through the rim tape and the hole in the wheel rim. Attach the locking ring on the first few threads, sufficient to hold the valve captive in its correct location.
11 Starting at the point furthest from the valve, push the tyre bead over the edge of the wheel rim until it is located in the central well. Continue to work around the tyre in this fashion until the whole of one side of the tyre is on the rim. It may be necessary to use a tyre lever during the final stages.
12 Make sure there is no pull on the tyre valve and again commencing with the area furthest from the valve, ease the other bead of the tyre over the edge of the rim. Finish with the area close to the valve, pushing the valve up into the tyre until the locking ring touches the rim. This will ensure that the inner tube is not trapped when the last section of bead is edged over the rim with a tyre lever.
13 Check that the inner tube is not trapped at any point. Reinflate the inner tube, and check that the tyre is seating correctly around the wheel rim. There should be a thin rib moulded around the wall of the

tyre on both sides, which should be an equal distance from the wheel rim at all points. If the tyre is unevenly located on the rim, try bouncing the wheel when the tyre is at the recommended pressure. It is probable that one of the heads has not pulled clear of the centre well.
14 Always run the tyres at the recommended pressures and never under or over inflate. The corect pressures are given in the Specifications Section of this Chapter.
15 Tyre replacement is aided by dusting the side walls, particularly in the vicinity of the beads, with a liberal coating of french chalk. Washing up liquid can also be used to good effect, but this has the disadvantage, where steel rims are used, of causing the inner surface of the wheel rim to rust.
16 Never replace the inner tube and the tyre without the rim tape in position. If this precaution is overlooked there is a good chance of the ends of the spoke nipples chafing the inner tube and causing a crop of punctures.
17 Never fit a tyre that has a damaged tread or sidewalls. Apart from legal aspects, there is a very great risk of a blowout, which can have very serious consequences on a two wheeled vehicle.
18 Tyre valves rarely give trouble, but it is always advisable to check whether the valve itself is leaking before removing the tyre. Do not forget to fit the dust cap, which forms an effective extra seal.

17.8 Oil seal can be levered out as shown or removed with bearing

17.9 Remove retaining nuts to release sprocket

17.10 Cush drive rubber must be renewed if deteriorated

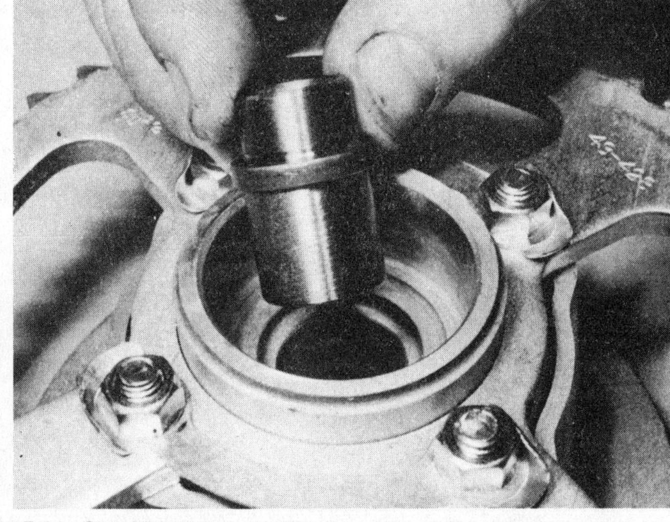
17.11a Shouldered spacer is fitted as shown into carrier

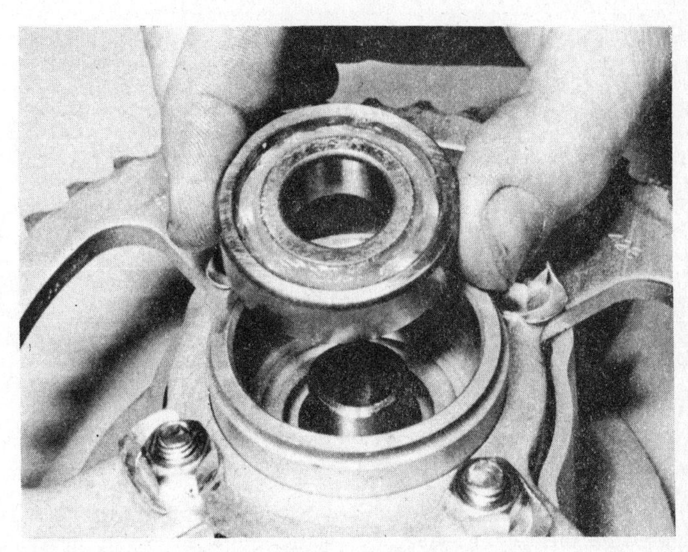
17.11b Pack bearing with clean grease before refitting

17.11c Oil seal must be renewed if damaged or worn ...

17.11d ... and is refitted as shown

19 Valve cores and caps

1 Valve cores seldom give trouble, but do not last indefinitely. Dirt under the seating will cause a puzzling 'slow-puncture'. Check that they are not leaking by applying spittle to the end of the valve and watching for air bubbles.
2 A valve cap is a safety device, and should always be fitted. Apart from keeping dirt out of the valve, it provides a second seal in case of valve failure, and may prevent an accident resulting from sudden deflation.

20 Wheel balancing

1 It is customary on all high performance machines to balance the wheels complete with tyre and tube. The out of balance forces which exist are eliminated and the handling of the machine is improved in consequence. A wheel which is badly out of balance produces through the steering a most unpleasant hammering effect at high speeds.
2 Some tyres have a balance mark on the sidewall, usually in the form of a coloured spot. This mark must be in line with tyre valve, when the tyre is fitted to the inner tube. Even then the wheel may require the addition of balance weights, to offset the weight of the tyre valve itself.
3 If the wheel is raised clear of the ground and is spun, it will probably come to rest with the tyre valve or the heaviest part downward and will always come to rest in the same position. Balance weights must be added to a point diametrically opposite the heavy spots until the wheel will come to rest in ANY position after it is spun.
4 Suzuki dealers can supply weights of 10-, 20-, or 30 grams for the GS125 ES model; these are clamped on to the side of the rim. Weights for the DR125 S and GS125 models are available from any motorcycle dealer or tyre-fitting centre; the range of weights available will vary considerably. This latter type of weight is slipped over a spoke and clamped firmly around the spoke nipple.

Chapter 6 Electrical system

Refer to Chapter 7 for information relating to DR125 SF, SH, SJ 'Raider', GN125 and GZ125 'Marauder' models

Contents

Specifications

Electrical system		
Voltage	12	
Earth	Negative	
Fuse rating		
GS125 ES, DR125 S	15A	
GS125	10A	
Battery	**GS125 ES**	**GS125, DR125 S**
Make	Yuasa	Yuasa
Type	YB7-A	YB4L-B
Capacity	8Ah	4Ah
Alternator		
No load voltage	70 volts or more @ 5000 rpm	50 volts or more @ 5000 rpm
Regulated voltage	13.5 - 16 volts @ 5000 rpm	13 - 16 volts @ 5000 rpm
Coil resistance	0.5 - 2.0 ohm	0.5 - 1.5 ohm
Starter motor		
Brush minimum length (two-brush starter)	9.0 mm (0.354 in)	
Commutator undercut minimum depth	0.2 mm (0.008 in)	
Starter relay resistance	3 - 4 ohm	
Bulbs	**GS125 ES, GS125**	**DR125 S**
Headlamp	12V, 35/35W	12V, 45/40W (45/45W, E model)
Parking lamp	12V, 3W	12V, 4W (3.4W, E model)
Stop/tail lamp	12V, 21/5W	12V, 21/5W
Turn signal lamp	12V, 21W	12V, 21W
Instrument/warning lamps - x5	12V, 3.4W	12V, 3.4W

1 General description

The 12 volt electrical system is powered by a crankshaft-mounted alternator located behind the crankcase left-hand outer cover. Output from the alternator is fed to a combined rectifier/regulator unit where it is converted from alternating current (ac) to direct current (dc) by the rectifier section, and the system voltage is regulated by the electronic voltage regulator.

2 Testing the electrical system

1 Simple continuity checks, for instance when testing switch units, wiring and connections, can be carried out using a battery and bulb arrangement to provide a test circuit. For most tests described in this Chapter, however, a pocket multimeter should be considered essential. A basic multimeter capable of measuring volts and ohms can be bought for a very reasonable sum and will prove an invaluable tool.

Note that separate volt and ohm meters may be used in place of the multimeter, provided those with the correct operating ranges are available.

2 Care must be taken when performing any electrical test, because some of the electrical components can be damaged if they are incorrectly connected or inadvertently shorted to earth. This is particularly so in the case of electronic components. Instructions regarding meter probe connections are given for each test, and these should be read carefully to preclude accidental damage occurring.

3 Where test equipment is not available, or the owner feels unsure of the procedure described, it is strongly recommended that professional assistance is sought. Errors made through carelessness or lack of experience can so easily lead to damage and need for expensive replacement parts.

4 A certain amount of preliminary dismantling will be necessary to gain acccess to the components to be tested. Normally, removal of the seat and side panels will be required, with the possible addition of the fuel tank and headlamp unit to expose the remaining components.

5 Because of the high cost of replacement electrical components it is recommended that test findings are double-checked by a Suzuki service agent before a suspect component is discarded.

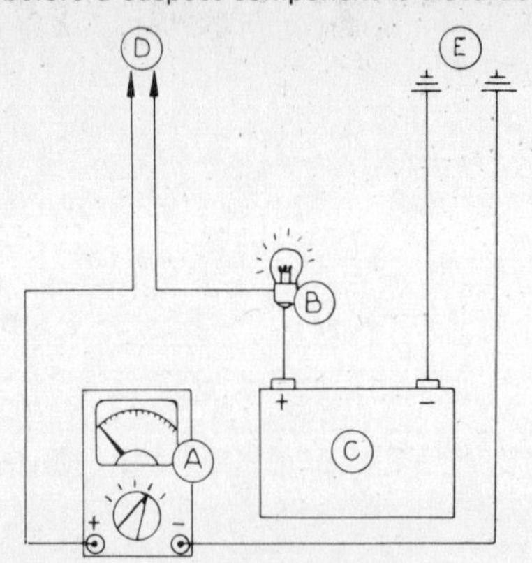

Fig. 6.1 Continuity test circuits

A Multimeter
B Bulb
C Battery
D Positive probe
E Negative probe

3 Wiring: layout and examination

1 The wiring harness is colour-coded and will correspond with the accompanying wiring diagram. When socket connectors are used, they are designed so that reconnection can be made in the correct position only.

2 Visual inspection will usually show whether there are any breaks or frayed outer coverings which will give rise to short circuits. Occasionally a wire may become trapped between two components, breaking the inner core but leaving the more resilient outer cover intact. This can give rise to mysterious intermittent or total circuit failure. Another source of trouble may be the snap connectors and sockets, where the connector has not been pushed fully home in the outer housing, or where corrosion has occurred.

3 Intermittent short circuits can often be traced to a chafed wire that passes through or is close to a metal component such as a frame member. Avoid tight bends in the lead or situations where a lead can become trapped between casings.

4 Charging system: checking the output

1 Remove the sidepanel to gain access to the battery terminals. Set a multimeter to the 0-20 volts dc scale and connect the red positive (+) probe to the positive battery terminal and the black negative (–) probe to the negative battery terminal.

2 Start the engine and raise the engine speed to 5000 rpm, or slightly more. Switch the headlamp on, to main beam, and note the meter reading. A reading of 13.5-16 volts (GS125 ES only) or 13-16 volts (GS125, DR125 S) will indicate that the system is functioning correctly.

3 If the voltage reading is significantly different, carry out the alternator test described in the following Section. If this test is satisfactory, then it may be assumed that the regulator/rectifier unit is defective and needs to be renewed. Before purchasing a replacement unit, it is well worth carrying out continuity checks on all of the relevant electrical leads and physically inspecting any connections for faults. Confirmation of a defective unit may be proved by carrying out the tests described in Section 6 of this Chapter.

4 On no account should the engine be run with the battery leads disconnected. This is because the resultant open voltage can destroy the regulator/rectifier unit diodes.

5 Alternator: testing

1 The alternator assembly can be tested in two ways: measure the output, with the engine running and measure the resistance values of the stator charging coils. In both tests it will be necessary to trace the main generator lead from the crankcase top up to the nearest connector block joining it to the main wiring loom.

2 In both tests, readings are taken between each pair of wires, making the meter connections as shown below:

GS125 ES	GS125, DR125 S
Yellow wire to Yellow wire	White/Green wire to White/Blue wire
Yellow wire to Yellow wire	White/Green wire to Yellow wire
Yellow wire to Yellow wire	White/Blue wire to Yellow wire

3 To test the alternator output, set a multimeter to the 0-100 volts ac scale, start the engine and raise the engine speed to 5000 rpm. Measure the outputs as shown above; if any one of the three readings obtained is less than 70 volts (GS125 ES models) or 50 volts (GS125, DR125 S models) the alternator is faulty. The results of this test can be confirmed by carrying out the test described below.

4 With the engine stopped, set the multimeter to the x1 ohm scale, check for the correct resistance value between each pair of terminals as shown above; the correct value is 0.5 - 2.0 ohm (GS125 ES models) or 0.5 - 1.5 ohm (GS125, DR125 S models). If the reading is significantly above or below the figure specified in any of the three tests, the coil concerned is faulty.

5 Check also between each lead has a good earth point on the engine. The alternator coils must be isolated from the stator core to function correctly, thus, if continuity is found in any test short-circuiting should be suspected.

6 If any test indicates that damage to the stator has occurred the stator should be removed from the machine and given a close visual inspection. It is possible that lack of continuity has been caused by a lead breakage somewhere between the wiring connectors and the stator. It may be possible to repair the damage or even to have a coil re-wired; always have your test results confirmed by a competent Suzuki dealer or auto-electrical expert, and ask his advice as to whether repairs are possible or if renewal of the complete stator assembly is required.

7 Removal and refitting of the alternator assembly is described in Sections 8 and 39 of Chapter 1.

6 Voltage regulator/rectifier unit: location and testing

1 The voltage regulator/rectifier unit is a heavily-finned sealed metal unit that is mounted immediately in front of the battery, on the left- or right-hand side of the machine (as appropriate). To test the unit, unplug the multi-pin block connector to expose the unit terminals (DR125 S models) or trace the unit lead back to the nearest connector and separate this (GS125 ES, GS125 models).

2 With a multimeter set to its x1 ohm range, refer to the accompanying tables, whichever is appropriate and measure the resistance between the combinations of leads shown. Adhere strictly to the sequence given in the table and take great care not to allow the probes of the meter to short to earth or against each other. If, during the test sequence, the resistance readings shown on the meter scale differ from those given in the table, then the regulator/rectifier unit may be assumed to be defective and should be renewed.

6.1a Location of voltage regulator/rectifier unit – all GS125 models ...

6.1b ... and DR125 S model

Unit: Ω

⊖ Probe of tester \ ⊕ Probe of tester	R	W/Bl	W/R	Y	B/W
R		OFF	OFF	OFF	OFF
W/Bl	7-8		OFF	OFF	OFF
W/R	7-8	OFF		OFF	OFF
Y	7-8	OFF	OFF		OFF
B/W	35-55	7-8	7-8	7-8	

Fig. 6.2 Regulator/rectifier test table – GS models

Wiring colour code

R	*Red*	*W/R*	*White and red*
Y	*Yellow*	*B/W*	*Black and white*
W/Bl	*White and blue*		

7 Battery: examination and maintenance

1 The battery is housed in a separate tray located behind the right-hand sidepanel on all GS125 models, and behind the left-hand sidepanel on DR125 S models. It is retained by a rubber band.

2 The transparent plastic case of the battery permits the upper and lower levels of the electrolyte to be observed without disturbing the battery by removing the side cover. Maintenance is normally limited to keeping the electrolyte level between the prescribed upper and lower limits and making sure that the vent tube is not blocked. The lead plates and their separates are also visible through the transparent case, a further guide to the general condition of the battery. If electrolyte level drops rapidly, suspect over-charging and check the system.

3 Unless acid is spilt, as may occur if the machine falls over, the electrolyte should always be topped up with distilled water to restore the correct level. If acid is spilt onto any part of the machine, it should be neutralised with an alkali such as washing soda or baking powder and washed away with plenty of water, otherwise serious corrosion will occur. Top up with sulphuric acid of the correct specific gravity (1.260 to 1.280) only when spillage has occurred. Check that the vent pipe is well clear of the frame or any of the other cycle parts.

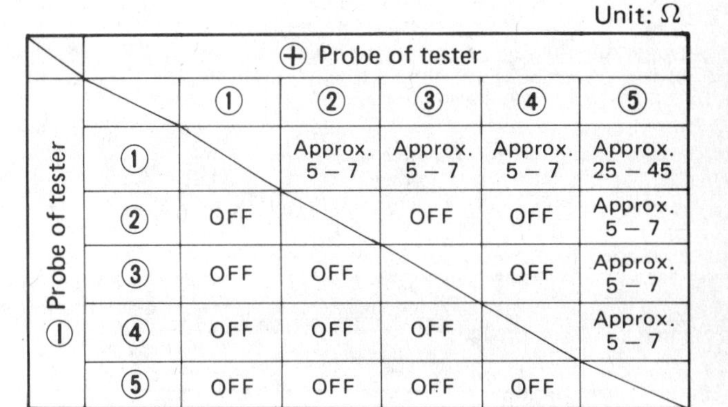

Unit: Ω

⊖ Probe of tester \ ⊕ Probe of tester	①	②	③	④	⑤
①		Approx. 5 – 7	Approx. 5 – 7	Approx. 5 – 7	Approx. 25 – 45
②	OFF		OFF	OFF	Approx. 5 – 7
③	OFF	OFF		OFF	Approx. 5 – 7
④	OFF	OFF	OFF		Approx. 5 – 7
⑤	OFF	OFF	OFF	OFF	

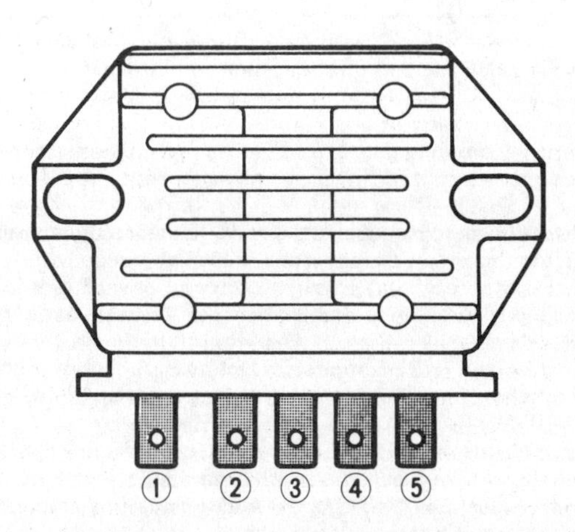

Fig. 6.3 Regulator/rectifier test table – DR model

4 It is seldom practicable to repair a cracked battery case because the acid present in the joint will prevent the formation of an effective seal. It is always best to renew a cracked battery, especially in view of the corrosion which will be caused if the acid continues to leak.

5 If the machine is not used for a period of time, it is advisable to remove the battery and give it a 'refresher' charge every six weeks or so from a battery charger. The battery will require recharging when the specific gravity falls below 1.260 (at 29°C - 68°F). The hydrometer reading should be taken at the top of the meniscus with the hydrometer vertical. If the battery is left discharged for too long, the

plates will sulphate. This is a grey deposit which will appear on the surface of the plates, and will inhibit recharging. If there is sediment on the bottom of the battery case, which touches the plates, the battery needs to be renewed. Prior to charging the battery refer to the following Section for correct charging rate and procedure. If charging from an external source with the battery on the machine, disconnect the leads, or the rectifier will be damaged.
6 Note that when moving or charging the battery, it is essential that the following basic safety precautions are taken:

a) Before charging check that the battery vent is clear or, where no vent is fitted, remove the combined vent/filler caps. If this precaution is not taken the gas pressure generated during charging may be sufficient to burst the battery case, with disastrous consequences.
b) Never expose a battery on charge to naked flames or sparks. The gas given off by the battery is highly explosive.
c) If charging the battery in an enclosed area, ensure that the area is well ventilated.
d) Always take great care to protect yourself against accidental spillage of the sulphuric acid contained within the battery. Eyeshields should be worn at all times. If the eyes become contaminated with acid they must be flushed with fresh water immediately and examined by a doctor as soon as possible. Similar attention should be given to a spillage of acid on the skin.

Note also that although, should an emergency arise, it is possible to charge the battery at a more rapid rate than that stated in the following Section, this will shorten the life of the battery and should therefore by avoided if at all possible.
7 Occasionally, check the condition of the battery terminals to ensure that corrosion is not taking place, and that the electrical connections are tight. If corrosion has occurred, it should be cleaned away by scraping with a knife and then using emery cloth to remove the final traces. Remake the electrical connections whilst the joint is still clean, then smear the assembly with petroleum jelly (NOT grease) to prevent recurrence of the corrosion. Badly corroded connections can have a high electrical resistance and may give the impression of complete battery failure.

8 Battery: charging procedure

1 Whilst the machine is used on the road it is unlikely that the battery will require attention other than routine maintenance because the generator will keep it fully charged. However, if the machine is used for a succession of short journeys only, mainly during the hours of darkness when the lights are in full use, it is possible that the output from the generator may fail to keep pace with the heavy electrical demand, especially if the machine is parked with the lights switched on. Under these circumstances it will be necessary to remove the battery from time to time to have it charged independently.
2 The normal maximum charging rate for any battery is 1/10 the rated capacity. Hence the charging rate for the battery fitted to GS125 ES models is 0.8 amp, while that of the 4 Ah battery fitted to the GS125 and DR125 S models is 0.4 amp. A slightly higher charge rate may be used in emergencies only, but this should never exceed 1 amp.
3 Ensure that the battery/charger connections are properly made, ie the charger positive (usually coloured red) lead to the battery positive (the red wire) lead, and the charger negative (usually coloured black or blue) lead to the battery negative (the black/white wire) lead. Refer to the previous Section for precautions to be taken during charging. It is especially important that the battery cell cover plugs are removed to eliminate any possibility of pressure building up in the battery and cracking its casing. Switch off the charger if the cells become overheated, ie over 45°C (117°F).
4 Charging is complete when the specific gravity of the electrolyte rises to 1.260 - 1.280 at 20°C (68°F). A rough guide to this state is when all cells are gassing freely. At the normal (slow) rate of charge this will taken between 3 - 15 hours, depending on the original state of charge of the battery.
5 If the higher rate of charge is used, never leave the battery charging for more than 1 hour as overheating and buckling of the plates will inevitably occur.

9 Fuse: location and renewal

1 The electrical system is protected by a single fuse of 10 or 15 amp rating, as appropriate. It is retained in a plastic casing set in the battery positive (+) terminal lead, and is clipped to a holder immediately in front of the battery. If the spare fuse is ever used, replace it with one of the correct rating as soon as possible.
2 Before renewing a fuse that has blown, check that no obvious short circuit has occurred, otherwise the replacement fuse will blow immediately it is inserted. It is always wise to check the electrical circuit thoroughly, to trace the fault and eliminate it.
3 When a fuse blows while the machine is running and no spare is available, a 'get you home' remedy is to remove the blown fuse and wrap it in silver paper before replacing it in the fuse holder. The silver paper will restore the electrical continuity by bridging the broken fuse wire. This expedient should never be used if there is evidence of short circuit or other major electrical faults, otherwise more serious damage will be caused. Replace the 'doctored' fuse at the earliest possible opportunity, to restore full circuit protection.

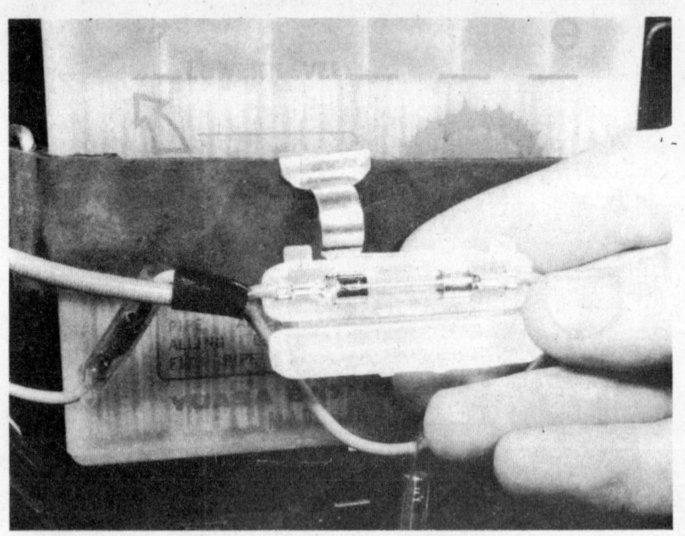

9.1 Fuse is in a plastic casing clipped in front of battery

10 Switches: general

1 While the switches should give little trouble, they can be tested using a multimeter set to the resistance function or a battery and bulb test circuit. Using the information given in the wiring diagram at the end of this Manual, check that full continuity exists in all switch positions and between the relevant pairs of wires. When checking a particular circuit follow a logical sequence to eliminate the switch concerned.
2 As a simple precaution always disconnect the battery before removing any of the switches, to prevent the possibility of a short circuit. Most troubles are caused by dirty contacts, but in the event of the breakage of some internal part, it will be necessary to renew the complete switch.
3 It should, however, be noted that if a switch is tested and found to be faulty, there is nothing to be lost by attempting a repair. It may be that worn contacts can be built up with solder, or that a broken wire terminal can be repaired, again using a soldering iron. The handlebar switches can all be dismantled to a greater or lesser extent. It is, however, up to the owner to decide if he has the skill to carry out this sort of work.
4 While none of the switches require routine maintenance of any sort, some regular attention will prolong their life a great deal. In the author's experience, the regular and constant application of WD40 or a similar water-dispersant spray not only prevents problems occurring due to waterlogged switches and the resulting corrosion, but also makes the switches much easier and more positive to use. Alternatively, the switch may be packed with a silicone-based grease to achieve the same result.

11 Starter motor: examination and renovation

Removal and refitting

1 To remove the starter motor from the engine unit, first disconnect the leads from the battery (negative lead first), to isolate the electrical system. Remove the rubber grommet and detach the heavy duty cable from the terminal on the starter motor body. The starter motor is secured to the crankcase by two bolts which pass through a flange in its end cap. Unscrew the two bolts and withdraw the motor.

2 Refitting is the reverse of the above; renew and smear with grease the sealing O-ring fitted around the motor left-hand end boss, and apply thread locking compound to the threads of the motor retaining bolts before refitting each one. Smear petroleum jelly or silicone grease over the starter motor terminal before refitting the rubber grommet.

Two-brush starter motor – GS125 ESZ, ESD, ESF and ESK

3 The end covers are secured to the motor body by two long screws which pass through the length of the motor. With these screws removed, the right-hand end cover can be separated from the motor body and the brush retaining plate exposed.

4 With the flat of a small screwdriver, lift the end of the spring clip bearing on the end of each brush and draw the brushes from their holders. Measure the length of each brush; if it is less than the service limit of 9.0 mm (0.35 in), renew the brush.

5 Before fitting the brushes, make sure that the commutator is clean, using a strip of glass paper. Never use emery cloth or 'wet-and dry' as the small abrasive fragments may embed themselves in the soft copper of the commutator and cause excessive wear of the brushes. Finish off the commutator with metal polish to give a smooth surface and finally wipe the segments over with a rag moistened with methylated spirit to ensure a grease free surface. Check that the mica insulators, which lie between the segments of the commutator, are undercut. If the amount of undercut is seen to be less than the service limit of 0.2 mm (0.008 in), then the armature should be withdrawn from the motor body and returned to an official Suzuki service agent or an experienced auto-electrician for recutting.

6 Fit the brushes in their holders and check that they slide quite freely. If the original brushes are being refitted, make sure that they are fitted in their original positions as they will have worn to the profile of the commutator.

7 If the motor has given indications of a more serious fault, withdraw the armature from the motor body, set a multimeter to its resistance function and carry out a check for equal resistance between the commutator segments and for good insulation between each commutator segment and the armature core. A fault in insulation or continuity will require renewal of the armature.

8 Before commencing reassembly of the motor, check the condition of the O-ring which forms a seal between the flanged end cap and the motor body and the condition of both the oil seal and O-ring located in the opposite end cap. If any of these items are seen to be damaged or deteriorating then they must be renewed. The seal can be removed from its location within the end cap by carefully easing it from position with the flat of a small screwdriver. Ensure that the new seal enters the end cap squarely and lubricate its outer surface and lip with a small amount of clean engine oil to ease fitting.

9 Check the condition of the shims, located one at either end of the armature, and commence reassembly of the starter motor whilst referring to the figure accompanying this text for the fitted positions of the components. It will be found that some difficulty will be experienced when refitting the right hand end cap and brush holder assembly. Care must be taken to locate correctly the brushes on the commutator and to align correctly the protrusion cast in the motor body with the corresponding notch in the end cap. A similar method of alignment is used for the opposing cap. Secure the assembly with the two long screws and check that the armature is able to rotate.

Four-brush starter motor – GS125 ESM, ESR and ESX

Note: *Reassembly of the starter motor requires a degree of patience due to the brush retaining method being somewhat awkward and the help of a couple more pairs of hands is advised.*

10 Remove the two screws from the left-hand end cap to allow both end caps and the motor body to separate; as the armature is withdrawn from the flanged end cap, the brushes will spring out of their holders.

11 One pair of brushes is attached to the brushholder and can be released once their retaining screws are removed, and the other pair are fixed to the plate beneath the main brushholder. No service limit is available for the brushes, and therefore comparison with a new set is the only means of determining whether they are worn; a set of new brushes measures 7 mm (0.28 in).

12 Refer to paragraphs 5 and 7 of this section for commutator and armature checks.

13 Check the condition of the O-rings around each end cap and the O-ring at the motor-to-engine join.

14 Begin reassembly by installing the brushholder in the flanged end cap. Working on one brush at a time, install the coil spring in the holder and slide the brush into position (make sure that the working surface faces the commutator). In the case of the remote pair of brushes, route their leads around the post and tighten their screws to secure the brushholder. Have an assistant hold the brush retracted in its holder whilst you work on the others. When all are held retracted in their holders lower the armature into position and relax tension on the brushes.

15 Slide the motor body down over the armature, aligning its mark with the end cap fork. Fit the shim on the other end of the armature and fit the end cap, together with its lockplate. Align the end cap correctly and install the two long screws to secure the assembly. Check that the armature is able to rotate.

12 Starter solenoid switch: testing

1 If the starter will not operate, first suspect a discharged battery. This can be checked by trying the horn or switching on the lights. If this check shows the battery to be in good shape, suspect the starter switch which should come into action with a pronounced click. It is located close to the battery, to which it is connected by a heavy duty cable. Before condemning the starter solenoid, carry out the following tests.

2 Disconnect the earth lead from the negative (–) terminal of the battery and move it well clear of the terminal. Remove the rubber cover from the starter solenoid and disconnect the low tension lead at the connector. Unscrew and remove the two nuts from the solenoid and disconnect the leads from the threaded terminals. Remove the starter solenoid from its retainer by sliding it out of position.

3 The solenoid coil should be tested by connecting a battery across the solenoid coil terminals. The low tension wire runs to one terminal, the second terminal is that to which the heavy duty cable from the battery is connected. On connecting the battery, a pronounced click should be heard as the contacts close. Maintain the power so that the contacts remain closed, and check for continuity across the two heavy duty terminals. This should be done with a multimeter set to the resistance function. If continuity does not exist, there is evidence that the starter relay has failed.

4 Alternatively, connect a multimeter set to the x1 ohm range between the low tension wire and the switch mounting bracket or a similar good earth point. If a reading of 3-4 ohms is obtained the switch is in good condition. If not, the switch is faulty and must be renewed.

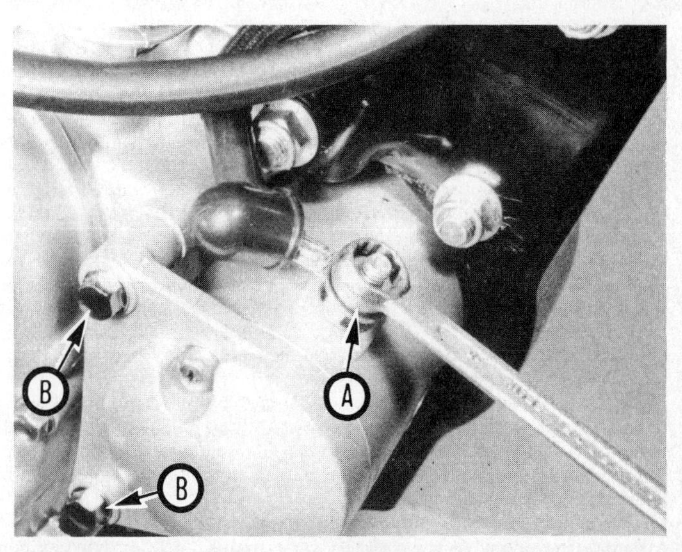

11.1a Cable connection (A) and mounting bolts (B) on 2-brush motor

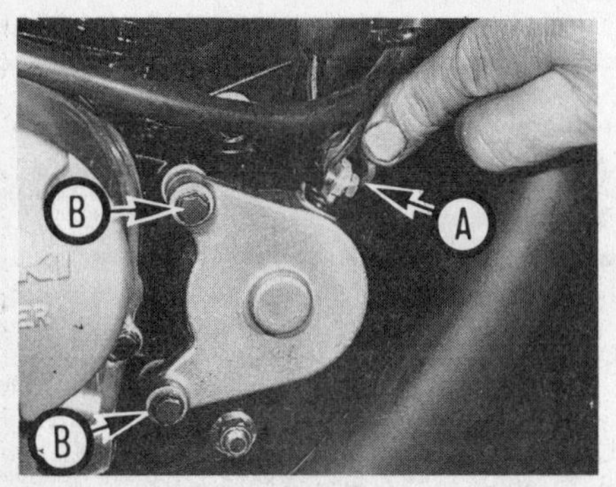

11.1b Cable connection (A) and mounting bolts (B) on 4-brush motor

11.1c Withdraw starter motor from engine

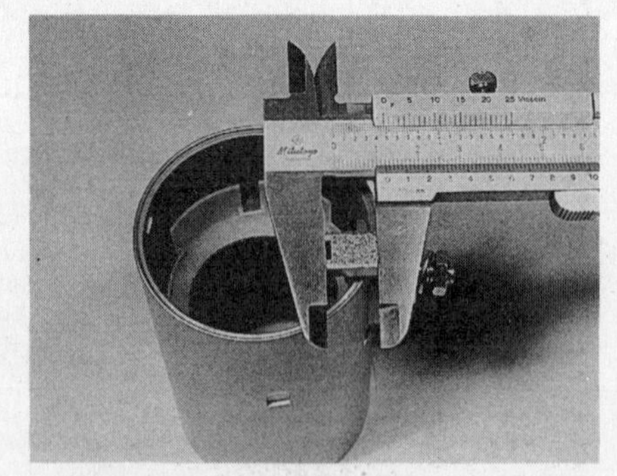
11.4 Measuring brush length

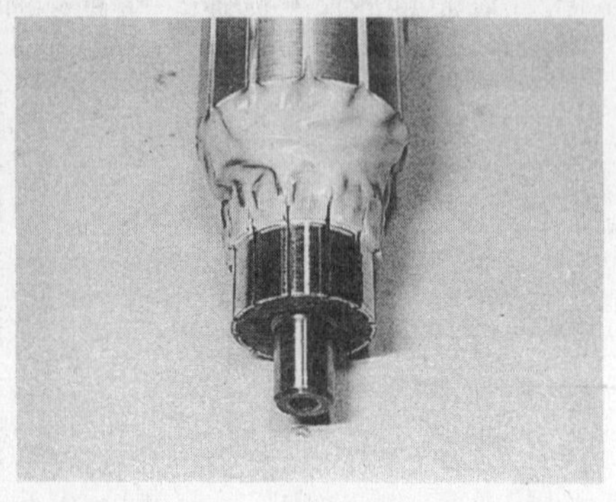
11.5 Examine and clean commutator as described in text

11.9 Notch in end cap must align with protrusion in motor body

11.10 Long screws pass through motor from left-hand side on 4-brush motor

11.15a Align motor body with locating marks on each end cap

11.15b Lockplate tanks must mesh with cutouts in left-hand end cover

12.1 Starter motor solenoid is mounted under seat

13 Neutral/gear position indicator assembly: testing

1 If a fault occurs in the circuit, first check that all connections are clean and securely fastened, then check that the bulbs and bulb holders are in good condition. Renew any defective component; the gear position indicator bulbs (GS125 ES models only) are a special type that are available only from Suzuki dealers. To test the switch, trace the lead from the crankcase top up to the nearest connector.
2 Separate the connector and check that continuity exists between the switch wire terminal and a good earth point when the gearbox is in the neutral position. On GS125 ES models, select first gear and check for continuity between the white/yellow wire terminal and earth, then select neutral and test between the blue wire terminal and earth, and so on, making a total of six tests in the different gear positions. If the switch appears to be faulty, remove it as described in Section 8 of Chapter 1, refitting being described in Section 39 of the same Chapter.
3 Before renewing the switch, check for continuity between the switch brass contact and the wire terminal to ensure that the fault is not due to a break in the wire which might easily be repaired. It is also worthwhile attempting to clean up the switch contacts using fine emery paper before buying unnecessarily a new component.

14 Turn signal relay: location and testing

1 The relay is a rectangular sealed black plastic box. It is mounted to the rear of the battery on all GS125 models and immediately behind the steering head on DR125 S models; it is rubber-mounted and is connected by a two-pin connector block.
2 If the turn signal lamps cease to function correctly, there may be any one of several possible faults responsible which should be checked before the relay is suspected. First check that the turn signal lamps are correctly mounted and that all the earth connections are clean and tight. Check that the bulbs are of the correct wattage and that corrosion has not developed on the bulbs or in their holders. Any such corrosion must be thoroughly cleaned off to ensure proper bulb contact. Also check that the turn signal switch is functioning correctly and that the wiring is in good order. Finally, ensure that the battery is fully charged.
3 Faults in any one or more of the above items will produce symptoms for which the turn signal relay may be blamed unfairly. If the fault persists even after the preliminary checks have been made, the relay must be at fault. Unfortunately the only practical method of testing the relay is to substitute a known good one.

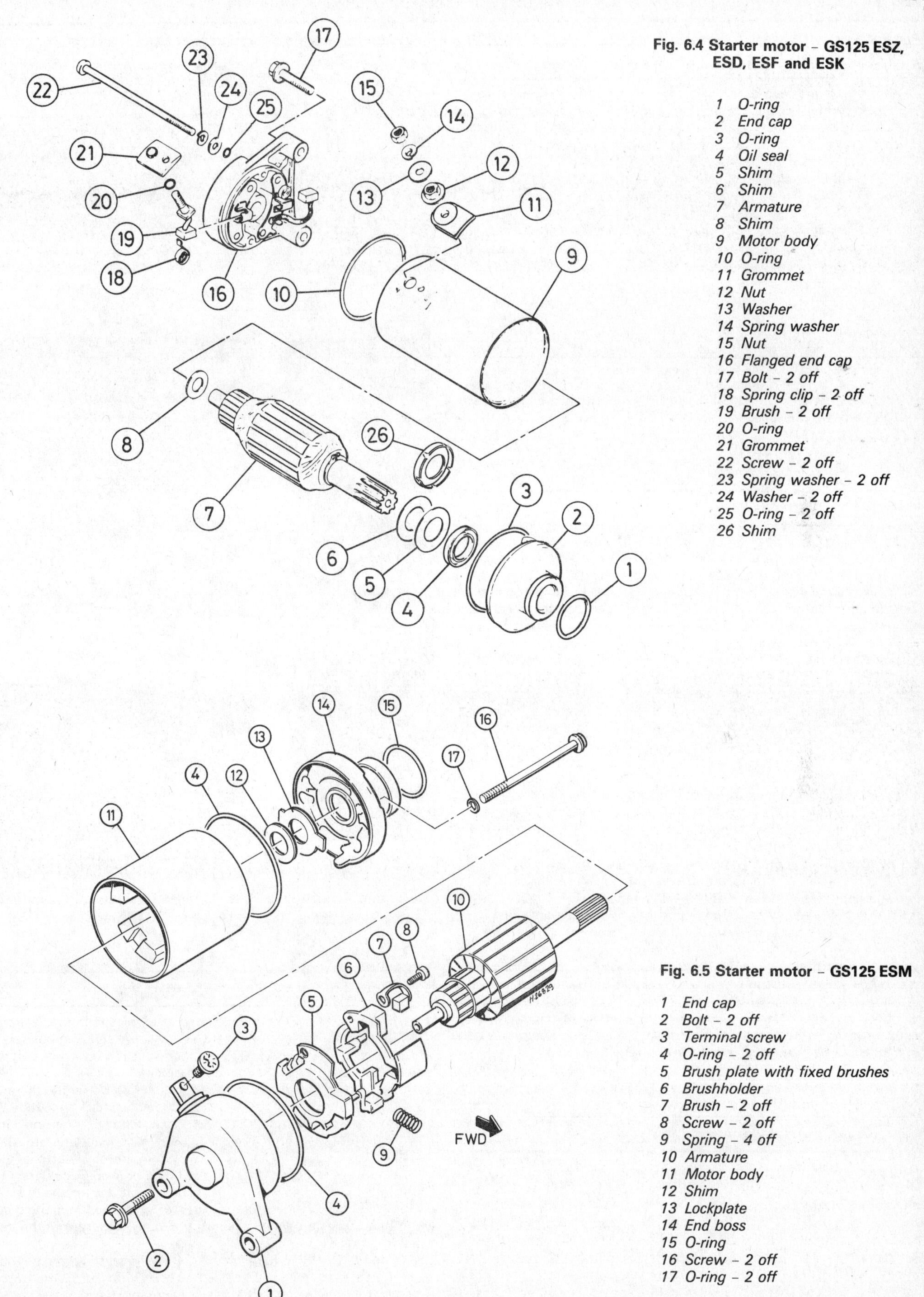

Fig. 6.4 Starter motor – GS125 ESZ, ESD, ESF and ESK

1 O-ring
2 End cap
3 O-ring
4 Oil seal
5 Shim
6 Shim
7 Armature
8 Shim
9 Motor body
10 O-ring
11 Grommet
12 Nut
13 Washer
14 Spring washer
15 Nut
16 Flanged end cap
17 Bolt – 2 off
18 Spring clip – 2 off
19 Brush – 2 off
20 O-ring
21 Grommet
22 Screw – 2 off
23 Spring washer – 2 off
24 Washer – 2 off
25 O-ring – 2 off
26 Shim

Fig. 6.5 Starter motor – GS125 ESM

1 End cap
2 Bolt – 2 off
3 Terminal screw
4 O-ring – 2 off
5 Brush plate with fixed brushes
6 Brushholder
7 Brush – 2 off
8 Screw – 2 off
9 Spring – 4 off
10 Armature
11 Motor body
12 Shim
13 Lockplate
14 End boss
15 O-ring
16 Screw – 2 off
17 O-ring – 2 off

15 Horn: location and testing

1 The horn is mounted on a flexible steel bracket to the frame front downtube. No maintenance is required other than regular cleaning to remove road dirt and occasional spraying with WD40 or a similar water dispersant lubricant to minimise internal corrosion.
2 Different types of horn may be fitted; if a screw and locknut is provided on the outside of the horn, the internal contacts may be adjusted to compensate for wear and to cure a weak or intermittent horn note. Slacken the locknut and rotate slowly the screw until the clearest and loudest note is obtained, then retighten the locknut. If no means of adjustment is provided on the horn fitted, it must be renewed.
3 If the horn fails to work, first check that power is reaching it by disconnecting the wires. Substitute a 12 volt bulb, switch on the ignition and press the horn button. If the bulb lights, the circuit is proved good and the horn is at fault; if the bulb does not light, there is a fault in the circuit which must be found and rectified.
4 To test the horn itself, connect a fully-charged 12 volt battery directly to the horn. If it does not sound, a gentle tap on the outside may free the internal contacts. If this fails, the horn must be renewed as repairs are not possible.

14.1 Location of turn signal relay – all GS125 models

16 Headlamp: bulb renewal and beam alignment

1 On GS125 ES models only, remove the headlamp fairing by unscrewing the three mounting bolts, two on each side, adjacent to the headlamp mounting brackets and one underneath the front of the fairing, in the centre. Unscrew the two screws which secure the headlamp rim to the shell on GS125 models, or the three screws which serve the same purpose on DR125 S models. Withdraw the headlamp rim and reflector unit.
2 Twist anti-clockwise the bulb holder and withdraw both bulb and holder. On DR125 SZ and D models, pull the bulb away from the holder to separate the two, while on all GS125 and DR125 SE models, the bulb is pushed into the holder and twisted anti-clockwise to release it. Refitting is a straightforward reversal of the above.
3 The parking lamp bulb is a conventional bayonet fitting in a bulbholder that is pressed into a rubber grommet set in the rear of the reflector unit.
4 Beam height on all models is effected by tilting the headlamp shell after the mounting bolts have been loosened slightly. On DR125 S models only, the horizontal alignment of the beam can be adjusted by altering the position of the screw which passes through the headlamp rim. The screw is fitted at the 4 o'clock position when viewed from the front of the machine. Turning the screw in a clockwise direction will move the beam direction over to the left-hand side.
5 In the UK, regulations stipulate that the headlamp must be arranged so that the light will not dazzle a person standing at a distance greater than 25 feet from the lamp, whose eye level is not less than 3 feet 6 inches above that plane. It is easy to approximate this setting by placing the machine 25 feet away from a wall, on a level road, and setting the dip beam height so that it is concentrated at the same height as the distance of the centre of the headlamp from the ground. The rider must be seated normally during this operation and also the pillion passenger, if one is carried regularly.

15.1 Spray horn regularly with water dispersant lubricant to prevent corrosion

16.1 Remove retaining screws to release headlamp rim

17 Instrument panel: bulb renewal

1 Remove the headlamp fairing or cowling (GS125 ES and DR125 S models only), then withdraw the headlamp shell and reflector unit to expose the base of the instrument panel. Release the cap nuts and screws (as applicable) to release the panel base, then pull out the relevant bulb holder which is a press fit in the panel.
2 The bulbs are of the capless type, which are pulled out of their respective holders.
3 Refitting is the reverse of the above. If sufficient care is taken, it will not be necessary to disconnect the instrument drive cables to release the panel.

16.2 Bulb is retained to rear of reflector unit by a separate bulb holder

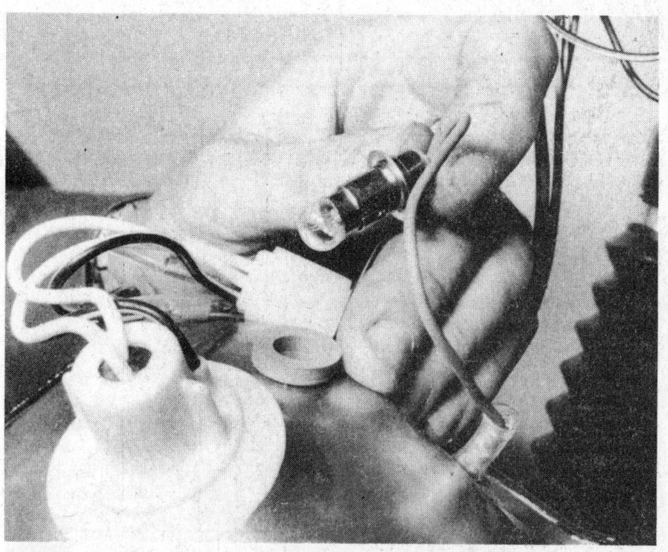

16.3 Parking lamp bulb is retained by rubber grommet

17.1a Remove cap nuts and screws to release instrument panel bottom half ...

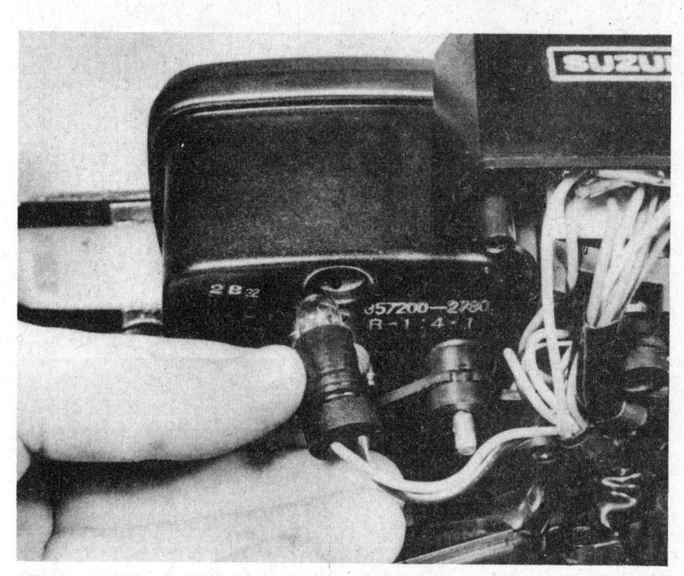

17.1b ... so that bulb holders can be withdrawn

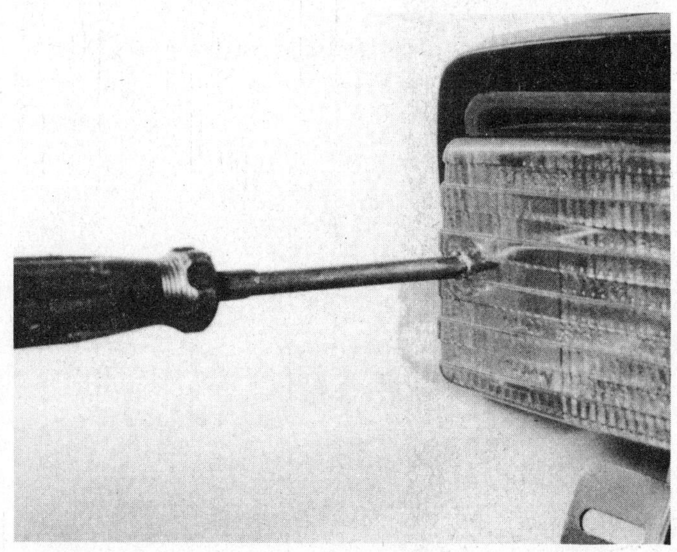

18.1a Remove screws to release tail lamp lens ...

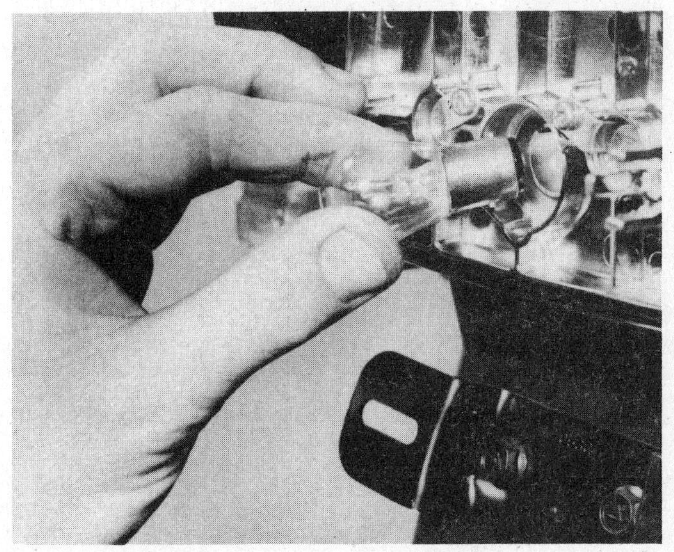

18.1b ... bulb has offset pins to prevent incorrect refitting

18 Stop/tail lamp: bulb renewal

1 Remove the two screws securing the lens and withdraw the lens, taking care not to tear the sealing gasket. Press the bulb in and twist it anti-clockwise to release it.
2 Remove all traces of moisture and corrosion from the interior of the lamp and the bulb holder, and renew the sealing gasket if it is damaged. Refit the bulb, noting that the pins are offset so that the bulb can be fitted only the correct way. Do not overtighten the lens retaining screws or the lens may crack.

19 Turn signal lamps: bulb renewal

1 Remove the two screws securing each lamp lens and withdraw the lens, taking care not to tear the sealing gasket. Press the bulb in and twist it anti-clockwise to release it.
2 Remove all traces of moisture and corrosion from the interior of the lamp and from the bulbholder, and renew the sealing gasket if it is damaged. Owners of DR125 S models may wish to improve the visibility of the turn signal lamps by lining the interior of the lamp with cooking foil.
3 Refitting is the reverse of the removal procedure. Do not overtighten the lens retaining screws or the lens may crack.

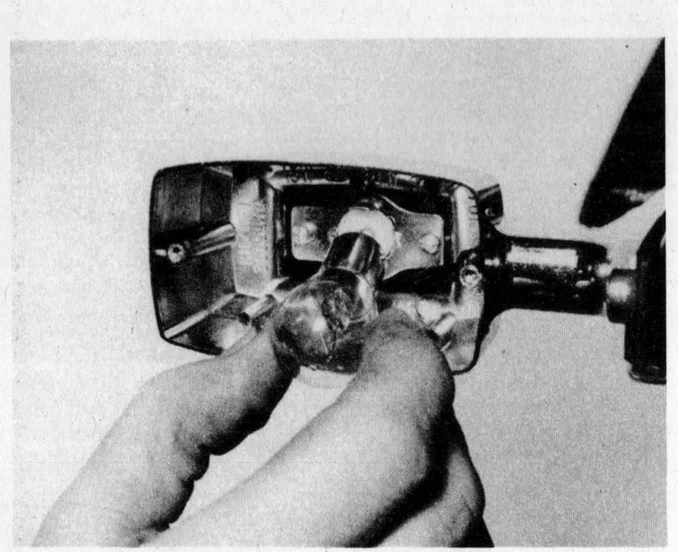

19.1a Turn signal bulb is a conventional bayonet fitting

19.1b Do not overtighten lens retaining screws

Chapter 7
The DR125 SF, SH, SJ 'Raider', GN125 and GZ125 'Marauder' models

Contents

Specifications

Specifications are given for the DR125SF, SH and SJ models only where different from that shown for the DR125SE model in Chapters 1 to 6. Specifications are given for the GN125 and the GZ125 models only where different from that shown for the GS125ES model in Chapters 1 to 6. Where a specification is not available the abbreviation N/Av will be found.

Model dimensions and weights

DR model:	
Overall length	2145 mm (84.5 in)
Overall width	845 mm (33.3 in)
Overall height	1190 mm (46.9 in)
Wheelbase	1390 mm (54.7 in)
Ground clearance	260 mm (10.2 in)
Dry weight	111 kg (245 lb)
GN model:	
Overall length	1945 mm (76.6 in)
Overall width	815 mm (32.1 in)
Overall height	1110 mm (43.7 in)
Wheelbase	1280 mm (50.4 in)
Ground clearance	175 mm (6.9 in)
Dry weight	107 kg (236 lb)
GZ model:	
Overall length	2160 mm (85.0 in)
Overall width	815 mm (32.1 in)
Overall height	1090 mm (42.9 in)
Wheelbase	1450 mm (57.1 in)
Ground clearance	140 mm (5.5 in)
Dry weight	125 kg (275 lb)

Routine maintenance

See Section 2 for service intervals

DR model:	
Tyre pressures – tyres cold:	
Front	22 psi
Rear	28 psi
Minimum compression pressure	128 psi (9 kg/cm2)
Front fork oil capacity – per leg	316 cc (11.12 fl oz)
Front fork oil level	170 mm (6.693 in)
Final drive chain free play	25 – 35 mm (1.0 – 1.4 in)
GZ model:	
Valve clearances (cold):	
Inlet	0.08 – 0.13 mm (0.003 – 0.005 in)
Exhaust	0.13 – 0.18 mm (0.005 – 0.007 in)
Tyre pressures – tyres cold:	
Front	25 psi
Rear	29 psi (solo), 33 psi (pillion)
Front fork oil capacity – per leg	369 cc (13.0 fl oz)
Front fork oil level	105 mm (4.1 in)
Final drive chain free play	5 – 15 mm (0.2 – 0.6 in)
GN model:	
Tyre pressures – tyres cold:	
Front	25 psi
Rear	29 psi (solo), 33 psi (pillion)
Front fork oil capacity – per leg	174 cc (6.13 fl oz)
Front fork oil level	166 mm (6.5 in)

Specifications relating to Chapter 1

Compression pressure

DR model:	
Nominal	156 – 213 psi (11 – 15 kg/cm2)
Service limit	128 psi (9 kg/cm2)
GZ model:	
Nominal	142 – 213 psi (10 – 15 kg/cm2)
Service limit	114 psi (8 kg/cm2)

Camshaft and rocker gear

DR model:	
Camshaft lobe height:	
Inlet	34.178 – 34.218 mm (1.3456 – 1.3472 in)
Service limit	33.878 mm (1.3338 in)
Exhaust	33.327 – 33.367 mm (1.3121 – 1.3137 in)
Service limit	33.027 mm (1.3003 in)
GZ model:	
Camshaft lobe height:	
Inlet and exhaust	33.120 – 33.160 mm (1.3039 – 1.3055 in)
Service limit	32.820 mm (1.2921 in)
GN model:	
Camshaft journal oil clearance	0.021 – 0.055 mm (0.0008 – 0.0022 in)
Camshaft journal OD	21.970 – 21.991 mm (0.8650 – 0.8658 in)

Valves, guides and springs

DR models:	
Valve lift:	
Inlet	8.0 mm (0.315 in)
Exhaust	6.8 mm (0.268 in)
Valve stem tip/collet groove minimum length	2.6 mm (0.102 in)
Valve spring minimum free length:	
Inner	36.0 mm (1.417 in)
Outer	39.3 mm (1.547 in)
Valve spring pressure:	
Inner	7.8 – 9.2 kg @ 32.5 mm (17.20 – 20.28 lb @ 1.28 in)
Outer	18.9 – 22.3 kg @ 36.0 mm (41.67 – 49.16 lb @ 10.42 in)
GZ models:	
Valve spring minimum free length:	
Inlet inner	35.1 mm (1.38 in)
Inlet outer	39.9 mm (1.57 in)
Exhaust inner	36.0 mm (1.42 in)
Exhaust outer	39.3 mm (1.55 in)
Valve spring pressure:	
Inlet inner	7.1 – 9.2 kg @ 32.5 mm (15.56 – 20.28 lb @ 1.28 in)
Inlet outer	17.3 – 21.3 kg @ 36.0 mm (38.14 – 46.96 lb @ 1.42 in)
Exhaust inner	7.8 – 9.2 kg @ 32.5 mm (17.20 – 20.28 lb @ 1.28 in)
Exhaust outer	18.9 – 22.3 kg @ 36.0 mm (41.67 – 49.16 lb @ 1.42 in)
Valve clearances (cold):	
Inlet	0.08 – 0.13 mm (0.003 – 0.005 in)
Exhaust	0.13 – 0.18 mm (0.005 – 0.007 in)

Clutch

DR model:	
Adjuster screw setting (see Ch. 1.37)	0 – 1/4 turn back
GZ model:	
Adjuster screw setting (see Ch. 1.37)	1/8 – 1/4 turn back

Gearbox

DR model:	
Gear ratios (number of teeth):	
1st	3.000:1 (33/11T)
2nd	1.857:1 (26/14T)
3rd	1.368:1 (26/19T)
4th	1.095:1 (23/21T)
5th	1.923:1 (24/26T)
6th	0.833:1 (20/24T)
GZ model:	
Gear ratios (number of teeth):	
1st	3.000:1 (33/11T)
2nd	1.857:1 (26/14T)
3rd	1.368:1 (26/19T)
4th	1.095:1 (23/21T)
5th	0.913:1 (21/23T)
6th	-

Final drive

DR model:	
Drive chain:	
Number of links	134
Free play	25 – 35 mm (1.0 – 1.4 in)
Reduction ratio	3.312:1 (53/16T)
GZ model:	
Drive chain:	
Number of links	132
Free play	5 – 15 mm (0.2 – 0.6 in)
Reduction ratio	3.470:1 (59/17T)
GN model:	
Drive chain:	
Number of links	114
Reduction ratio	3.000:1 (42/14T)

Torque settings

DR models	Revised torque wrench settings are not available
GZ models:	
Cylinder head cover bolt	1.0 kgf m (7.0 lbf ft)
Cylinder head nut (8 mm)	2.7 kgf m (19.5 lbf ft)
Cylinder head nut (6 mm)	1.0 kgf m (7.0 lbf ft)
Camshaft sprocket bolts	1.1 kgf m (8.0 lbf ft)
Valve clearance adjuster lock nut	1.4 kgf m (10.0 lbf ft)
Rocker arm shaft bolt	0.9 kgf m (6.5 lbf ft)
Cam chain tensioner adjuster bolt	0.7 kgf m (5.0 lbf ft)
Cylinder barrel retaining nuts	1.0 kgf m (7.0 lbf ft)
Alternator rotor nut	5.5 kgf m (40.0 lbf ft)
Primary drive gear/oil pump drive gear nut	5.0 kgf m (36.0 lbf ft)
Clutch centre retaining nut	4.0 kgf m (29.0 lbf ft)
Selector arm stopper bolt	1.9 kgf m (13.5 lbf ft)
Engine oil drain plug	2.8 kgf m (20.0 lbf ft)
Gearbox sprocket nut	9.0 kgf m (65.0 lbf ft)
Engine mounting bolt nuts	4.1 kgf m (29.5 lbf ft)
Exhaust pipe/cylinder head bolts	1.1 kgf m (8.0 lbf ft)
Silencer mounting bolt	1.1 kgf m (8.0 lbf ft)
Starter clutch bolts	1.0 kgf m (7.0 lbf ft)

Specifications relating to Chapter 2

Fuel tank capacity

DR model:	
Overall	13.0 lit (2.9 gal)
GZ model:	
Overall	14.0 lit (3.1 gal)
GN model:	
Overall	10.0 lit (2.2 gal)

Carburettor

DR model:	
Make and type	Mikuni VM24SS
Bore	24 mm
ID number	44A00
Main jet	100
Main air jet	1.4 mm
Jet needle	5JP15
Clip position – grooves from top	3rd
Needle jet	N-8
Throttle valve cutaway	1.5
Pilot jet	30
Pilot air jet	1.1 mm
By-pass	0.9 mm
Pilot outlet	0.7 mm
Valve seat	2.0 mm
Starter jet	35
Pilot mixture screw – turns out	5/8
Float height	24.5 ± 1.0 mm (0.97 ± 0.04 in)
GZ model:	
Make and type	Mikuni BS26SS
Bore	26 mm
ID number	12F0
Main jet	110
Jet needle	5D97-54
Clip position – grooves from top	4th
Needle jet	P-0
Pilot jet	12.5
Pilot mixture screw – turns out	1 5/8
Float height	17.1 ± 0.5 mm (0.67 ± 0.02 in)
GN model:	
Make and type	Mikuni BS26SS
Bore	26 mm
ID number	05311
Main jet	102.5
Jet needle	4DZ35
Clip position – grooves from top	2nd
Needle jet	P-2
Pilot jet	35
Pilot air jet	180
Valve seat	2.0 mm
Starter jet	25
Pilot mixture screw – turns out	1 7/8
Float height	21.4 ± 0.5 mm (0.84 ± 0.02 in)

Specifications relating to Chapter 3

Ignition system

DR model:	
Ignition timing:	
Up to 1950 rpm	13° BTDC
Above 3800 rpm	37° BTDC

Flywheel generator

DR model:	
Pulser coil resistance (blue/yellow to green/white)	94 – 140 ohm
Source coil resistance (brown to pink)	97 – 146 ohm
GZ model:	
Pulser coil resistance (blue to green)	150 – 300 ohm
GN model:	
Pulser coil resistance (black/blue to green)	90 – 120 ohm

Ignition HT coil

DR model:	
Primary winding resistance (white/blue to black/white)	0 – 0.5 ohm
Secondary winding resistance (plug cap to white/blue)	13 – 15 K ohm
GZ model:	
Primary winding resistance (negative terminal to positive terminal)	3 – 5 ohm
Secondary winding resistance (positive terminal toplug cap)	17 – 28 K ohm
GN model:	
Primary winding resistance (white to orange/white)	2 – 5 ohm
Secondary winding resistance (orange/white toplug cap)	15 – 25 K ohm

Spark plug

GN and GZ models:

Type	NGK DR8EA or ND X24ESR-U

Specifications relating to Chapter 4

Front forks

DR model:	
Wheel travel	205 mm (8.1 in)
Fork spring minimum free length	511 mm (20.118 in)
Fork oil level	170 mm (6.693 in)
Fork oil quantity – per leg	316 cc (11.12 fl oz)
GZ model:	
Wheel travel	120 mm (4.7 in)
Fork spring minimum free length	301 mm (11.9 in)
Fork oil level	105 mm (4.1 in)
Fork oil quantity – per leg	369 cc (13.0 fl oz)
Fork oil type	SAE 10
GN model:	
Fork spring minimum free length	386 mm (15.2 in)
Fork oil level	166 mm (6.5 in)
Fork oil quantity – per leg	174 cc (6.13 fl oz)
Fork oil type	SAE 10

Rear suspension

DR model:	
Wheel travel	200 mm (7.9 in)
Suspension unit spring set length	226 mm (8.898 in)
Swinging arm and suspension linkage dimensions	N/Av
GZ model:	
Wheel travel	90 mm (3.5 in)
Swinging arm pivot bolt max runout	0.3 mm (0.01 in)

Torque wrench settings:

DR models	Revised torque wrench settings are not available
GZ models:	
Handlebar clamp bolt	1.6 kgf m (11.5 lbf ft)
Handlebar holder nut	4.5 kgf m (32.5 lbf ft)
Steering stem top bolt	6.5 kgf m (47.0 lbf ft)
Front fork top yoke pinch bolt	2.3 kgf m (16.5 lbf ft)
Front fork lower yoke pinch bolt	3.3 kgf m (24.0 lbf ft)
Front fork top bolt	2.3 kgf m (16.5 lbf ft)
Front fork damper rod Allen bolt	2.0 kgf m (14.5 lbf ft)
Rider's footrest bolt	2.6 kgf m (19.0 lbf ft)
Swinging arm pivot nut	7.2 kgf m (52.0 lbf ft)
Torque arm nut (front and rear)	1.3 kgf m (9.5 lbf ft)
Rear suspension unit mounting bolt or nut	2.9 kgf m (21.0 lbf ft)

Specifications relating to Chapter 5

Brakes

DR model:	
Type – front	Hydraulic disc (refer to information for GS125ES in Chapter 5)
Brake disc thickness:	
Standard	3.3 – 3.7 mm (0.13 – 0.15 in)
Service limit	3.0 mm (0.12 in)
GZ model:	
Type – front	Hydraulic disc (refer to information for GS125ES in Chapter 5)
Brake disc thickness:	
Standard	5.0 ± 0.2 mm (0.20 ± 0.01 in)
Service limit	4.5 mm (0.18 in)

Wheels

GN model:	
Type	Wire spoked
Size	N/Av
GZ model:	
Type	Wire spoked
Size:	
Front	16 x MT2.50
Rear	15 x MT3.00

Tyres

	Front	Rear
DR model:		
Size	80/80 – 21 45P	100/80 – 18 53P
Pressures – solo and with pillion	22 psi	28 psi
GZ model:		
Size	110/90 – 16 59P	130/90 – 15M/C 66P
Pressures	25 psi	29 psi (solo), 33 psi (with pillion)
GN model:		
Size	2.75 – 18 4PR	3.50 – 16 4PR
Pressures	25 psi	29 psi (solo), 33 psi (with pillion)

Torque wrench settings

DR model	Revised torque wrench settings are not available
GZ model:	
Front wheel axle pinch bolt	2.3 kgf m (16.5 lbf ft)
Front wheel axle	6.5 kgf m (47.0 lbf ft)
Rear wheel axle nut	7.8 kgf m (56.5 lbf ft)
Front brake caliper mounting bolts	3.9 kgf m (28.0 lbf ft)
Front brake pad retaining bolt	1.8 kgf m (13.0 lbf ft)
Front brake hose union bolt	2.3 kgf m (16.5 lbf ft)
Front brake caliper bleed nipple	0.75 kgf m (5.5 lbf ft)
Front brake master cylinder clamp bolts	1.0 kgf m (7.0 lbf ft)
Front brake disc bolts	2.3 kgf m (16.5 lbf ft)
Rear sprocket nuts	5.0 kgf m (36.0 lbf ft)
Rear brake cam lever nut	1.0 kgf m (7.0 lbf ft)

Specifications relating to Chapter 6

Battery

GZ model:	
Type	Maintenance-free
Specification	FTX7L-BS
Capacity	12V 21.6 kC (6 Ah)/10HR

Fuse rating

GZ model	20A

Alternator

DR model:	
No load voltage	58 volts or more @ 5000 rpm
Regulated voltage	13.5 – 15.5 volts @ 5000 rpm
Coil resistance (yellow to yellow)	0.1 to 1.0 ohm
GZ model:	
No load voltage	45 volts or more @ 5000 rpm
Regulated voltage	13.5 – 15.0 volts @ 5000 rpm
Coil resistance (yellow to yellow)	0.3 to 1.5 ohm

Starter motor

GZ model:	
Brush minimum length	N/Av
Relay resistance	3 – 6 ohms

Bulbs

DR model:	
Headlamp	12V, 45/40W
Parking lamp	12V, 4W
Main beam warning lamp	12V, 1.7W
GZ model:	
Headlamp	12V, 60/55W
Parking lamp	12V, 4W
Main beam warning lamp	12V, 1.7W
Stop/tail lamp	12V, 21/5W
Turn signal lamp	12V, 21W
Instrument and high beam warning lamps	12V, 1.7W
Neutral and turn signal warning lamps	12V, 3.4W
GN model:	
Headlamp	12V, 45/45W
Parking lamp	12V, 3.4W
Stop/tail lamp	12V, 21/5W
Turn signal lamp	12V, 21W
Instrument and all warning lamps	12V, 3.4W

1 Introduction

The preceding Chapters of this manual cover the Suzuki GS125 and GS125ES models, also the early DR125S models imported into the UK from January 1982 to May 1986. This Chapter describes the later DR125S models imported from June 1985 onwards, the GN125 which was introduced into the UK in August 1993 and also the GZ125 which was introduced in the UK in July 1998.

Model changes

The DR125S, commonly known as the Raider model can be identified easily by its hydraulically-operated disc front brake. Apart from the modifications described later in this Chapter, the Raider models feature heavily revised styling, including the fitting of a larger-capacity fuel tank with a safety seat, a larger headlamp/instrument cowling, revised mudguards with an enduro-type toolbag on the rear, and revised side panels, turn signal lamps and tail lamp.

Engine response is improved by the fitting of a larger carburettor and a side-branch exhaust pipe. Vibration is reduced by partially rubber-mounting the engine and exhaust system, but chiefly by fitting rubber mountings to the handlebars and footrests. The instruments are simplified by the removal of the tachometer.

The principal alterations to the DR model, however, are to the cycle parts. The front forks feature longer travel and separate bushes that can be renewed to compensate for wear. The frame rear section is revised and the rear suspension, although still called 'Full Floater', is completely altered. The single suspension unit is now bolted directly to the frame at its upper end (and is therefore no longer 'fully-floating') and is attached to the swinging arm via a cast alloy bellcrank. Needle roller bearings with steel inner sleeves are fitted to the swinging arm/bellcrank pivots; note that the bellcrank is now fitted underneath the swinging arm to reduce the overall height of the assembly and that it completely replaces the rocker and link arm of the original system.

The front end of the bellcrank (or 'rear cushion lever' as Suzuki describe it) is connected to the frame by way of a pivot bolt supported by a spherical bearing. At the rear of the bellcrank is a plain forked end to which the suspension unit lower mounting is attached. Between these two lies a larger bore containing a needle roller race and a hard steel spacer. The spacer is connected to an extension of the swinging arm via a pivot bolt which passes through an eccentric bore. It is the spacer, which is allowed to rotate as the suspension moves, which provides the rising-rate effect. It does so by effectively altering the distance between the bellcrank to swinging arm bolt and the suspension unit lower mounting, and thus changing the leverage ratio.

The hydraulic disc front brake is of a similar design to that fitted to the GS125ES models.

The GZ125 Marauder shares its name with the larger capacity VZ800 Marauder. In true cruiser style it has a low-slung seat, wire spoked wheels, forward mounted footrests, deep flared mudguards and a single instrument pod.

The GN125 shares many components with the GS125ES covered in the preceding chapters of this manual. Use of wire spoked wheels and chrome plating on the mudguards, exhaust, headlamp shell, instruments, rear shock absorbers and grab rail give the machine a definite 'classic' look.

Model identification

As with the earlier models, modified versions have appeared during their life. Since there are only slight cosmetic variations between each of these versions they can be identified only by their frame numbers. Given below are each version's full model designation, the initial frame number with which its production run commenced and the approximate date of import into the UK.

Model	Frame number	Dates
DR125SF	SF43B – 100001 on	June 1985 to February 1987
DR125SH	SF43B – 103042 on	February 1987 to March 1988
DR125SJ	SF43B – 104350 on	March 1988 to November 1993
GN125R	NF41A – 145362 on	August 1993 to November 1996
GN125V	NF41A – 232733 on	November 1996 to August 1997
GN125W	NF41A – 248475 on	August 1997 to November 1998
GN125X	NF41A – 268499 on	November 1998 on
GZ125W	JS1AP111200100001 on	July 1998 to January 1999
GZ125X	JS1AP111200100469 on	January 1999 on

Note that this Chapter only describes those differences in the later models which require additional information or a modification in working procedure; in all other respects the machines are similar to the earlier models. If working on one of these later models, therefore, check with this Chapter first to establish whether there is any alteration in the information or procedure required; if not the task is substantially the same as that described in the relevant part of Chapters 1 to 6. For DR125S models, refer back to the DR125S information in previous chapters unless listed here. For GN and GZ models, refer back to the GS125ES information in previous chapters unless listed here.

2 Routine maintenance: revised service intervals

Owners should note that the main service intervals have been considerably revised as shown below. Carry out the tasks as described in the Routine maintenance section at the front of this Manual, with reference as necessary to this Chapter or to Chapters 1 to 6 for further information. Service intervals are as follows:

Pre-ride check – as described in Routine maintenance

Monthly or every 600 miles (1000 km)

Check, adjust and lubricate the final drive chain. Reduce interval if the machine is used off-road.

Three monthly, or every 1000 – 1250 miles (1600 – 2000 km)

Additional engine/transmission oil change – even though the service intervals have been reduced, the extra oil change is still advisable at approximately twice the manufacturer's recommended frequency.

Every 2500 miles (4000 km)

Change the engine oil and filter
Check the valve clearances
Check the spark plug
Adjust the cam chain tension
Clean the air filter element
Check the tightness of all nuts, bolts and other fasteners
Check the fuel feed pipe and carburettor settings
Check the battery (not GZ125)
Check the clutch adjustment
Check the brakes and brake hose – as described for GS125ES models
Check the tyres
Check the steering
Check the front forks

Every 5000 miles (8000 km)

Repeat all the previous maintenance operations, then carry out the following:
Renew the spark plug
Clean the fuel tap filter bowl (not GZ125)

Additional routine maintenance items

The fuel feed pipe, front brake hose and front brake hydraulic fluid should all be renewed at least at the intervals given in Routine maintenance for the GS125ES models.

Replace the air cleaner element every 7500 miles (12 000 km)

Service intervals – time

Suzuki have set certain time intervals, to be applied for those machines which are used infrequently; the service is to be carried out at the mileage interval or at the time interval, whichever comes first. These intervals are as follows:
2500 mile check, or every 20 months
5000 mile check, or every 40 months

In the author's opinion these time intervals are too infrequent and owners are recommended to use the six-monthly/annual time intervals instead, as described in Routine maintenance.

3.6a Pads are retained by two bolts and a double lock washer – DR model

3.6b Remove the bolts and lift away the fixed pad . . .

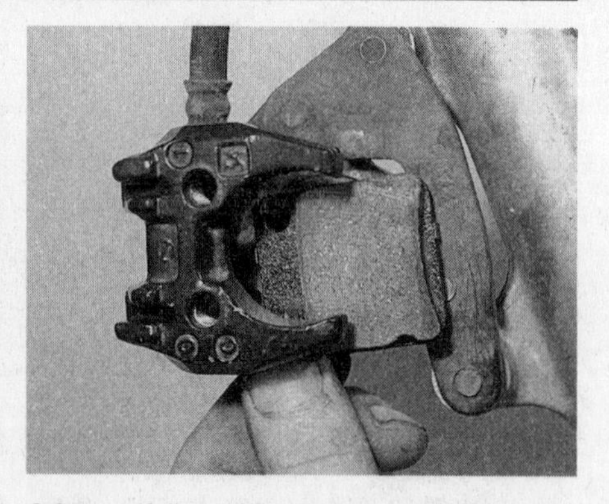

3.6c . . . followed by the moving pad – DR model

Observant owners may have noted that the revised manufacturer's schedule above does not include such important tasks as cleaning the oil pump filter gauze, changing the front fork oil, and greasing the rear suspension, steering head and wheel bearings. These tasks **must not** be overlooked; owners are recommended to carry them out at similar intervals to those given in Routine maintenance.

3 Routine maintenance: new procedures

1 When carrying out maintenance operations, owners should note the revised information which may be applicable, as given in the Specifications Section of this Chapter.

Front brake

2 All front brake components should be checked at the intervals given in Routine maintenance or Section 2 of this Chapter, as appropriate; in all cases procedures are similar to those described for GS125ES models.

3 The caliper and master cylinder designs differ somewhat in appearance, but the operating principle remains the same. It is worth noting that single-piston, or 'floating' calipers are particularly prone to mechanical wear of the axle pins on which the caliper body moves, or to sticking, due to corrosion of the same parts. This will be more evident on a machine used regularly off-road, which indicates a need for regular cleaning and inspection. Mud trapped in and around the caliper should be carefully removed when cleaning the machine after trail riding. **Do not** aim a hose or pressure washer directly at the caliper; dirt will be forced past its seals and into the working surfaces.

4 When checking the pads it is recommended that the caliper is detached from the fork leg to permit thorough cleaning of the piston area; if new pads are required the piston will have to be pushed back into the caliper bore to accept them and it is important that no dirt is allowed to enter the dust seal. Check the condition of the piston dust seal and look for indications of hydraulic fluid seepage. Check the condition of the dust seal on the caliper axle pin. If any seal appears to be worn or perished it is best to strip and overhaul the caliper, fitting new seals on reassembly. This can be regarded as preventative maintenance, and may save the cost of renewing a prematurely worn caliper.

5 Check the amount of friction material remaining on each brake pad by looking closely at them from above, standing astride the machine. Pads must usually be renewed when the friction material only (not including the metal backing) is worn to a thickness of 1.0 – 1.5 mm (0.04 – 0.06 in) or less at any point; on these machines this limit is indicated by either a red-painted groove cut around the periphery of each pad's friction material or by a deep notch cut in the 'upper' and 'lower' edges of the friction material, depending on the type of pads fitted. If either pad is so worn that the red-painted groove is in contact with the disc at any point, or if the deep notches have been worn away (as applicable), then both pads must be renewed immediately. If the wear limit marks cannot be seen, the pads must be removed for thorough cleaning and checking.

6 To renew the pads on DR models, flatten back the raised tabs of the retaining bolt lockwasher (working 'through' the front wheel), then slacken the retaining bolts and withdraw the caliper assembly from the fork leg. Unscrew the retaining bolts and withdraw the pads, noting the anti-squeal shim fitted behind the 'moving' pad (that fitted next to the piston) (**see illustrations**). Since the pads are identical it is permissible to swap them over to even out wear, if required, but note that the reason for such uneven wear should be found and rectified first. Note that the anti-squeal shim must be fitted so that its arrow points in the direction of disc rotation.

7 To renew the pads on GZ models, loosen the brake pad retaining bolt, which is located just above the brake hose union. Remove the two caliper mounting bolts which are located on the front fork and withdraw the caliper from the disc. Fully unscrew the pad retaining bolt

3.6d Note that the moving pad has anti squeal shim fitted to the backing metal as shown – DR model

3.6e Anti rattle pad spring fits between caliper jaw as shown – DR model

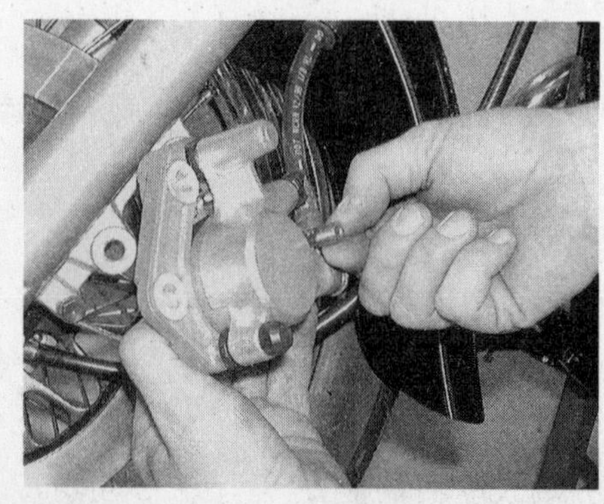

3.7 Remove pad retaining bolt to free the pads – GZ model

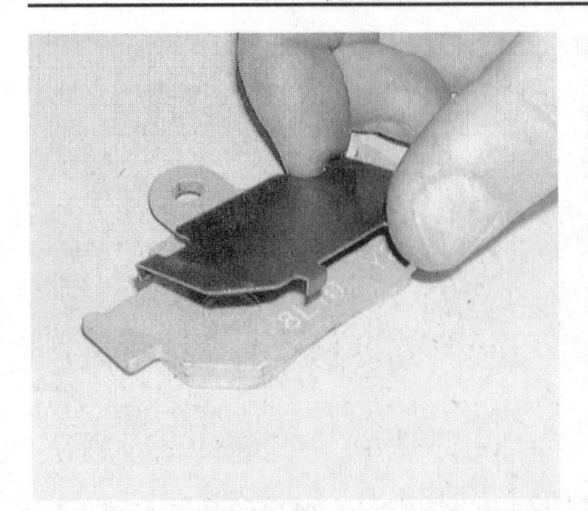
3.9a Anti squeal shim fits on the back of the brake pad - GZ model

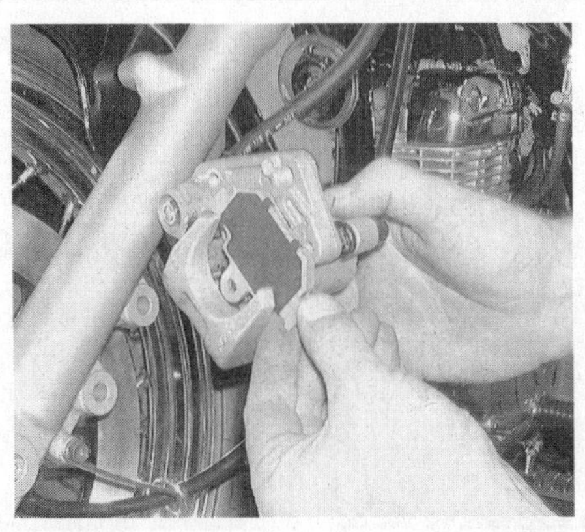
3.9b Fitting the brake pads back into the caliper - GZ model

3.10 Snail cam chain adjusters are fitted - DR model

and remove the brake pads **(see illustration)**. **Note:** *Do not operate the brake lever during this procedure.* Always renew the pads as a set.

8 On DR models, apply a thin smear of brake caliper grease to the caliper axle bolt and support pin, also to the edge of each pad's metal backing and to the plain shanks of the pad retaining bolts; be careful to use only the thinnest smear of grease and to keep it away from the disc or friction material. Tighten securely all retaining bolts to the specified torque (where applicable). Lock each pad retaining bolt by bending up against one of its flats an unused lock washer tab.

9 On GZ models, install the new pads making sure that the anti-squeal shims are fitted to the rear of the pads; secure the pads with the pad retaining bolt and tighten it to the specified torque **(see illustrations)**. Install the caliper on the disc and tighten the caliper mounting bolts to the specified torque. Once you have fitted the brake pads, pump the brake lever a few times to bring the pads back into contact with the disc, then check the brake fluid level.

Fig. 7.1 Air filter - DR model

1 Casing lid
2 Wing bolt
3 Washer
4 Element
5 Supporting frame
6 Filter inner
7 Casing
8 Screw - 3 off
9 Spring washer - 3 off
10 Spring clamp
11 Carburettor mounting stub
12 Clamp
13 Breather pipe
14 Clamp
15 Clip
16 Drain pipe

Final drive chain adjustment - DR model

10 The adjustment procedure is essentially as described in Routine maintenance but note the revised chain free play specification. On these later models, snail cam chain adjusters are fitted **(see illustration)**. To adjust the chain slacken the spindle nut and rotate each adjuster anti-clockwise (looking at it from the side of the machine) to push the wheel backwards and tighten the chain. To ensure accurate wheel alignment ensure that the same notch in each adjuster is aligned with its respective stopper pin; the notches are numbered to facilitate this. Tighten the spindle nut to the specified torque and re-check the setting. All other chain maintenance procedures are as described for the early models.

Cleaning the air filter - DR model

11 Remove the two screws which secure each side panel, then carefully detach both panels. Remove the two bolts, one on each side at the rear, which secure the seat, then lift it up at the rear and withdraw it. Unclip the lid of the air filter casing and withdraw it **(see illustrations)**. The element can then be released by unscrewing the wing bolt which secures it to the casing (see Fig. 7.1). The element

3.11a Air filter casing lid is clipped into position - DR model

3.11b Element assembly is secured by a wing bolt – DR model

3.12a Air filter cover screws – GZ model

3.12b Removing the air filter element – GZ model

4.3 Cylinder head cap covers – GZ model

itself can be cleaned as described for the earlier models in Routine maintenance.

Cleaning the air filter – GZ model

12 Remove the rider's seat and the left frame cover as described in Section 23. Remove the four screws which hold the filter cover in place and remove the filter **(see illustrations)**. Clean the filter by using compressed air. **Note:** *Always apply the compressed air to the outside of the filter. If it is applied to the inside of the filter the dirt blocks the filter pores, which restricts the air flow.* When cleaning the air filter remove the air cleaner drain hose plug, to enable any water to drain away.

13 Refitting is reverse of removal, noting that the triangular marks on the cover and casing must be aligned.

Adjusting the steering head bearings – DR and GZ models

14 The modified steering head bearings, once set up as described later in this Chapter, are adjusted using the method described in Routine maintenance (for GZ models see section 16 of this chapter).

Rear suspension – DR model

15 The manufacturer's recommended interval for checking and lubricating the rear suspension remains unchanged at 3000 miles (5000 km), as shown in Routine maintenance. However in view of the number of bearings and pivots involved and their exposed position, the author suggests the assembly be stripped for examination and lubrication fairly frequently. The exact interval must depend upon the use to which the machine is put; if used solely for road work, the rate of wear is unlikely to be very great, but frequent off-road use is likely to cause accelerated deterioration of the bearings if lubrication is neglected. As a guide, the used example shown in the photographs had no signs of abuse, or even heavy use, yet there were definite indications of wear in the suspension pivots and bearings after 6000 miles.

4 Engine modifications: general

Tachometer drive – DR and GZ models

1 Note that the tachometer has been deleted and with it the various components of its drive mechanism. The task of removing and refitting the cylinder head cover is therefore simplified.

Cylinder head fasteners – DR model

2 With reference to Fig. 1.21, note that the cap nut on the front (exhaust side) right-hand cylinder head stud is replaced by a standard-pattern flanged nut. The arrangement of steel and copper washers remains unchanged.

Cylinder head side covers – GZ model

3 The engine has been fitted with decorative chrome covers on both sides of the cylinder head. The covers are fixed to their mounting brackets by two bolts **(see illustration)**. If removing the cylinder head cover (rocker cover), note how the brackets are secured to each side of the cover.

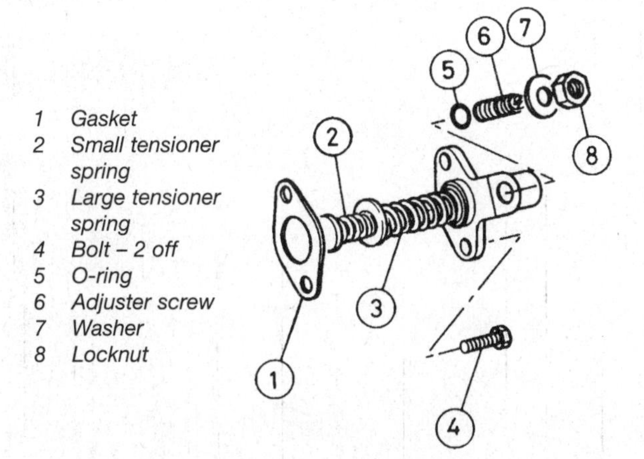

Fig. 7.2 Cam chain tensioner assembly – DR model

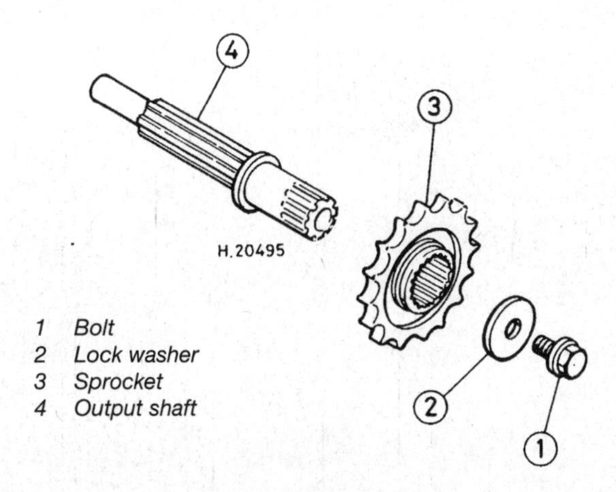

Fig. 7.3 Gearbox sprocket retainer – DR model

Cam chain tensioner – DR model

4 The cam chain tensioner assembly has been modified to incorporate a double plunger with two separate springs (see Fig. 7.2). If dismantling the unit, note the correct position of all components as a guide to reassembly. Note that the two springs are listed separately and can thus be renewed if they are thought to be faulty. Adjustment is carried out as described in *Routine maintenance* at the beginning of this manual.

Gearbox sprocket retainer – DR model

5 The gearbox sprocket is now retained by a single bolt threaded into the output shaft end. If the bolt is disturbed at any time its threads and those of the shaft should be cleaned carefully, and it should be tightened securely with a drop of thread locking compound such as Loctite 242; note that a torque wrench setting is not available. If the large washer is of the Belleville type (ie conical) it must be refitted with its convex surface outwards (see Fig. 7.3).

Engine mountings – DR model

6 The crankcase lower rear mounting is modified to accept bonded rubber bushes, to reduce the level of engine vibration transmitted to the frame. These bushes should not require attention during the interval between engine overhauls, but should be checked for wear each time the engine unit is stripped. If renewal is necessary, it is advisable to use a drawbolt arrangement to extract the old bushes and to fit the new ones. Failing this, use a long drift passed through from one side to drive out the opposite bush. When fitting the new bushes, ensure that they enter their bore squarely.

6.1 Fuel tap – GZ model

5 Fuel tank: removal and reassembly – DR model

Note that the fuel tank is retained by two bolts at the rear, as described for the GS models in Chapter 2.

6 Fuel tap: general – GZ model

The fuel tap fitted to this model is similar to the GS model apart from it not having a filter bowl. To clean the internal filter follow the procedure covered in Chapter 2 **(see illustration)**.

7 Carburettor: general – DR model

The later models use a Mikuni carburettor which is similar in design to that fitted to the GS models. Servicing procedures are as described in Chapter 2 and Routine maintenance for these models, noting the different settings given in the Specifications Section of this Chapter. Refer to Fig. 7.4 for details of the modifications to the throttle valve and choke assemblies.

8 Carburettor: removal, overhaul and refitting – GZ model

1 Remove the fuel tank. Slacken the screws of the clamps which secure the carburettor to the inlet stub and air filter hose and manoeuvre it free from the left-hand side of the machine. Disconnect the wiring from the throttle position sensor.
2 Pull back the rubber cap from the end of the choke plunger retaining nut and unscrew the nut. Withdraw the cable and choke plunger completely. Slacken the locknuts of the throttle cable adjuster to create slack in the throttle cable, then disconnect the cable nipple from the pulley.
3 Remove the two retaining screws from the carburettor top cap and remove the cap (take care that the spring does not fly out of the carburettor). Remove the spring and then carefully remove the piston valve along with the diaphragm. Lift the spring seat and jet needle out of the piston (see Fig. 7.5).
4 Turn the carburettor over and remove the float chamber by withdrawing the two retaining screws.
5 Remove the screw and pivot pin which hold the float assembly in place and lift out the float together with the needle valve. Unscrew and remove the main jet and the jet holder (noting its O-ring), followed by the adjacent pilot jet. The needle jet is a press fit in the carburettor body.
6 Remove the pilot air jet followed by the pilot screw (see Fig. 7.5). **Note:** *Before removing the pilot screw, it's setting must be determined*

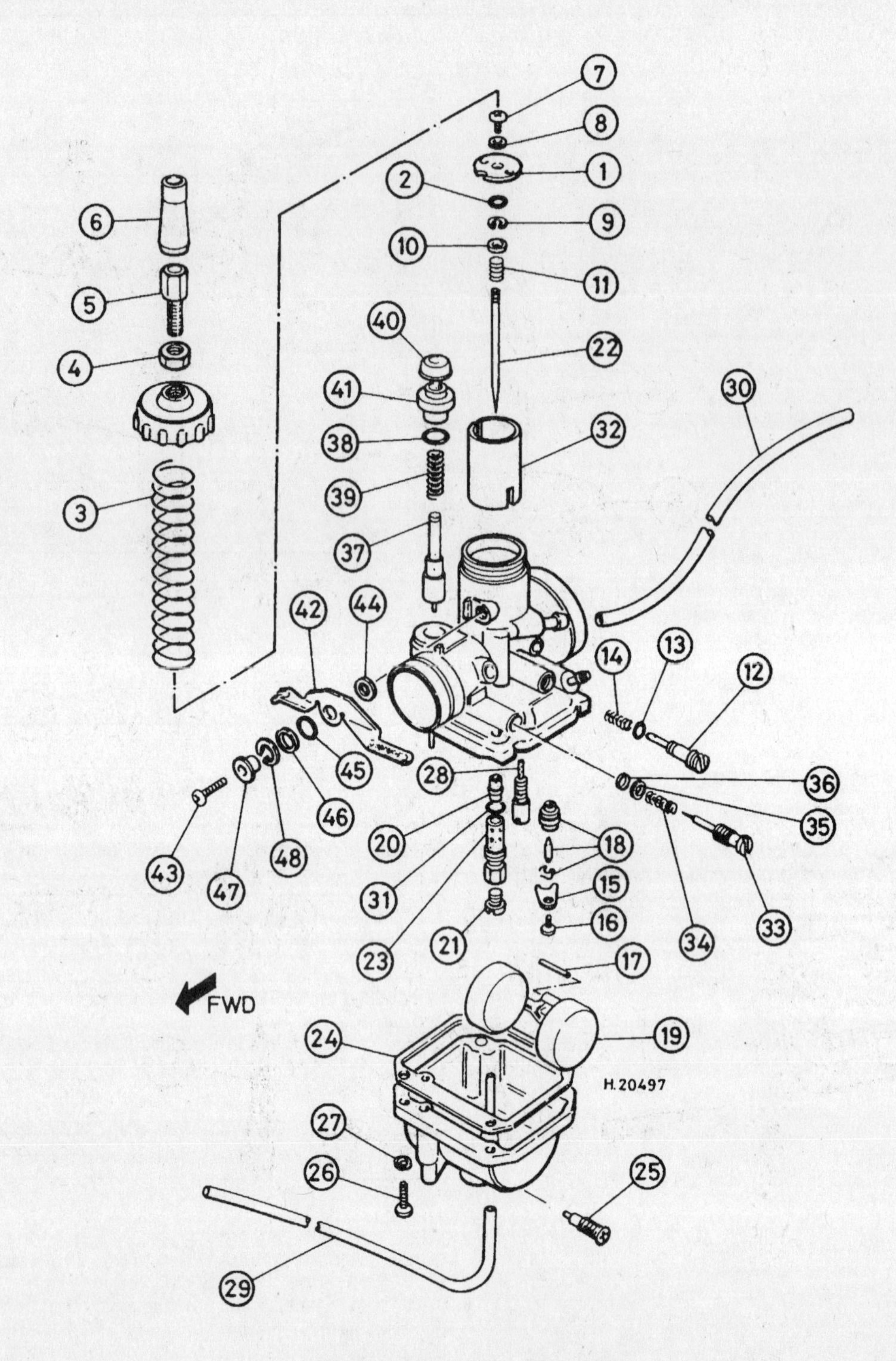

Fig. 7.4 Carburettor – DR model

1 Retaining plate
2 Ring
3 Return spring
4 Lock nut
5 Adjusting screw
6 Rubber cap
7 Screw
8 Spring washer
9 E-clip
10 Spacer
11 Spring
12 Throttle stop screw
13 O-ring
14 Spring
15 Retaining plate
16 Screw
17 Pivot pin
18 Float needle valve assembly
19 Float
20 Needle jet
21 Main jet
22 Jet needle
23 Main jet holder
24 Gasket
25 Drain screw
26 Screw – 4 off
27 Spring washer – 4 off
28 Pilot jet
29 Overflow pipe
30 Breather pipe
31 O-ring
32 Throttle valve
33 Pilot mixture screw
34 Spring
35 Washer
36 O-ring
37 Choke plunger
38 O-ring
39 Spring
40 Cap
41 Retainer
42 Choke lever
43 Screw
44 Washer
45 Ring
46 Washer
47 Ring
48 Spring washer

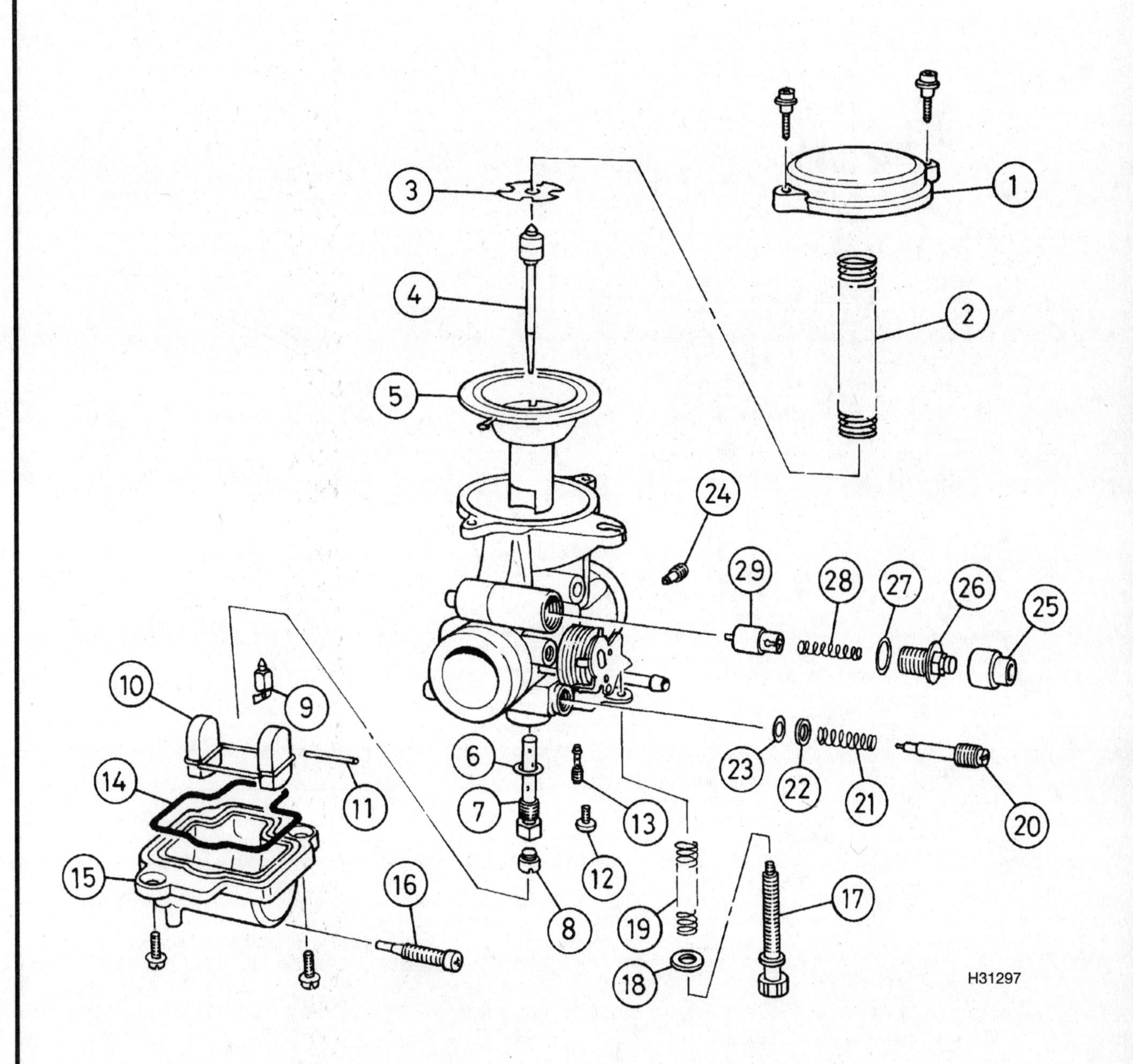

Fig. 7.5 Carburettor – GZ model

1 *Cap*
2 *Spring*
3 *Spring seat*
4 *Jet needle*
5 *Piston valve/diaphragm*
6 *O-ring*
7 *Main jet holder*
8 *Main jet*
9 *Float needle valve*
10 *Float*
11 *Pivot pin*
12 *Float pivot pin screw*
13 *Pilot jet*
14 *Sealing ring*
15 *Float chamber*
16 *Drain screw*
17 *Throttle stop screw*
18 *Washer*
19 *Spring*
20 *Pilot screw*
21 *Spring*
22 *Washer*
23 *O-ring*
24 *Pilot air jet*
25 *Cap*
26 *Choke plunger retaining nut*
27 *Washer*
28 *Spring*
29 *Choke plunger*

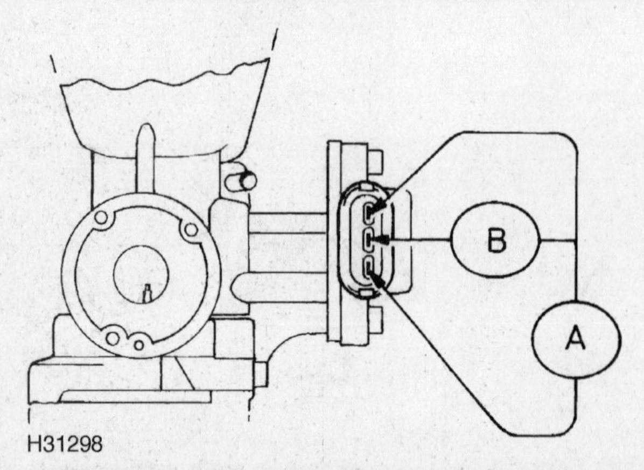

Fig. 7.6 Throttle position sensor test connections (see text) – GZ model

Fig. 7.7 Throttle position sensor test result graph – GZ model

so that it can be returned to the same position on installation. Slowly turn the pilot screw clockwise and count the number of turns until it is lightly seated, record the setting, then unscrew the pilot screw.

7 Clean all the carburettor parts with an aerosol type carburettor cleaner and dry the components with compressed air. Never use wire to clean jets or passageways, as the wire may cause damage. Always reassemble the carburettor with new gaskets and seals and note the following points.

8 When reassembling you will need to check the float height. Refer to Section 8 in Chapter 2 noting the float height specification at the beginning of this chapter.

9 If the pilot screw was disturbed, assemble its components correctly (see Fig. 7.5) and screw it into position until it seats lightly. Now back it out of the number of turns previously recorded.

10 If the throttle stop screw (tickover screw) has been removed, turn the screw until the throttle valve's bottom end is aligned with the foremost by-pass hole in the venturi.

11 Do not disturb the throttle position sensor unless absolutely necessary. If it has been removed, or a new sensor is being fitted, it must be set up correctly. Secure the sensor with its retaining screws but do not tighten them at this stage. Using an ohmmeter, measure and record the exact resistance (A) between the outer two terminals (see Fig. 7.6); the resistance reading should be between 3.5 – 6.5 K ohm. Now measure the resistance (B) between the two terminals shown in Fig. 7.6 whilst holding the throttle valve fully open by turning the throttle pulley by hand. The resistance measured should be 79.6% of that measured in test A. Refer to the graph in Fig. 7.7, noting the example of 5 K ohm in test A and 3.98 K ohm in test B. If the sensor needs adjusting, slacken its screws and rotate its body by a small amount as necessary, then tighten its screws. Check the resistance again and readjust if necessary.

12 When fitting the piston valve, align the projection on the diaphragm with the groove of the carburettor body.

9 Exhaust system: general – DR model

1 The side branch exhaust pipe does not affect servicing procedures except that its junctions must be checked frequently and re-painted to prevent the formation of rust-traps.

2 The silencer mountings are reduced to two, the front of which is rubber-mounted. This apart, the general comments made in Chapter 2.10 apply equally to these later models.

10 CDI/ignitor: general

1 The ignition system was revised for the later models and features a new CDI/ignitor unit (see Fig. 7.8 for DR model) with associated changes to the wiring. Unfortunately, no revised test information was available at the time of writing, so all that can be done in the event of a fault is to check by substituting a new ignitor unit, or to eliminate all other possible causes by testing the source and pulser coil resistances and checking that the ignition HT coil is functioning normally.

2 On DR models the CDI unit is mounted on the rear mudguard under

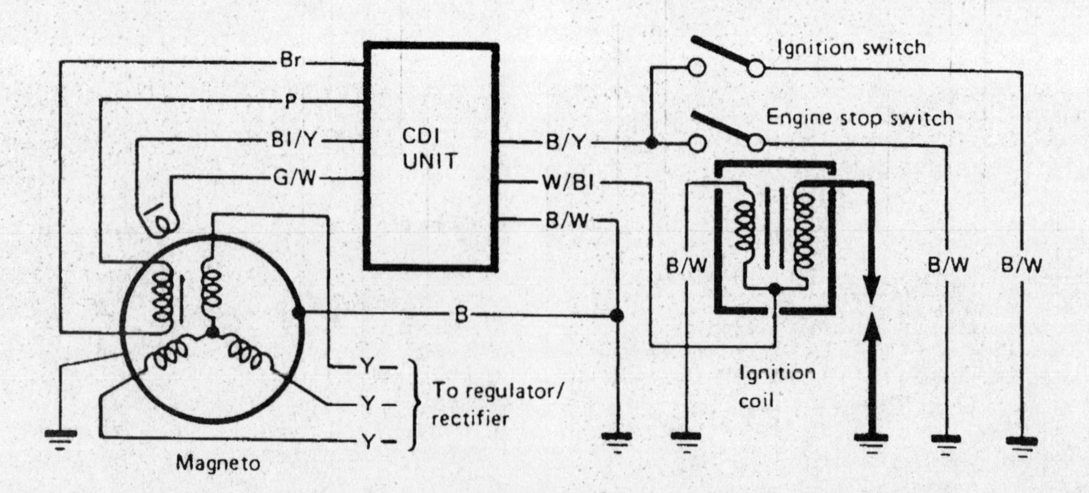

Fig 7.8 Ignition system circuit diagram – DR model

B Black *Bl* Blue *Br* Brown *G* Green *P* Pink *W* White *Y* Yellow

10.2a CDI unit location – DR model

the seat **(see illustration)**. On GN models the ignitor unit is located directly under the seat **(see illustration)**. On GZ models, remove the seat and luggage box to access the ignitor unit **(see illustration)**.

11 Ignition coil: testing - GZ model

1 The ignition coil location is under the fuel tank. It can be tested in the same way as described in Chapter 3 section 5, but note the following points regarding the test connections. Disconnect the black/yellow and black wires from the coil terminals (note their location first) and pull the plug cap off the spark plug.
2 To check the primary windings, connect the meter between the two wire terminals of the coil and note the reading **(see illustration)**.
3 To check the secondary windings, connect the meter between the positive (+ve) wire terminal (black/yellow wire terminal) and the spark plug cap and note the reading **(see illustration)**.
4 Refer to the specifications at the beginning of this chapter for expected test results.

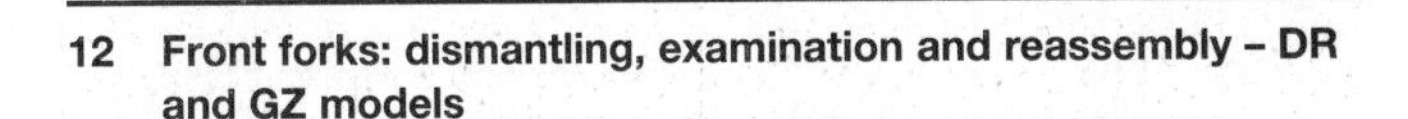

12 Front forks: dismantling, examination and reassembly - DR and GZ models

Refer to Fig. 7.9 and the accompanying photographs relating to the DR model.

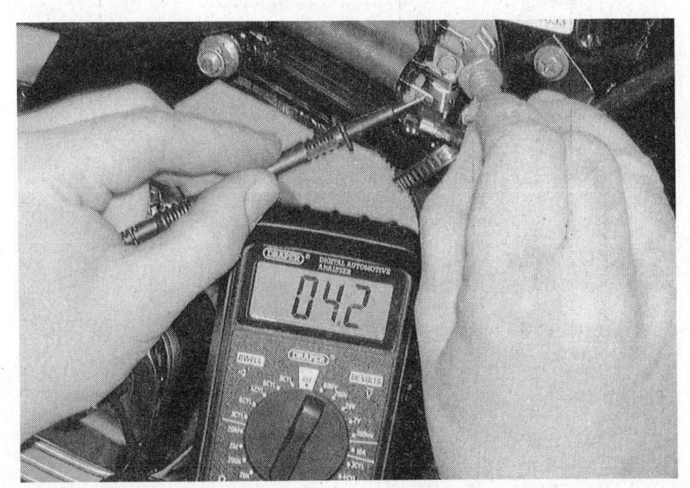

11.2 Measuring the ignition coil primary windings – GZ model

10.2b Ignitor unit location – GN model

10.2c Ignitor unit location – GZ model

1 On DR models dismantle the fork as described in Chapter 4, Section 3, paragraphs 1 to 6.
2 On GZ models, unscrew the fork top bolt and take care as its last threads come free as it is under spring tension. Withdraw the spacer, washer and fork spring. Refer to Chapter 4, Section 3, paragraphs 4 to 6.

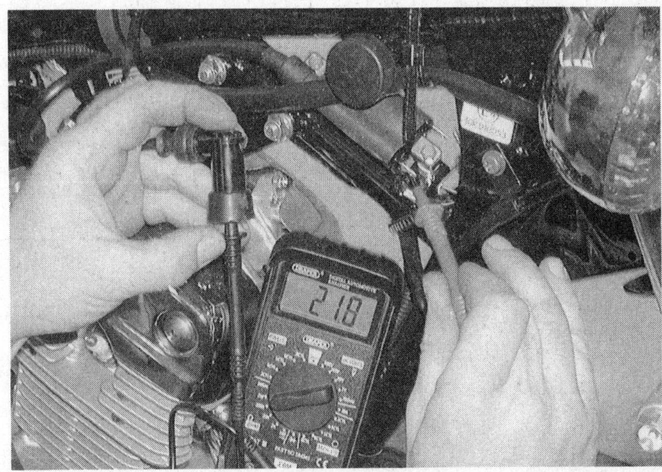

11.3 Measuring the ignition coil secondary windings – GZ model

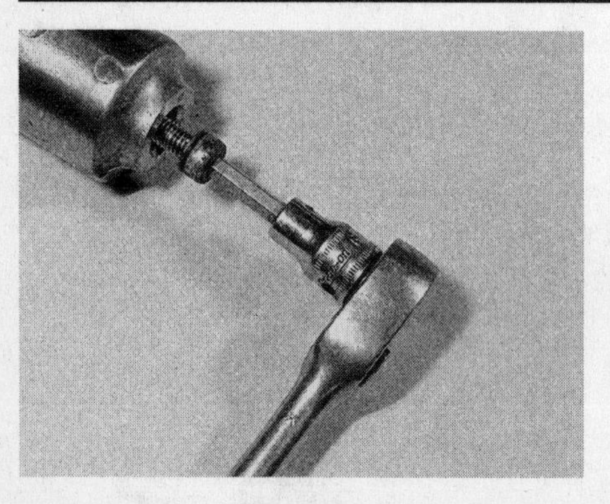

12.3a Remove the damper rod bolt using an Allen key

12.3b Prise out the dust seal and slide it off the stanchion

12.3c Use screwdriver to remove circlip from lower leg

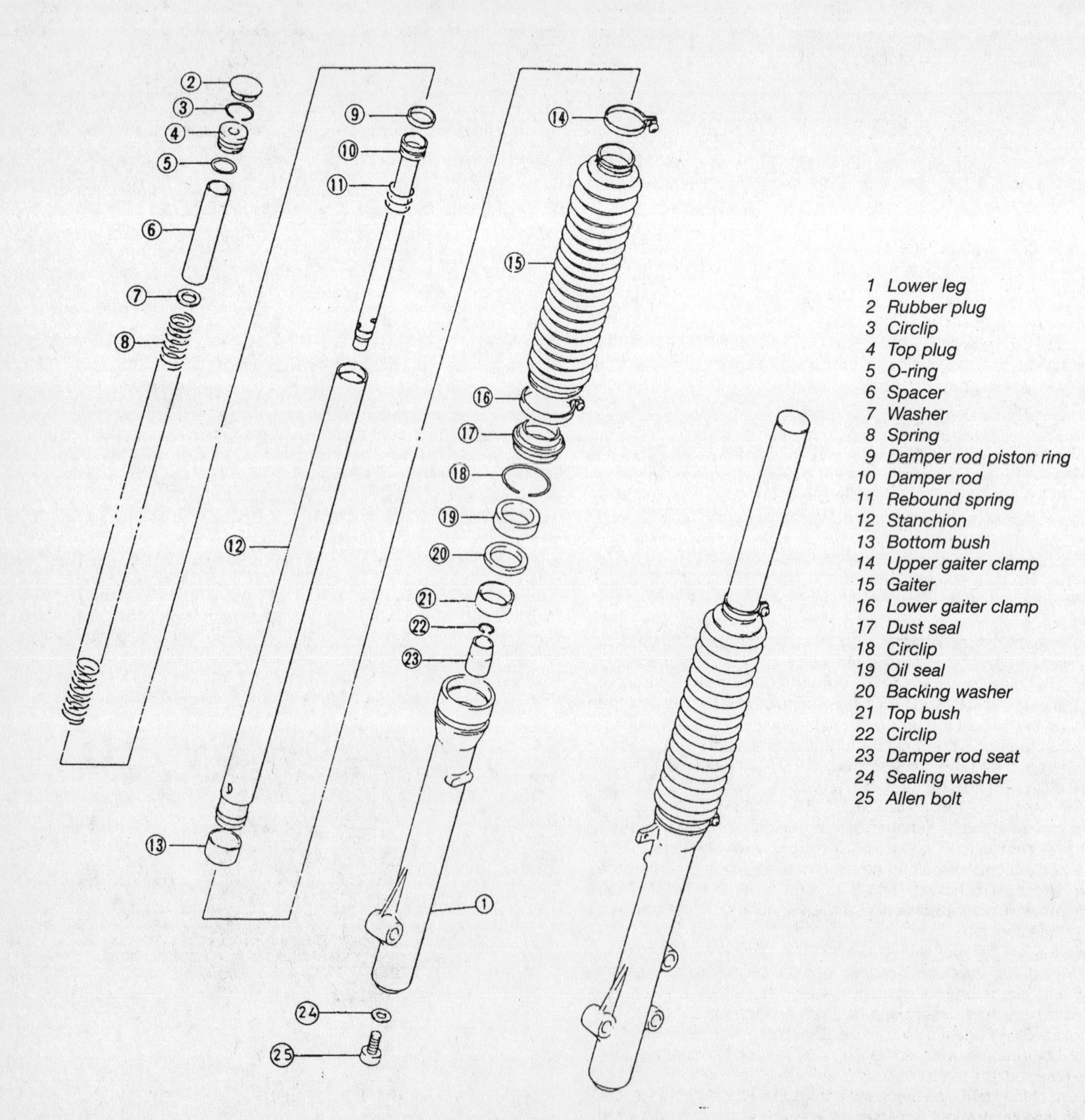

Fig. 7.9 Front forks – DR model

12.6 Bottom bush is split to allow its removal

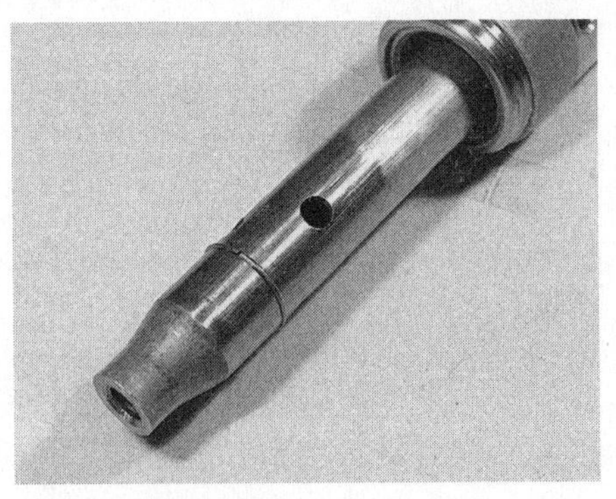

12.7a Insert damper rod and rebound spring into stanchion. Check that circlip is in place

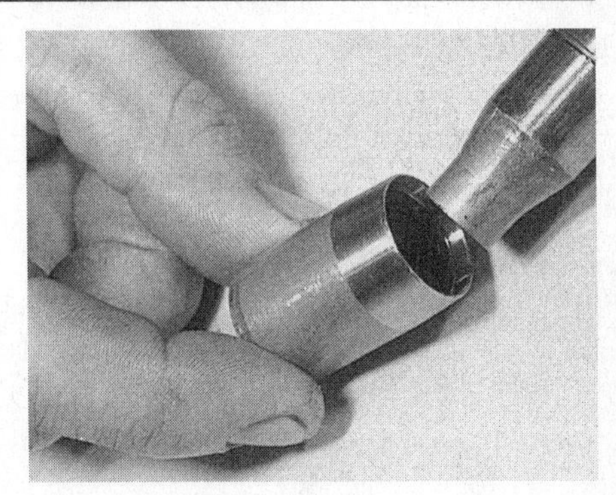

12.7b Fit damper rod seat over end of rod . . .

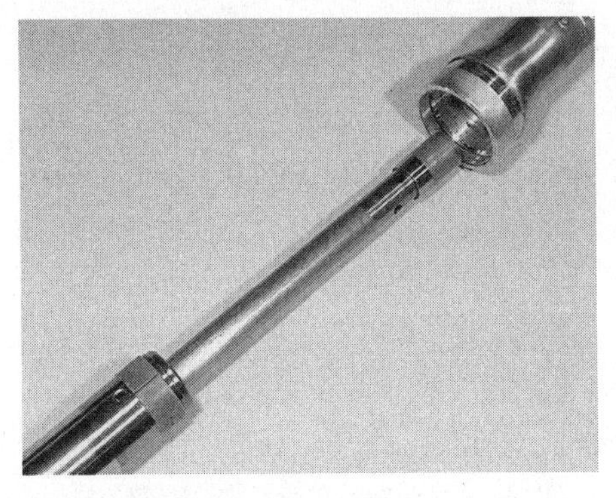

12.7c . . . and slide assembled stanchion into lower leg

12.7d Top bush must now be fitted into its recess . . .

12.7e . . . using backing washer to spread the load on the bush

3 On all models, once the damper rod bolt has been slackened and removed, the procedure is changed somewhat **(see illustration)**. Remove the dust seal from the top of the lower leg, working it out of position with a small screwdriver **(see illustration)**. Once the dust seal comes free of the lower leg it can be slid off and placed to one side. Below the dust seal lie the oil seal and top bush, and these are retained by a wire circlip. The circlip should be prised out of its groove using a small screwdriver **(see illustration)**.

4 If the stanchion is pulled outwards it will be noted that it cannot be removed completely. This is because the fork bushes meet. With luck, this will provide a means of extracting the top bush from its recess. Push the stanchion in by about six inches or so, then pull it sharply outwards. The lower bush will strike the underside of the top bush, forcing it out of the lower leg. This process should be repeated until the top bush and the oil seal are driven out of the lower leg. The stanchion assembly can then be withdrawn for further examination.

5 Unfortunately, the above method may not always work; sometimes the lower bush will pull through the top bush, leaving the latter in position in the top of the lower leg. If this happens, remove the stanchion and use a slide-hammer bearing extractor to drive out the top bush. Alternatively, it is possible to modify an old screwdriver, forming a hooked end which can be used to work the bush out of the lower leg recess. Whichever method is chosen, be warned that the bush may prove difficult to dislodge and care must be taken to avoid damaging the lower leg.

6 Examine the fork components as described in Section 4 of Chapter 4, bearing in mind that the bushes, rather than the sliding surfaces of the lower leg and stanchion, are subject to wear. No information on the nominal or service limit dimensions of these bushes are available, but wear should be self-evident during inspection. It goes without saying that if either or both bushes were damaged or scored during removal, they should be renewed as a matter of course. The lower bush can be removed from the end of the stanchion by opening the split with a screwdriver and sliding it off **(see illustration)**. When fitting the new bush, do not spread it more than is absolutely necessary to ease it over the stanchion end.

7 Reassemble the fork leg by reversing the dismantling sequence, noting the following points **(see illustrations)**. Check that the lower bush is lubricated with fork oil and that it is located correctly on the stanchion end. Slide the stanchion into the lower leg and secure the damper rod bolt in the normal way, tightening it to 1.5 – 2.5 kgf m (11 – 18 lbf ft) on DR models and 2.0 kgf m (14 lbf ft) on GZ models. The top bush should be fitted next, together with the backing washer **(see illustrations)**. Ideally, the bush should be tapped home using a tubular drift made from a length of tubing. If making up a tool of this type, check that there are no sharp edges on its inner face which might score the stanchion. It is a good idea temporarily to wrap the stanchion with PVC or masking tape to protect it. Alternatively, use a normal drift to tap the bush home, working around the top edge of the bush and making sure that it enters its bore squarely.

8 Check that the bush is fully home and that the washer is in place, then fit a new oil seal **(see illustration)**. Press this home and secure

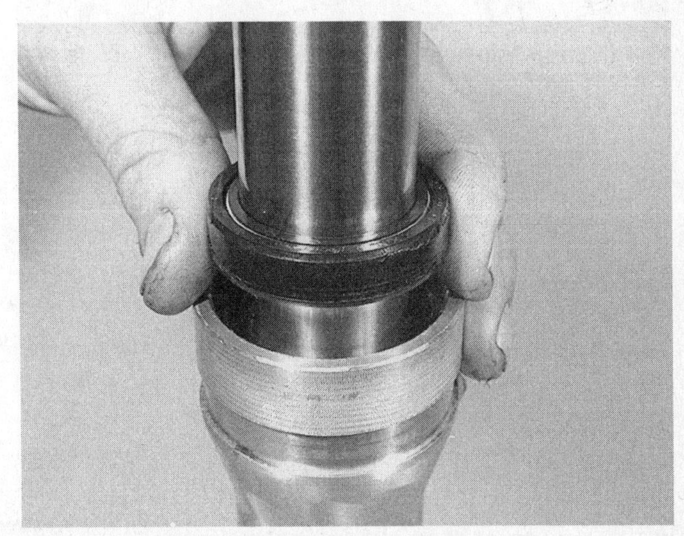

12.8a Fit the fork oil seal, keeping it square to the bone . . .

12.8b . . . and secure it with the wire circlip

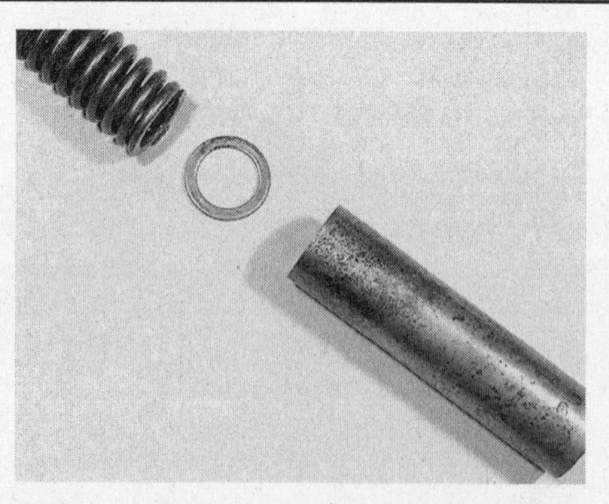
12.10a Do not omit to fit the washer and tubular spacer

12.10b Refit top plug and retaining circlip – DR model

the assembly with the wire circlip. Fit a new dust seal, and on DR models then fit the fork gaiter loosely in place.

9 Add the specified amount of the correct grade of fork oil, then check and adjust the oil level as described in Chapter 4 (or Routine maintenance).

10 Install the fork spring, noting that its tightly-wound coils face upwards on DR models and downwards on GZ models. Install the washer and the spacer **(see illustration)**. On DR models fit the top plug and circlip, then position the gaiter correctly and secure its retaining clips **(see illustration)**. On GZ models, screw the top bolt into the stanchion, noting that final tightening can be left until the fork is clamped in the yokes.

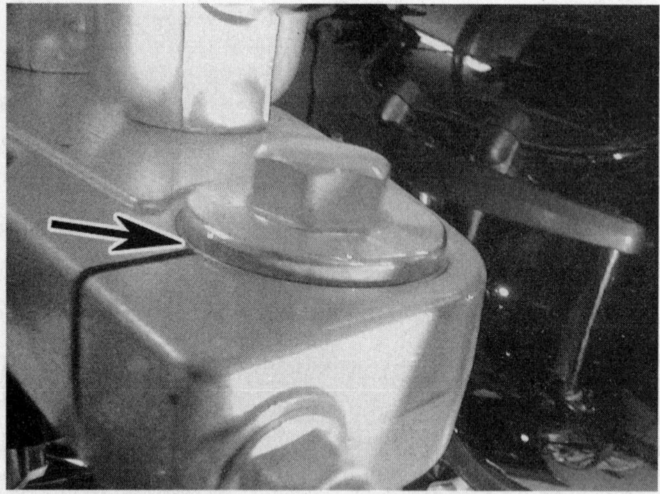
13.3 Align the upper surface of the stanchion with the upper surface of the upper yoke – GZ model

13 Front forks: removal and refitting – GZ model

1 Remove the front wheel (see Section 25) and unbolt the brake caliper from the fork leg; secure the brake caliper to the frame making sure not to bend or kink the brake hose. Remove the speedometer cable guide from the left-hand fork leg. Remove the four mudguard mounting bolts and remove the mudguard together with the brake hose guide.

2 To remove the front fork, loosen the upper and lower yoke clamp bolts and gently pull the fork legs downwards and out of the yokes. Note that if you are going to dismantle the forks, slacken the fork top bolts before the forks are removed from the yokes.

3 Install the forks into the yokes and make sure that the upper surface of the stanchion and the upper surface of the upper yoke are aligned **(see illustration)**. Tighten the yoke clamp bolts, and if dismantled the fork cap bolts, to the specified torque setting. Refit the mudguard and tighten its bolts temporarily at this stage. Refit the brake caliper and the front wheel. Push down on the handlebars to move the forks up and down several times and tighten the mudguard bolts securely.

14 Front forks: modifications – GN model

Removal and refitting

1 The procedure for removal and refitting of the front forks is very similar to that for the GS125 models. Refer to Chapter 4, Section 2, noting that the forks are retained in the upper and lower yokes by pinch bolts **(see illustration)**; there is no need to remove the fork top bolts, although you are advised to slacken them at this stage if you are going to dismantle the forks. There is also no need to disturb the handlebars.

Dismantling and reassembly

2 Follow the procedure in Chapter 4, Section 3, noting that the fork

14.1a Loosen the upper clamp bolt . . .

14.1b . . . and the lower clamp bolt, then . . .

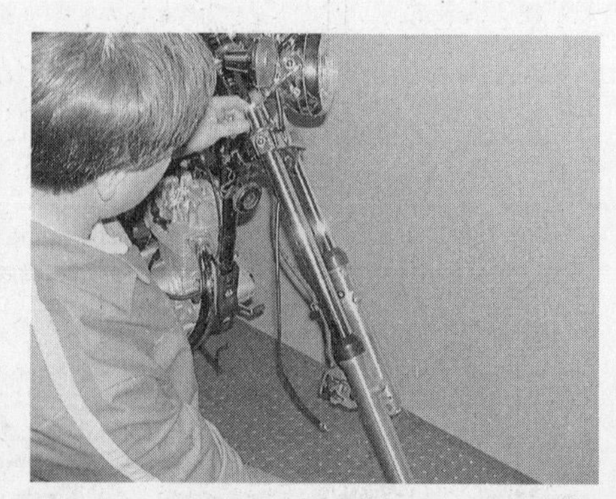
14.1c . . . slide out the fork – GN model

top bolt is unscrewed from the stanchion to free the spacer and fork spring; note that the top bolt will be under spring pressure. Note also that the gaiters fitted to early models are replaced by dust seals which are simply prised off the fork lower leg **(see illustration)**.

15 Steering head bearings: removal, examination and refitting – DR model

1 The steering head bottom bearing is of the taper-roller type, while the top bearing remains the cup and cone type described in Chapter 4. Refer to Fig. 7.10 for details.
2 Dismantling is basically as described in Chapter 4, Section 5, noting the differences described in this Chapter concerning the headlamp assembly and instrument panel. When the yokes themselves are withdrawn the bottom bearing inner race will remain in position on the steering stem, leaving only the steel balls of the top race to be collected.
3 The top bearing components are checked as described in Section 6 of Chapter 4, noting that it contains 18 No 8 (5/16 in) uncaged steel balls. The bottom bearing is checked in a similar way and must be renewed as a complete assembly if any signs of wear or damage are discovered; check the bearing track of the outer race and the rollers themselves with particular care. If the bearing is to be renewed the outer race can be removed and refitted using the method described in Chapter 4; note however that great care must be exercised to keep the race absolutely square in its housing on refitting.
4 The inner race can be prised gently off its locating shoulder by using two slim steel wedges and then tapped carefully up the stem until it can be withdrawn. On refitting, obtain a length of tubing as long as the stem and of the same diameter as the bearing inner race's upper end. Clean any burrs or dirt from the bearing and stem, lightly grease both components and slide the bearing as far as possible down the stem; ensure that the bearing is fitted correctly, with its wider, sealed, side downwards. Tap it fully into place on its locating shoulder, being

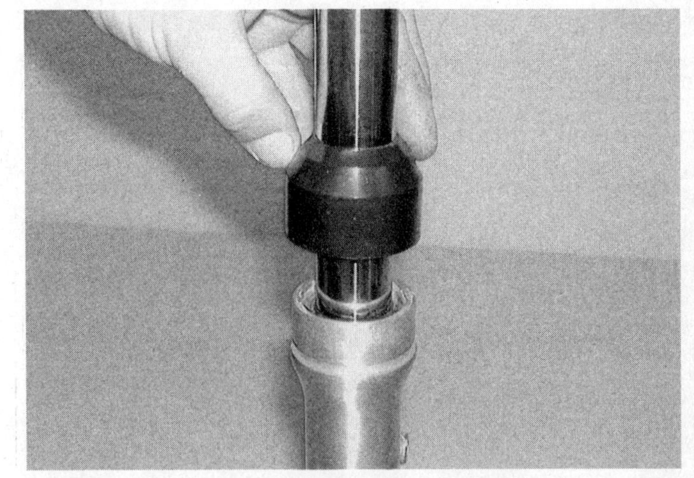

14.2 Dust seal can be prised off lower leg – GN model

1 Bottom yoke/steering stem
2 Lower bearing
3 Top yoke
4 Bolt
5 Washer
6 Upper handlebar clamp – 2 off
7 Lower handlebar clamp - 2 off
8 Bolt – 4 off
9 Spring washer – 4 off
10 Damping rubber – 2 off
11 Damping rubber – 2 off
12 Bolt – 2 off
13 Spacer – 2 off
14 Nut – 2 off
15 Split pin – 2 off
16 Cable guide
17 Bolt – 2 off
18 Spring washer – 2 off
19 Cable guide
20 Spring washer
21 Screw
22 Bolt – 2 off
23 Spring washer – 2 off
24 Washer
25 Cable guide
26 Headlamp bracket – 2 off
27 Damper – 4 off
28 Upper bearing cup
29 Upper bearing cone
30 Ball – 18 off
31 Dust cover
32 Adjuster nut

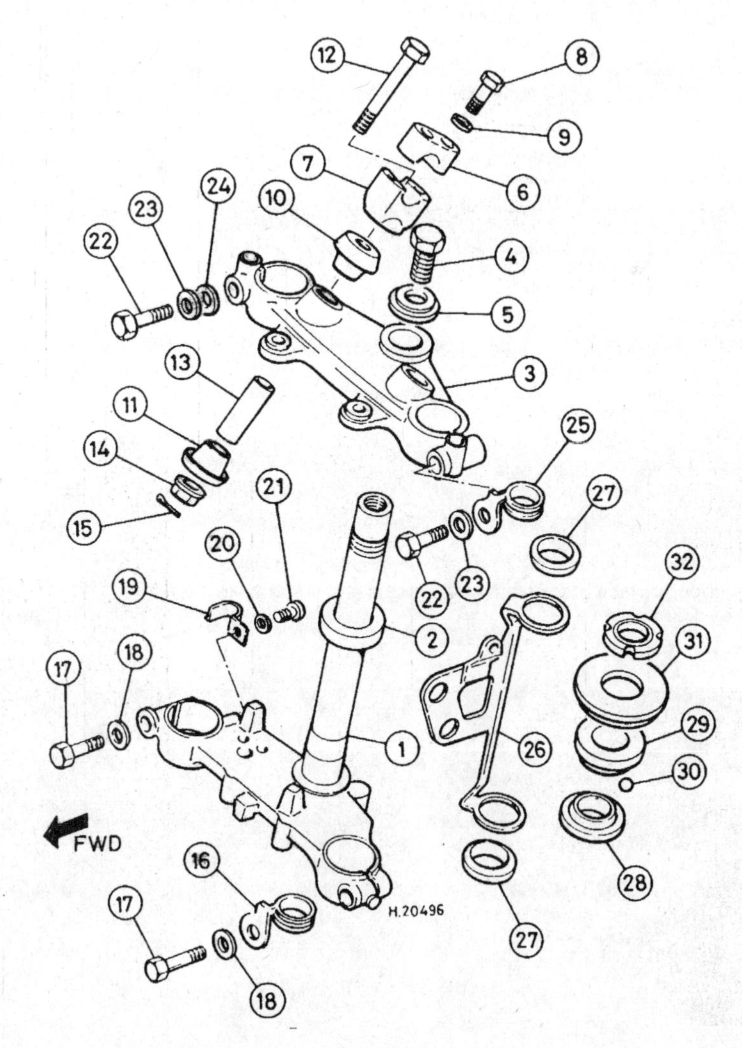

Fig. 7.10 Steering head assembly – DR model

careful to keep it square to the stem at all times. Using the tubing as a drift, tap the bearing down until it seats firmly on the machined surface of the bottom yoke.

5 ***Note:*** *taper roller steering head bearings, if badly adjusted or improperly installed, can give rise to many faults in the machine's handling and stability. If you are at all in doubt as to your ability to carry out this task take the machine to an authorised Suzuki dealer for the work to be done by an expert using the correct service tools.*

6 On refitting the yokes, pack both bearings with high-quality molybdenum disulphide-based grease and smear grease over the full length of the steering stem and inside the steering head. Do not forget to grease the bottom bearing outer race and the top bearing cup.

7 Insert the bottom yoke into the steering head and hold it in place while the eighteen balls of the top bearing are refitted, followed by the top cone, dust excluder and bearing adjustment ring.

8 After dismantling, taper roller bearings must be set up by tightening the adjuster ring firmly to preload the bearing. This is normally done by applying a specified torque wrench setting, similar to that given in Chapter 4, Section 5, paragraph 11; note however that in the case of these later models no torque wrench settings are available from the manufacturer, so some care will be required.

9 Settle the bearings by turning the bottom yoke five or six times from lock to lock, then release pressure by turning the adjusting ring back through 1/4 – 1/2 turn; in this position the bottom yoke should be completely free to move smoothly and easily from lock to lock with no trace of free play or stiffness. Refit the top yoke, using the fork stanchions to ensure that it is correctly aligned, and tighten the steering stem top bolt firmly. Do not overtighten the bolt; while specific torque wrench settings are not provided for these models, the correct setting should be close to that specified in Chapter 4.

10 Check that the steering stem assembly is free to move under its own weight from left to right and that there is no trace of free play. When the setting is correct, refit the forks and front wheel. Re-check the bearing adjustment before taking the machine out on the road.

16 Steering head bearings: removal, examination and refitting – GZ model

1 Remove the front wheel (Section 25).

2 Remove the front forks (Section 13).

3 Remove the handlebar clamp bolts caps to gain access to the bolts. Remove the bolts and lift the handlebars out of their locations on the upper yoke. There is no need to disconnect the wiring, cables or brake hose; rest the handlebars on the fuel tank cushioned by rag.

4 To remove the headlight undo the two retaining screws, lift the headlight out of the shell and carefully disconnect the wiring plug from the back of the bulb and pull out the parking lamp bulbholder **(see illustrations 36.2b and 36.2c)**. Undo the mounting nut from beneath the lower yoke and remove the headlight housing complete. Remove the two retaining screws on the steering stem lower cover and remove the cover.

5 Remove the front brake hose/speedometer cable guide by removing the retaining bolt. Disconnect the speedometer cable and remove the speedometer, by removing the two nuts which are located either side of the speedometer and remove the bolts **(see illustration)**.

6 Remove the two bolts (one on either side of the steering stem) to remove the turn signal brackets, then trace and disconnect their wiring.

7 Remove the two nuts from the base of the upper yoke and remove the handlebar holders. Remove the steering stem top bolt, then trace and disconnect the ignition switch wiring at the connector and lift the upper yoke off the steering head. Remove the steering stem nut from the stem using a C-spanner, whilst supporting the lower yoke/steering stem to prevent it falling to the ground.

8 Remove the lower yoke/steering stem from the steering head. Remove the dust seal, the upper bearing inner race and balls from the top of the steering head. Remove the lower bearing balls from the steering stem. Check all components for signs of distortion, wear or damaged and replace if necessary. If the ignition switch requires renewal, note that it is secured to the upper yoke with special bolts – refer to a Suzuki dealer for removal details.

9 Refer to Chapter 4, Section 6 for details of bearing renewal, disregarding the bearing size information.

10 Refitting is a reverse of the removal procedure noting the following points:

11 Make sure that you grease the bearings and their races well. Tighten the steering stem nut (special tool maybe required) to 4.5 kgf m (32.5 lbf ft). Turn the steering stem lower bracket about five or six times to the left and right and then loosen the steering stem nut 1/4 – 1/2 of a turn. **Note:** *make sure that the steering is smooth and turns easily in both directions.*

12 Refit the upper yoke. Refit the front forks and tighten the lower yoke clamp bolts. Tighten the steering stem head bolt to the specified torque.

13 Make sure that the upper surface of the front fork stanchion and the upper surface of the upper yoke are aligned and then tighten the upper and lower yoke clamp bolts to the specified torque.

14 Refit the handlebar holders and tighten their nuts temporarily. Refit the handlebars, make sure that the punch mark on the handlebars aligns with the handlebar clamp **(see illustration)** and tighten the handlebar clamp bolts to the specified torque. Make sure that the gap between the handlebars clamps are even. Tighten the handlebar holder nuts to the specified torque. Refit the caps to the handlebar clamp bolts.

15 When refitting the turn signals, make sure that you insert the projection on the indicator bracket into the hole of the steering stem upper bracket.

16 While holding the front forks, move them back and forth to ensure that the steering is not loose. Make sure the steering moves freely from side to side under its own weight. If the steering appears to be stiff or slack, slacken off the fork upper yoke pinch bolts and steering stem top bolt and readjust the steering stem nut.

16.5 Remove the two nuts to remove the speedometer – GZ model

16.14 Align the handlebar clamp with the handlebar punch mark – GZ model

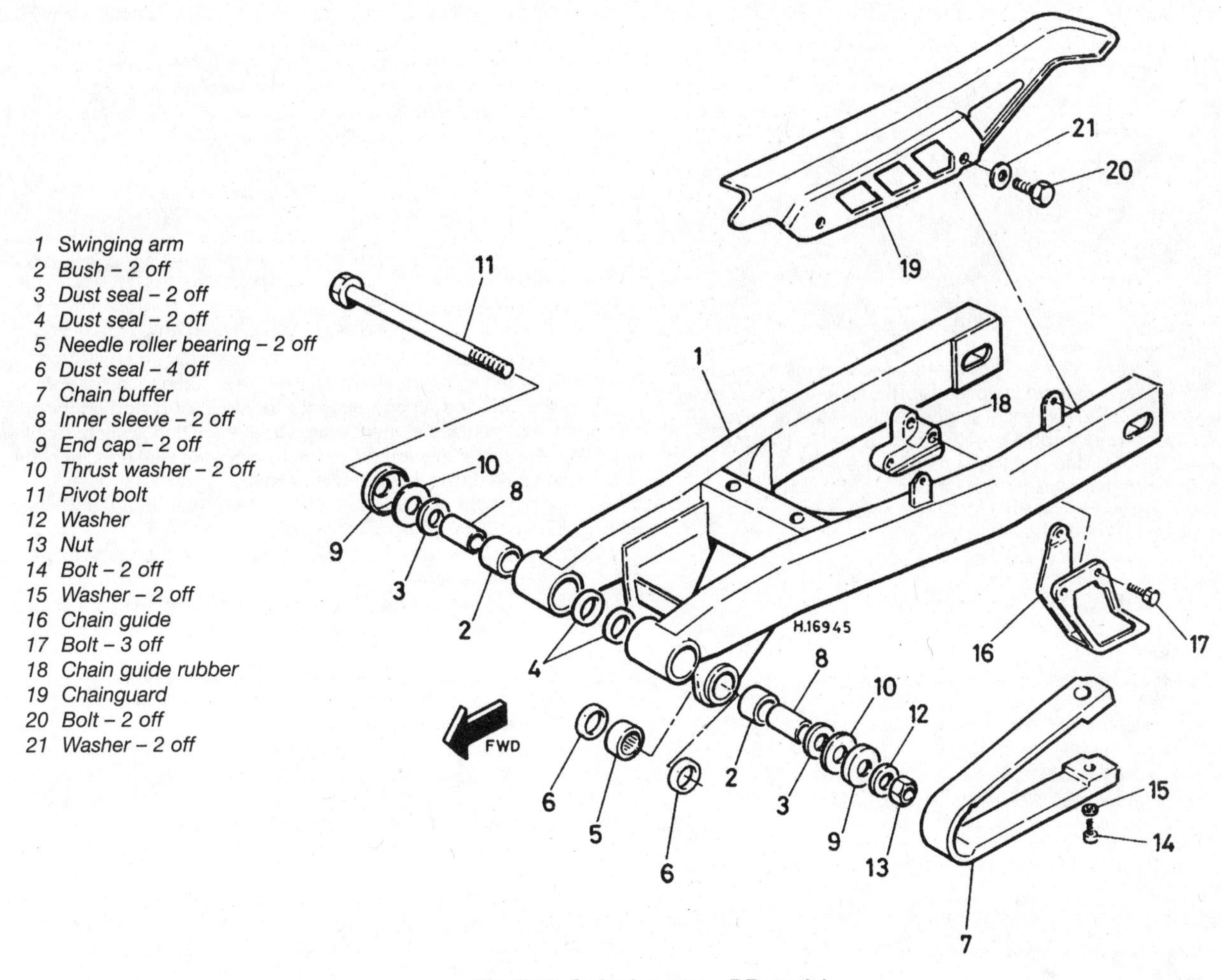

Fig. 7.11 Swinging arm – DR model

17 Swinging arm and suspension linkage: removal and refitting – DR model

Refer to Figs. 7.11 and 7.12.

1 Given that the link arm and rocker of the original suspension system is replaced by a single bellcrank, the basic principles of the procedure described in Chapter 4, Section 8 also apply to the later models; either the swinging arm pivot bolt and the swinging arm/bellcrank bolt can be removed so that the swinging arm can be withdrawn alone, or the swinging arm, bellcrank and suspension unit can be removed in turn after their respective retaining bolts have been unscrewed **(see illustrations)**.

2 As described in Chapter 4 it is recommended that the complete assembly is removed at regular intervals so that it can be cleaned, checked for wear and greased thoroughly before reassembly.

3 The assembly is installed by reversing the removal sequence, having ensured that all bearings have been greased and that the various dust seals are in place. Do not omit to fit the end caps and thrust washers which fit on the outer face of each of the swinging arm pivot bosses. The swinging arm is a tight fit between the frame plates and it will be necessary to lift the bosses up and forward of the mounting plates, fit

17.1a Remove suspension unit lower mounting bolt and swinging arm to bellcrank bolt

17.1b Remove swinging arm pivot bolt and tap pivot bolt out

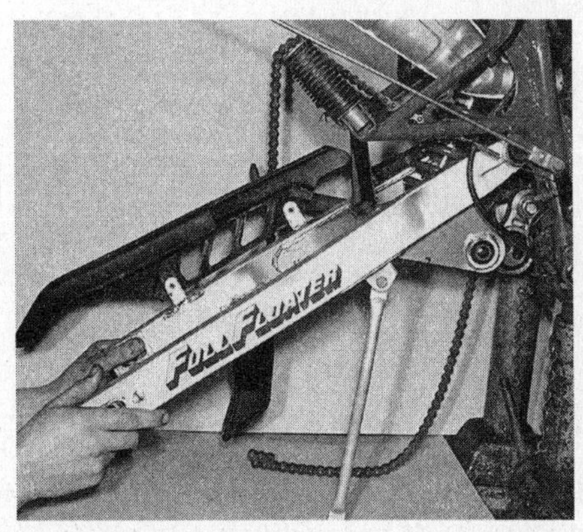

17.1c Swinging arm can now be manoeuvred out of frame

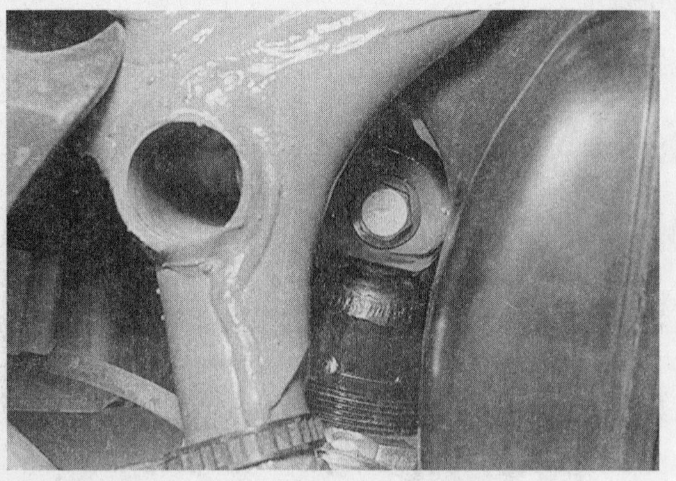

17.1d Suspension unit top mounting bolt can be just reached

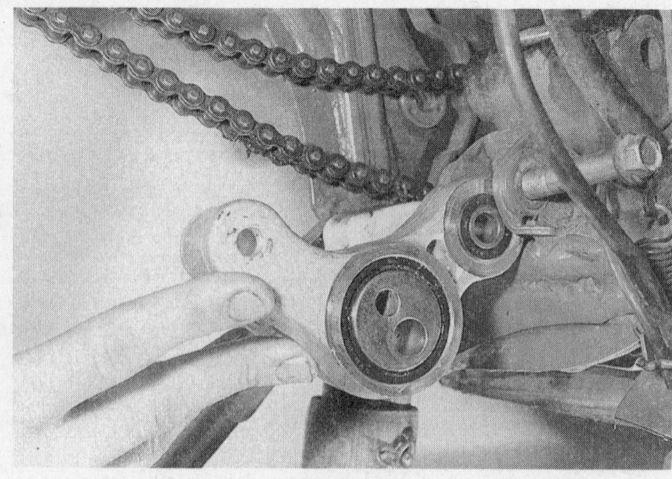

17.1e Release bellcrank to frame bolt and remove the bellcrank

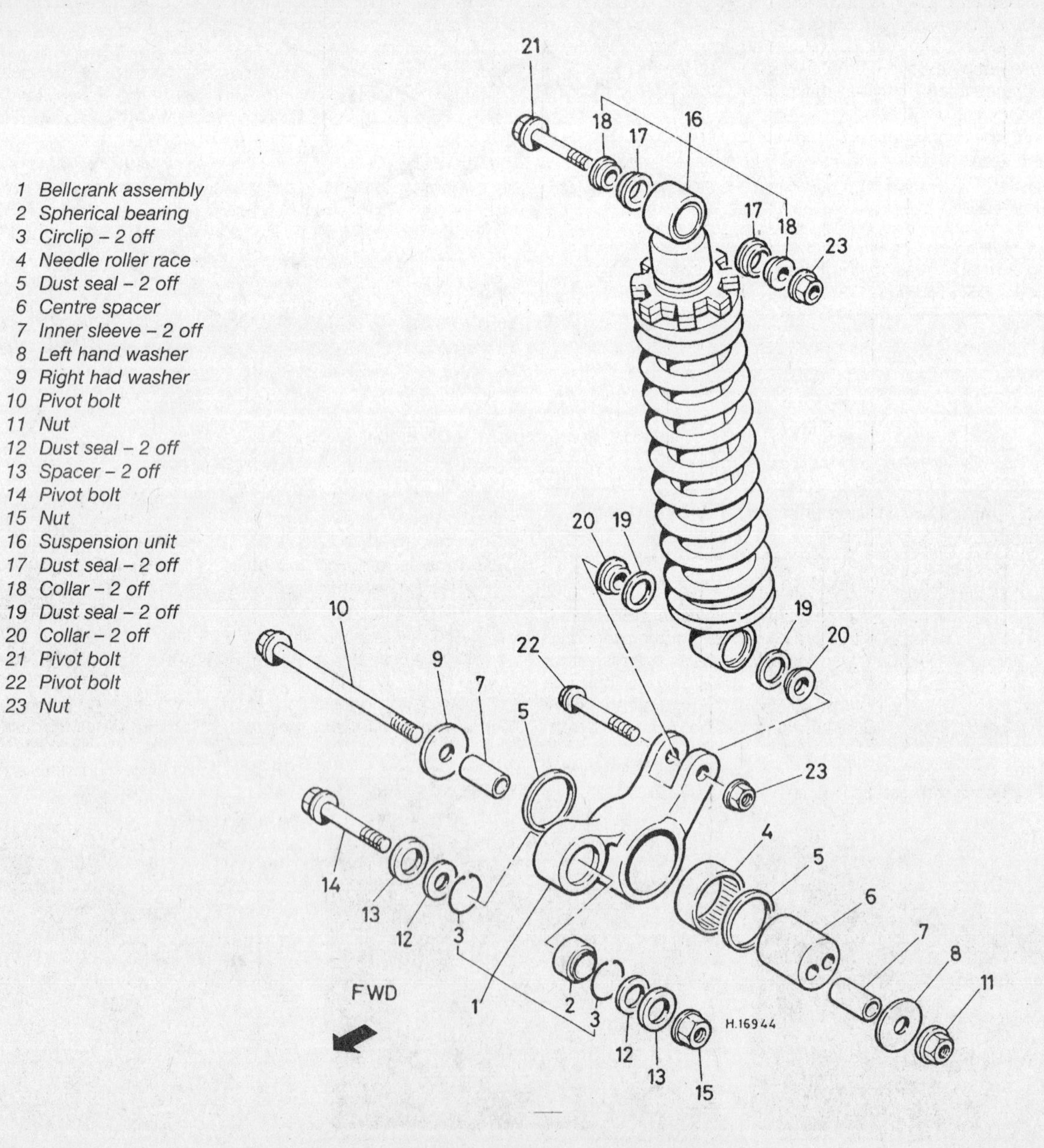

Fig. 7.12 Rear suspension linkage – DR model

17.3a Lift bosses above the forward of frame mounting holes, then fit dust seals and lower assembly into position

17.3b Refit and tighten the bellcrank bolts

the end caps and thrust washers, and then lower the assembly into position **(see illustration)**. When fitting the swinging arm to bellcrank pivot bolt, it may be necessary to turn the spacer slightly until the bolt holes align. Tighten securely all mounting nuts and bolts; note that revised torque wrench settings are not available **(see illustration)**. Do not forget to grease thoroughly all pivot bolts as well as the bores through which they are to be fitted.

4 Although no precise instructions are available, it would appear that the bellcrank centre spacer eccentric bore should be in the lower of its two possible positions on reassembly. Refer to the accompanying photographs for details.

18 Swinging arm and suspension linkage: examination and renovation - DR model

Refer to Figs. 7.11 and 7.12.

1 The swinging arm pivots on two bushes of synthetic material with steel inner sleeves which actually bear on the pivot bolt. Dust seals are fitted on each side of the pivot bearings and thrust washers are fitted inside the large end caps to eliminate endfloat. These are checked and renewed, if worn, as described in Chapter 4, Section 9.

2 The swinging arm to bellcrank pivot is fitted with two needle roller bearings, one on each side **(see illustration)**. Again dust seals are fitted on each side of each bearing and a hardened steel inner sleeve is fitted to each **(see illustrations)**. These are also checked and renewed as described in Chapter 4.

3 The bellcrank itself uses a spherical bearing at its front (frame) mounting; these bearings are used to ensure correct alignment between the various components. Check that the dust seals are intact (it is probably a sound precaution to renew these anyway) and that the bearing itself moves smoothly and easily. Wash out any old grease with solvent, then grease the bearing carefully before refitting the dust seals. If the bearing is worn or damaged, it can be driven out after the retaining circlips have been removed **(see illustration)**. Support the bellcrank on a large diameter socket, then use a smaller socket as a drift **(see illustrations)**.

4 The bellcrank large centre bearing uses uncaged needle rollers, so care must be taken not to lose the rollers during dismantling or cleaning

18.2a Check the caged needle roller bearings and lubricate with grease

18.2b Grease and fit the inner sleeve . . .

18.2c . . . and fit new dust seals as shown

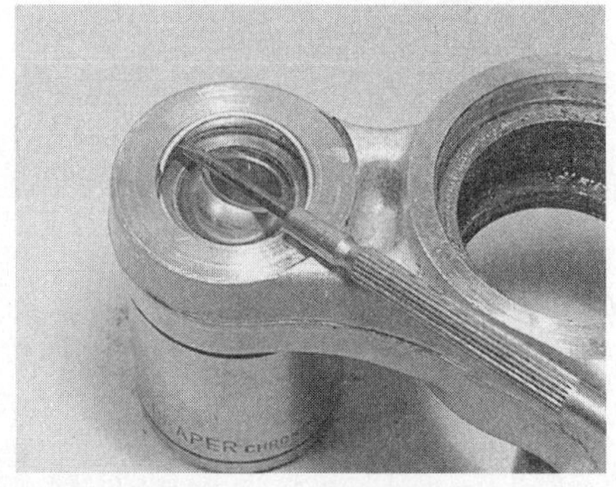

18.3a Prise out the seal and retaining circlips

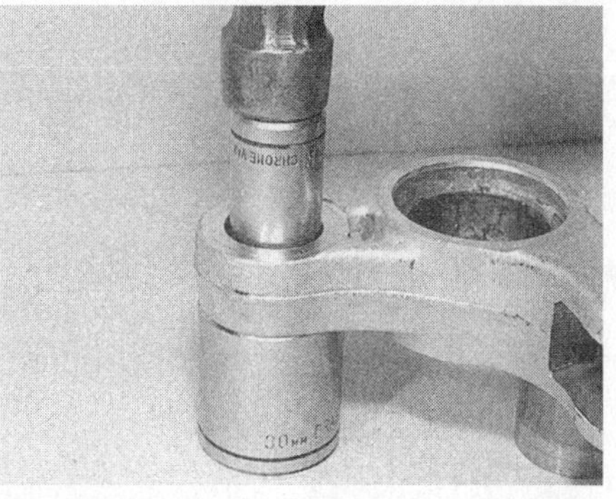

18.3b Use sockets as shown to drive out spherical bearing

18.3c When fitting a new bearing, make sure that it enters the bore squarely

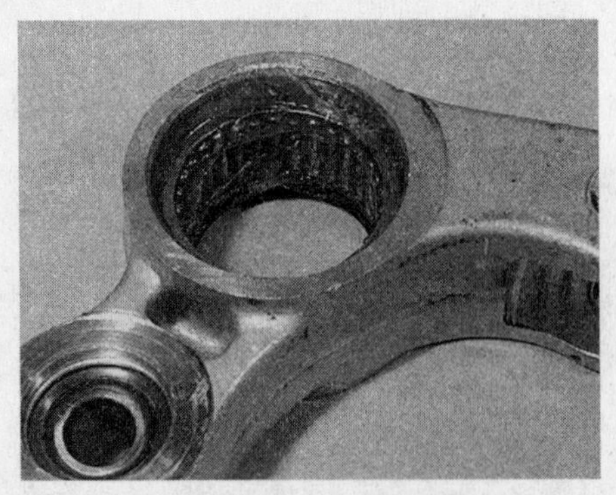
18.4a Large centre bearing uses uncaged rollers – be careful to avoid losing them

18.4b Dust seals press into place in recesses

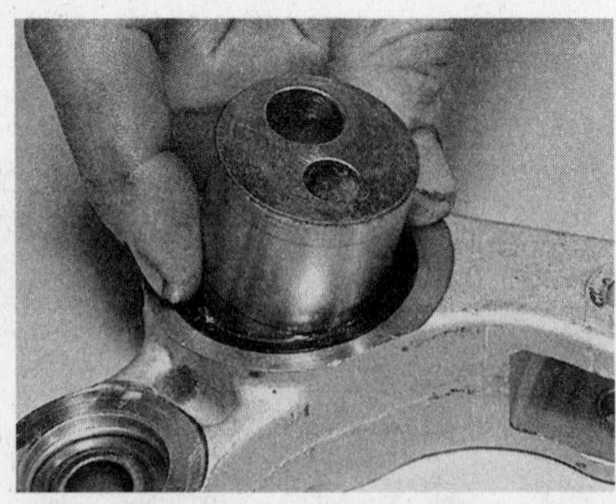
18.4c Centre spacer can now be fitted. Note eccentric bore

(see illustration). When refitting the rollers, grease the outer race and use this to hold the rollers in place. It is preferable to renew the large dust seals each time the assembly is dismantled **(see illustration)**. Note that the bearing rollers are not available separately; the complete bearing must be renewed if it is worn or damaged. If renewal is necessary the bearing outer race can be easily driven out provided sockets or drifts of the required size are available. On refitting be very careful to keep the race absolutely square to the bellcrank housing at all times and to tap it in very gently; if the outer race is distorted through careless work, the complete bearing must be renewed **(see illustration)**.

19 Swinging arm: general – GZ model

1 The procedure for removal and refitting of the swinging arm is very similar to that for the GS125 covered in Chapter 4 apart from the following differences (see Fig. 7.13):

2 Once you have removed the chainguard, you need to remove the rear brake cable and it's holder from the swinging arm.

3 To gain access to the pivot shaft remove the chrome end caps (one on each side of the bike).

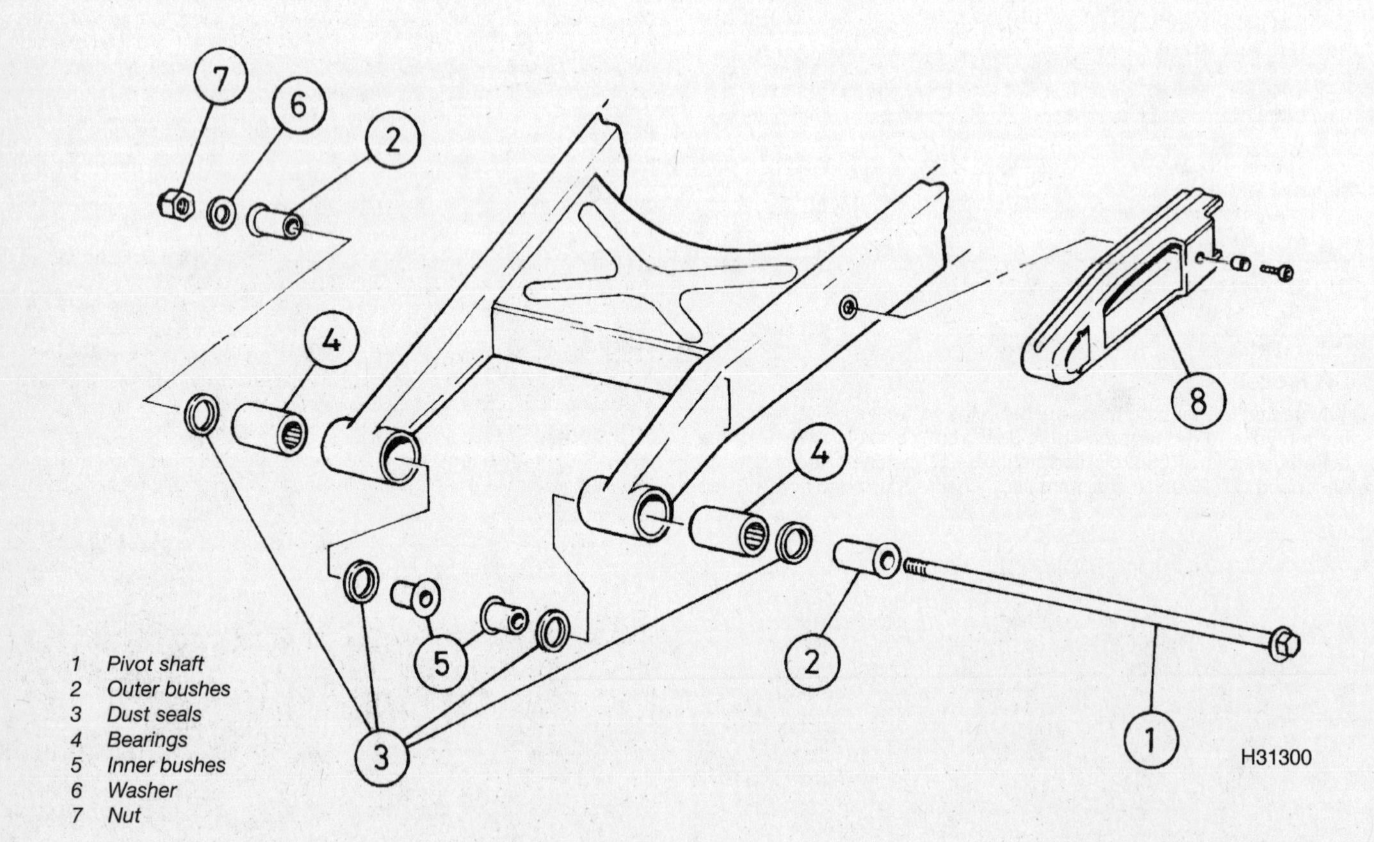

Fig. 7.13 Swinging arm pivot assembly – GZ model

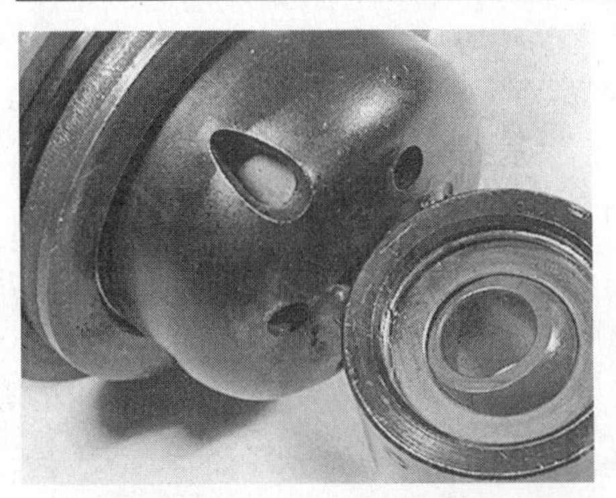
20.1 Suspension unit mounting eyes are also fitted with spherical bearings – DR model

20.3 Use C-spanner to adjust the suspension unit spring length – DR model

21.4 Side stand interlock switch – GZ model

20 Rear suspension unit: modifications – DR model

1 The suspension unit is fitted with spherical bearings at both mounting eyes **(see illustration)**. These, however, are not fitted as separate parts; the complete unit must be renewed if either bearing is found to be worn. The dust seal and collar on each side of each bearing can be renewed if required, as described in Chapter 4. Bearing life will be greatly extended if the unit is kept as clean as possible, with the seals renewed whenever necessary, and if they are cleaned and repacked with grease whenever the suspension is dismantled
2 The original cam-type method of spring preload adjustment is replaced by an adjuster and locknut running on a fine thread on the unit body. This arrangement provides a wide range of adjustment. The standard setting for the unit is to set the adjuster to give an installed spring length of 226 mm (8.898 in). The spring preload can be varied from this setting to suit rider weight and preference, or to suit various surfaces.
3 Adjustment requires the use of a C-spanner of suitable size, and is best carried out with the unit removed; access to the adjuster and locknut is very limited, and it is essential that the threads on the body are completely clean if damage is to be avoided **(see illustration)**. To alter the spring length, slacken the locknut, then move the adjuster nut to obtain the desired spring length. Tighten the locknut to secure the adjuster, and grease the threads prior to installing the unit. This will help prevent corrosion and make subsequent adjustment easier.

21 Footrests, stands and handlebars: modifications – DR and GZ models

DR model

1 Note that rubber mountings are set in the footrest assemblies and in the top yoke to further protect the rider from vibration. Whenever the general checks of controls and tightness of fasteners are made, as described in Routine maintenance, check the condition of the mounting rubbers and renew them if they appear unduly sloppy or if the rubber itself is cracked or perished.
2 The construction of each mounting is self-explanatory on close examination (refer also to Fig. 7.10 for details of the handlebar mountings); ensure that the mounting components are correctly refitted and securely fastened on reassembly. The split pins securing the handlebar mounting retaining nuts should be renewed whenever they are disturbed.
3 With reference to Chapter 4, Section 11, always remove the rubber mounting before heating either part of the footrest assembly during the course of repairs.

GZ model

4 This model is not fitted with a centre stand. The side stand however, is fitted with an interlock switch **(see illustration)**, which is designed to prevent the machine being ridden with the side stand down. To check the operation of the switch, sit on the machine with the stand in the up position. Pull in the clutch lever and start the engine. Now engage first gear and with the clutch lever still held in, lower the side stand; if the switch is working correctly, the engine should stop.

22 Instrument panel: modifications – DR and GZ models

DR model

1 The DR models are fitted with a simplified instrument panel assembly. The tachometer fitted to the early models is omitted, and the panel reduced in size as a result. Like earlier versions, the assembly is held by a rubber-mounted bracket, the removal and refitting of which is self-explanatory. For details, refer to the accompanying line drawing (Fig. 7.14). Note that the speedometer is now driven by a separate gearbox as described in Chapter 4, Section 14 for the disc brake models (**see illustrations**).

GZ model

2 The GZ models are also fitted with a simplified instrument panel. Details of removal can be found in Section 16 of this Chapter.

22.1a Revised instrument panel – DR model

22.1b Panel is retained by rubber bushed studs and domed nuts – DR model

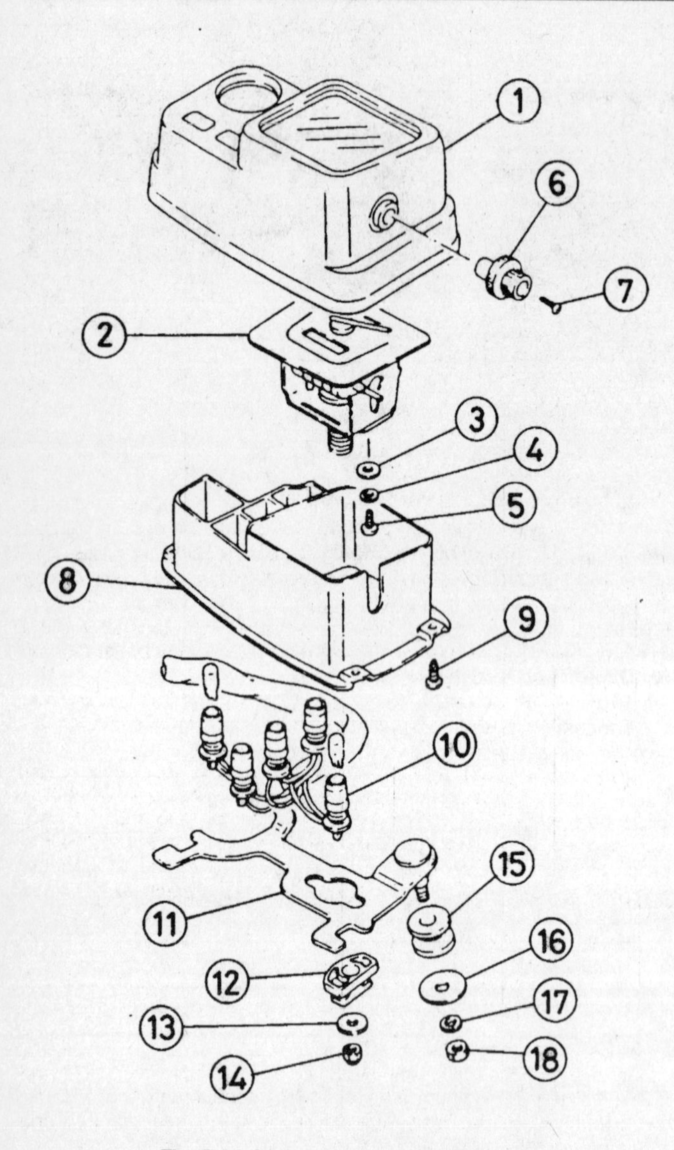

Fig. 7.14 Instrument panel – DR model

1 Top cover	*10 Bulbs and wiring*
2 Speedometer	*11 Mounting bracket*
3 Washer – 2 off	*12 Grommet – 2 off*
4 Spring washer – 2 off	*13 Washer – 2 off*
5 Screw – 2 off	*14 Nut – 2 off*
6 Reset knob	*15 Grommet – 2 off*
7 Screw	*16 Washer – 2 off*
8 Housing	*17 Spring washer – 2 off*
9 Screw – 3 off	*18 Nut – 2 off*

23 Seat and side covers: removal and refitting – GZ models

1 Remove the rider's seat by using the ignition key in the lock in the right-hand side cover **(see illustration)**. To remove the pillion seat, remove the two retaining screws positioned at the front of the seat, followed by the retaining bolt at the rear of the seat. Refitting is the reverse of removal, noting that the tab at the front of the rider's seat should engage under the fuel tank bracket before pushing the seat down at the rear to latch it.

2 To remove either side cover, first remove the rider's seat, then remove the single retaining screw and ease the pegs on the inside of the cover out of their grommets in the frame **(see illustration)**. When removing the right-hand cover, note that the seat lock cable must be disconnected. Refitting is the reverse of removal.

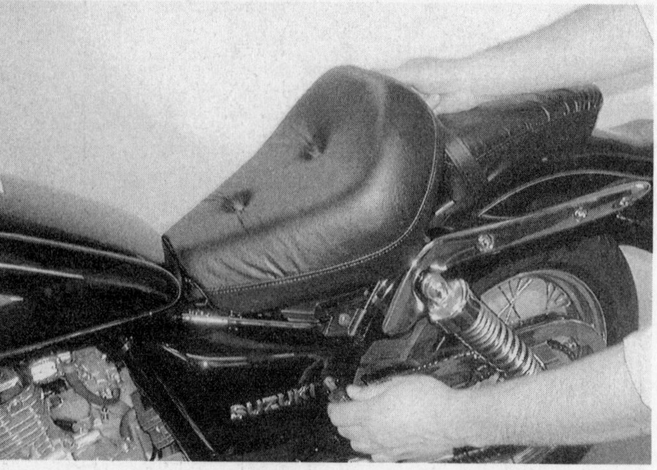

23.1 Operate key in seat lock to unlatch rider's seat – GZ model

23.2 Side covers are retained by a single screw and grommets – GZ model

24.1 Remove countersunk screw and withdraw speedometer cable – DR model

24 Front wheel: general – DR model

The front wheel is identical in layout to that shown in Chapter 5 for the GS125 ES models. Removal and refitting is as described in Section 4 of that Chapter for the disc brake models, noting that the speedometer cable is retained by a countersunk screw **(see illustration)** and that the spindle nut is now of the self-locking type;

25.1a Loosen the axle pinch bolt – GZ model

25.1b Removing the axle bolt – GZ model

the split pin is no longer fitted. Position the speedometer gearbox as shown in the accompanying photograph, so that the cable has a smooth straight run; the gearbox may butt against a lug on the fork lower leg to ensure this.

25 Front wheel: removal and refitting – GZ model

1 Place a support or jack under the engine so that the weight of the machine is off the front wheel. Loosen the Allen headed pinch bolt located at the bottom of the right-hand fork **(see illustration)**, then loosen the axle bolt. Remove the axle bolt **(see illustration)** and remove the wheel, disengaging the speedometer drive unit as the wheel is withdrawn from the forks.

25.2 Remove the six disc bolts with the use of an Allen key – GZ model

2 Remove the dust cover and spacer which are located on the opposite side to the brake disc. If required, remove the brake disc by removing the six bolts **(see illustration)**.

3 When refitting the brake disc make sure that it is free of dust and grease. Apply a thread locking compound to the disc bolts and tighten them to the specified torque setting.

4 Fit the spacer (head against the wheel hub) and dust cover to the right-hand side of the wheel. Move the wheel into position and refit the speedometer drive unit, making sure that the two lugs on the speedometer drive fit neatly into the wheel recesses. Slide the axle into place and thread it into the left-hand fork. Tighten the axle to the specified torque setting, then tighten the axle pinch bolt to the specified torque setting. Ensure that the speedometer cable is engaged in the wire cable guide.

26 Speedometer drive unit: removal and refitting – GZ model

1 The speedometer drive unit is located on the front wheel left-hand side. Unscrew the knurled nut to allow the speedometer cable to be withdrawn from the drive unit. Remove the wheel to free the speedometer drive unit (see Section 25). Turn the gear and check that it turns smoothly together with the pinion.

2 When refitting the unit apply some grease to the gear and align the lugs on the gear with the recesses in the front wheel **(see illustration)**. Position the drive unit so that the cable will be able to engage the wire cable guide attached to the fork leg. Refit the wheel.

3 The speedometer cable is retained to the drive unit by a knurled nut **(see illustration)**.

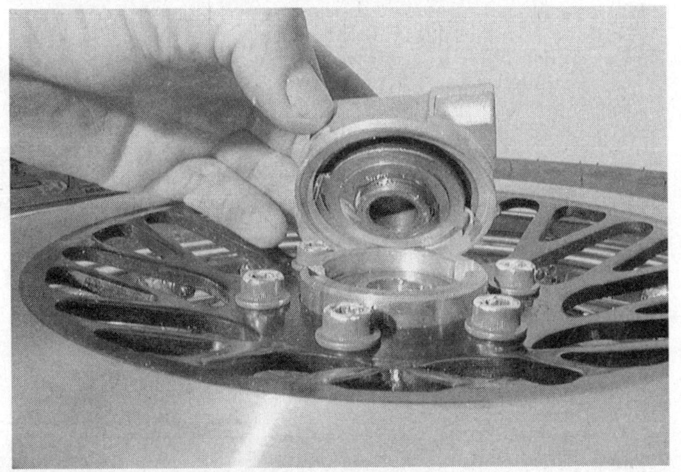
26.2 Align the speedo lugs with the recesses in the wheel – GZ model

26.3 Refit the speedo cable – GZ model

1 Pad retaining bolt – 2 off
2 Tab washer
3 Brake pads
4 Shim
5 Dust seal
6 Washer
7 Axle bolt
8 Mounting bracket
9 Dust seal
10 Piston
11 Fluid seal
12 Bolt – 2 off
13 Support pin
14 Boot
15 Cap
16 Bleed nipple
17 Anti-rattle spring
18 Washer
19 Cap

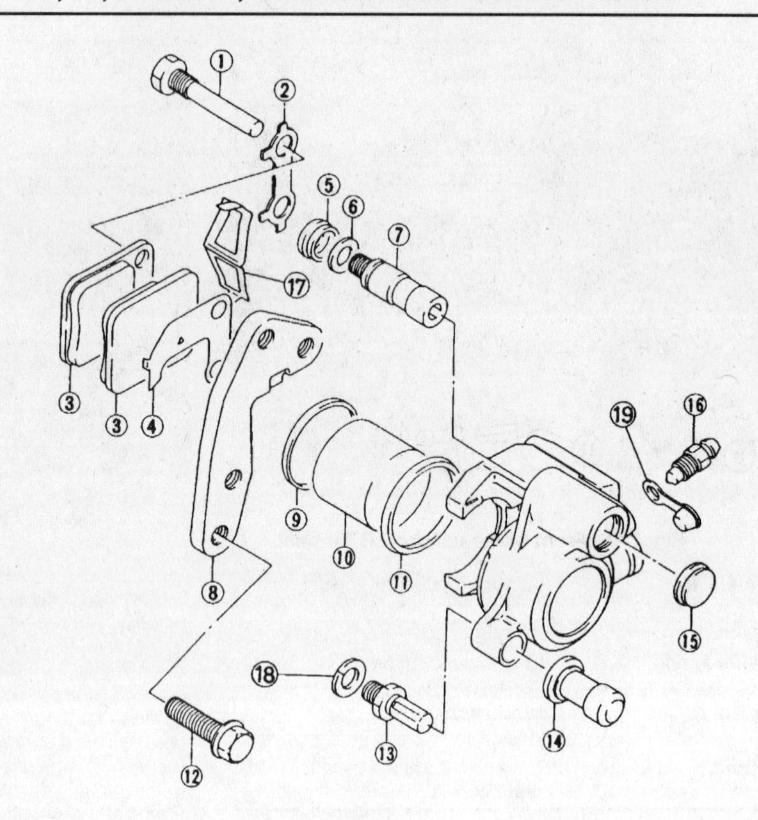

Fig. 7.15 Front brake caliper – DR model

27 Front brake: general – DR model

On DR 125 models the front brake is as previously mentioned in this Chapter is similar to that described for the GS125 ES models in the preceding Chapter, noting the differences shown in Fig. 7.15 and the revised information given in the Specifications Section of this Chapter. Note also that the brake disc is retained by four countersunk Allen bolts **(see illustration)** and that no torque wrench settings are available from the manufacturer. For all servicing procedures not given in this Chapter, refer to the information given for the GS125 ES models.

28 Brake caliper: removal, overhaul and refitting – GZ model

1 Disconnect the brake hose from the caliper by removing the union bolt. Allow the brake fluid to drain into a container.
2 Slacken the brake pad retaining bolt and remove the caliper by removing the two mounting bolts located in the front fork leg **(see illustrations)**. Remove the brake pads (see Section 3) and separate the mounting bracket from the caliper (see Fig. 7.16). Remove the pad spring from the caliper.
3 Place a clean rag over the piston to prevent it from popping out of the caliper. Using compressed air, force out the piston. Remove the dust seal and the piston seal (make note of its position before removing it). Do not reuse the seals, always renew them.
4 Inspect all the caliper parts for signs of damage, eg nicks and scratches. Clean the piston, piston bore and new seals with brake fluid but do not wipe off the fluid once applied. Ensure that the piston seal is fitted with its chamfer in the correct direction (see Fig. 7.17).
5 Refitting is reversal of removal but make sure that you push the piston all the way into the caliper. This enables you to insert the brake pads with ease. Tighten all the bolts to the specified torque. Bleed the brake system to remove any air (see Chapter 5, Section 8). Before sliding the mounting bracket into the caliper, lubricate the axle pins with brake caliper grease.

27.1 Disc is held direct to wheel hub by countersunk Allen bolts – DR model

28.2a Remove the caliper mounting bolts . . .

28.2b . . . and withdraw the caliper – GZ model

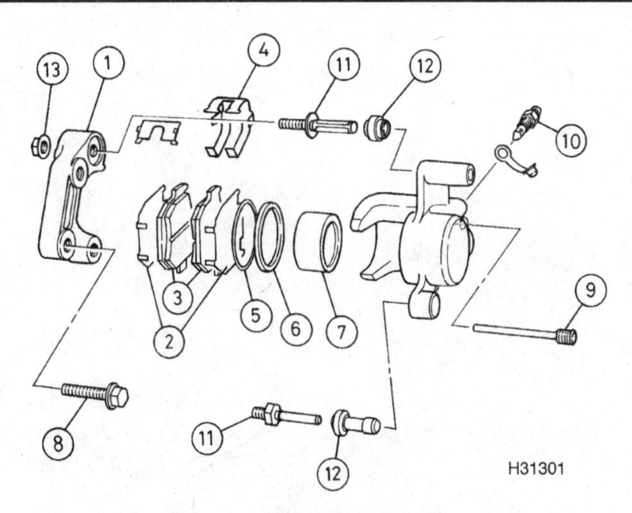

Fig. 7.16 Front brake caliper – GZ model

1 *Mounting bracket*
2 *Shims*
3 *Brake pads*
4 *Pad anti-rattle spring*
5 *Dust seal*
6 *Piston seal*
7 *Piston*
8 *Brake caliper mounting bolt*
9 *Brake pad retaining bolt*
10 *Air bleed nipple*
11 *Axle bolts*
12 *Dust boots*
13 *Nut*

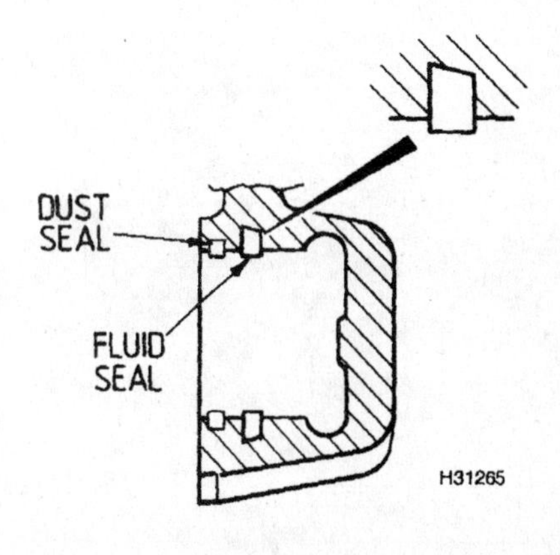

Fig. 7.17 Ensure the caliper piston (fluid) seal is fitted correctly – GZ model

29 Master cylinder: removal, inspection and refitting – GZ model

Follow the procedure in Chapter 5, Section 9, noting the following differences:

a) The front brake switch wires are simply disconnected from the switch; there is no need to withdraw the headlamp for access to the wiring connectors.

b) The fluid reservoir cover, plate and diaphragm are retained by two screws ***(see illustration)****.*

c) Refer to the torque wrench settings at the beginning of this Chapter.

30 Rear wheel: modifications – DR model

The rear wheel is generally similar to previous versions, the only significant change being the use of bolts, rather than studs and nuts, to retain the sprocket. The revised swinging arm includes snail type

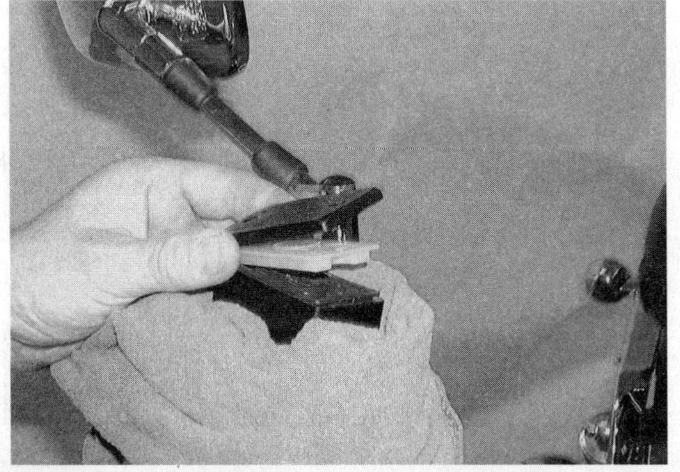

29.1 Removing the master cylinder cap and diaphragm – GZ model

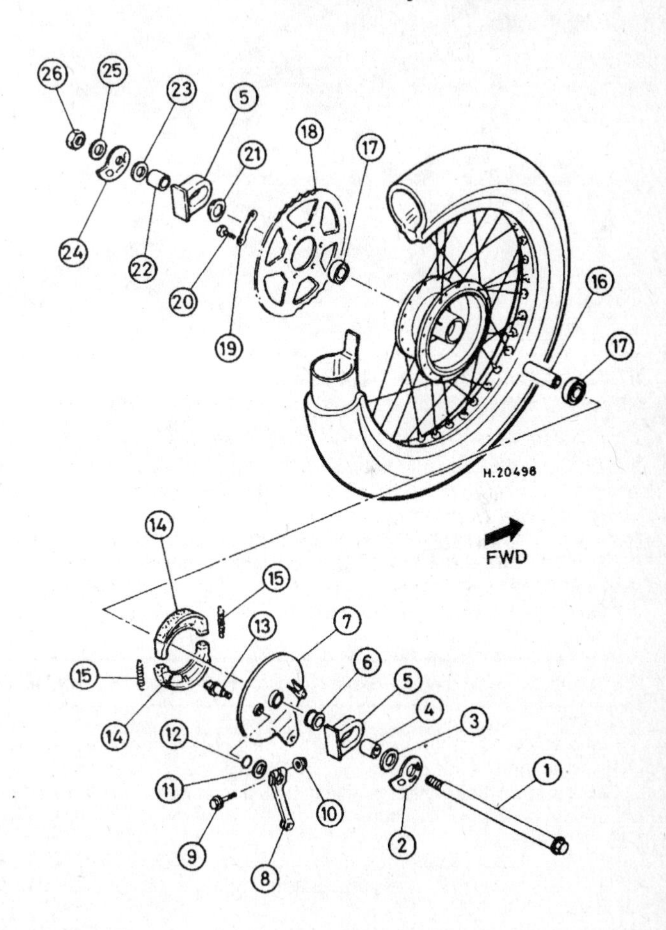

Fig. 7.18 Rear wheel – DR model

1 *Wheel spindle*
2 *Snail cam chain adjuster*
3 *Washer*
4 *Spacer*
5 *Spindle carrying block – 2 off*
6 *Spacer*
7 *Brake backplate*
8 *Operating lever*
9 *Bolt*
10 *Nut*
11 *Washer*
12 *Dust seal*
13 *Operating cam*
14 *Brake shoe – 2 off*
15 *Return spring – 2 off*
16 *Spacer*
17 *Bearing – 2 off*
18 *Sprocket*
19 *Tab washer – 3 off*
20 *Bolt – 6 off*
21 *Spacer*
22 *Spacer*
23 *Washer*
24 *Snail cam chain adjuster*
25 *Washer*
26 *Nut*

30.1 Note spindle-carrying block and spacer fitted inside swinging arm ends – DR model

31.1 Remove the rear brake adjusting nut – GZ model

31.2 Remove the rear axle bolt – GZ model

31.3a Remove the rear brake panel from the right side of the wheel . . .

31.3b . . . then lift out the sprocket and carrier from the left side

31.4 Separating the sprocket from the carrier – GZ model

chain adjusters in place of the drawbolt type fitted previously; note that these are handed and must be refitted with the marks and numbers facing outwards. Do not forget the washers fitted inside the snail cams and note the modified spindle-carrying blocks and spacers that fit inside the swinging arm fork ends **(see illustration)**; refer to Fig. 7.18 for details. Note also that the spindle nut is now of the self-locking type, the split pin being no longer fitted.

31 Rear wheel and rear brake: removal, overhaul and refitting – GZ model

Removal

1 Remove the safety clip and fully unscrew the rear brake adjusting nut from the end of the brake cable (**see illustration**). Remove the split pin (always replace with a new one) and the torque arm bolt, washer and nut to disconnect the torque arm from the brake backplate. Remove the axle nut.

2 Place a support under the rear of the engine to raise the rear wheel off the ground. Loosen the drive chain adjusting nuts on both sides of the rear wheel. Slide out the axle bolt (**see illustration**). Carefully remove the chain from the sprocket and remove the wheel.

3 Remove the brake backplate **(see illustration)**. Lift the rear wheel sprocket along with the sprocket carrier from the rear wheel **(see illustration)**.

4 Remove the four nuts on the back of the chain guide ring to enable you to separate the sprocket from the carrier (**see illustration**).

Overhaul

5 Refer to Chapter 5, Section 12 for brake shoe and drum inspection.

6 Refer to Chapter 5, Section 16 for wheel bearing and sprocket carrier bearing renewal procedures.

7 Check the cush drive rubber damper blocks in the wheel for any signs of wear or damage and renew them if necessary **(see illustration)**. The sprocket teeth must also be checked for signs of wear and if necessary the sprocket renewed, together with the gearbox sprocket and drive chain.

Refitting

8 Refitting is the reverse of removal. Note the torque settings given in the specifications at the beginning of this Chapter for the brake cam lever nut, sprocket nuts, torque arm nut and axle nut.

9 Adjust the drive chain tension as described in Routine maintenance at the front of this manual, noting the chain free play specification in this chapter's specifications.

31.7 Remove the sprocket damper rubber blocks – GZ model

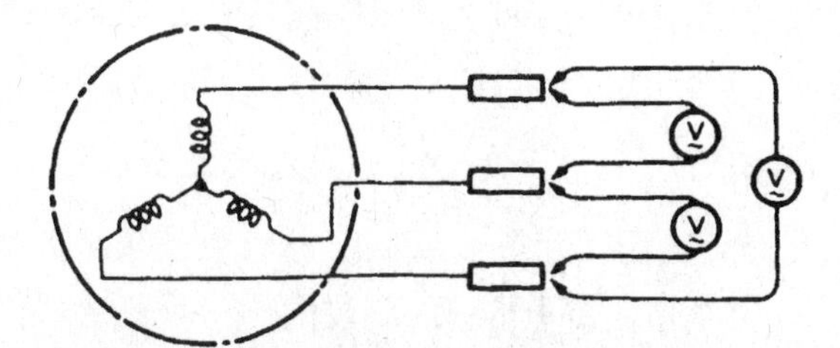

Fig. 7.19 Alternator no load test - DR model

32 Battery: general - GZ model

1 The battery fitted to this model is a sealed for life and is maintenance free, therefore requiring no regular check of its electrolyte level. The battery is located behind the right-hand frame cover.
2 The state of charge of the battery can be checked by measuring its voltage at the terminals. Using a DC voltmeter, connect its positive probe to the battery positive terminal and its negative probe to the battery negative terminal. Battery voltage should be above 12.5V.
3 If voltage is below 12.5V, Suzuki advise charging the battery at 0.7A for 5 to 10 hours or at 3A for 1 hour. If voltage is still below 12.5V after charging, charge once again, but if still below 12.5V renew the battery.

33 Alternator: testing - DR model

Noting the different information given in the Specifications Section of this Chapter, the alternator coils and output are tested as described in Chapter 6 for the GS125 ES models. The accompanying illustration (Fig. 7.19) shows details of the no-load voltage output test, which also applies to the earlier models.

34 Regulator/rectifier unit: testing - DR and GZ models

DR model

1 The regulator/rectifier unit is tested in much the same way as described in Chapter 6, Section 6. It should be noted, however, that the unit fitted to these models differs from earlier types **(see illustration)**. When making the various tests, set the multimeter to the kilo ohms scale, then carry out the test sequence in Fig.7.20.

GZ model

2 The test procedure recommended by Suzuki requires the use of a multimeter with a diode test facility, ideally their own multi circuit tester (pt no. 09900-25008). You should be able to obtain an indication of the unit's condition using another test meter, but note that the actual test values in the table cannot be guaranteed.
3 The regulator/rectifier is mounted on the left-hand side of the machine, near the carburettor. Disconnect the battery negative (-ve) terminal before carrying out any work on the unit. Disconnect the two- and three-pin wire connectors, then remove its two mounting bolts and remove the regulator/rectifier unit for testing.
4 Using a multimeter set to the diode test function, check the

34.1 Regulator/rectifier unit is mounted adjacent to the battery on DR model

Unit: Approx. kΩ

⊖ Probe of tester to:	⊕ Probe of tester to:					
	Y	Y	Y	B/W	O/B	R
Y		OFF	OFF	OFF	OFF	3.2
Y	OFF		OFF	OFF	OFF	3.2
Y	OFF	OFF		OFF	OFF	3.2
B/W	3.2	3.2	3.2		0.5	8.7
O/B	3.9	3.9	3.9	0.5		10
R	OFF	OFF	OFF	OFF	OFF	

Fig. 7.20 Regulator/rectifier test table - DR model

B/W Black with white tracer *O/B Orange with black tracer*
R Red *Y Yellow*

regulator/rectifier unit by applying its probes across the terminals shown in the accompanying table (Fig. 7.21). In each case, you are taking two readings, by reversing the meter probes. If the unit does not produce the expected results it is probably faulty, but do have your findings confirmed by a Suzuki dealer before purchasing a new unit.

35 Fuse: location and renewal - DR and GZ models

1 Later models are fitted with a flat-blade fuse in place of the conventional cartridge type used on previous models. Whilst this does not have any material effect on the function of the fuse, it does mean that only the correct replacement fuse can be used in the event of a failure (see Fig. 7.22). It follows that a replacement fuse of the correct

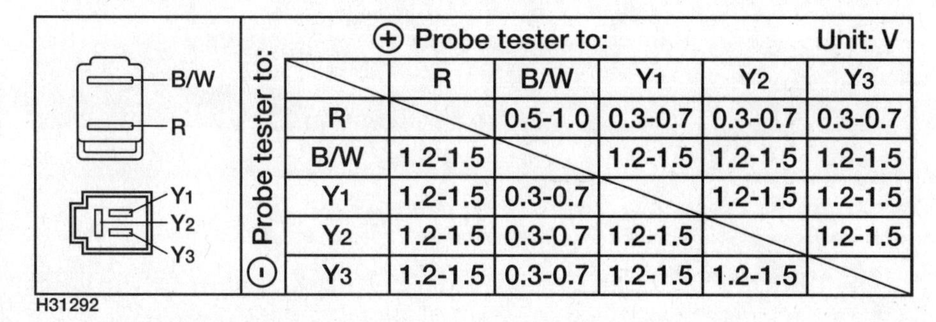

⊕ Probe tester to: Unit: V

⊖ Probe tester to:	R	B/W	Y1	Y2	Y3
R		0.5-1.0	0.3-0.7	0.3-0.7	0.3-0.7
B/W	1.2-1.5		1.2-1.5	1.2-1.5	1.2-1.5
Y1	1.2-1.5	0.3-0.7		1.2-1.5	1.2-1.5
Y2	1.2-1.5	0.3-0.7	1.2-1.5		1.2-1.5
Y3	1.2-1.5	0.3-0.7	1.2-1.5	1.2-1.5	

Fig. 7.21 Regulator/rectifier terminals and test table - GZ model

R Red Y Yellow B/W Black with white tracer

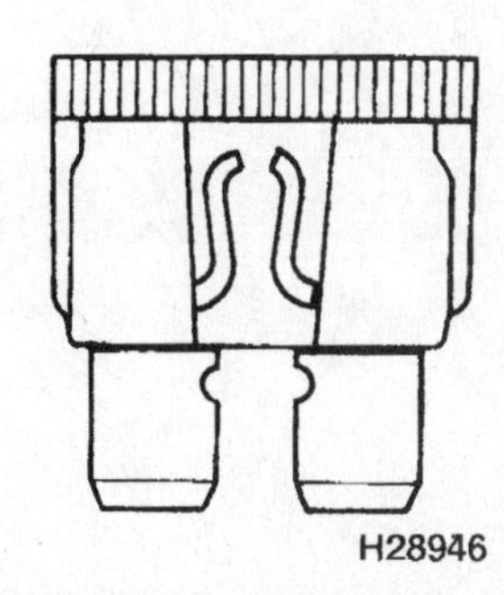

Fig. 7.22 A blown fuse can be identified by a break in its element

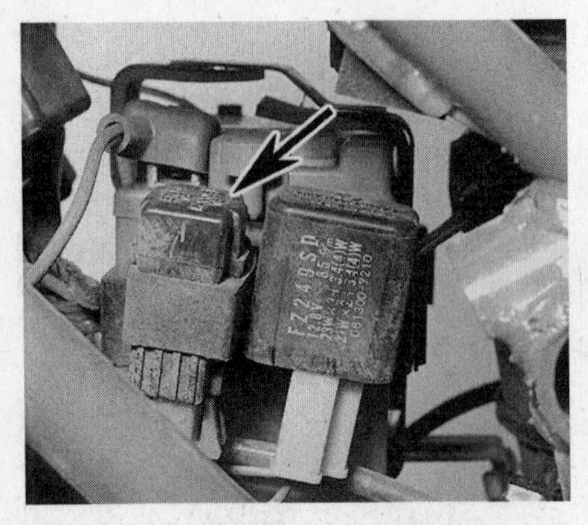

35.2a Fuse holder location – DR model

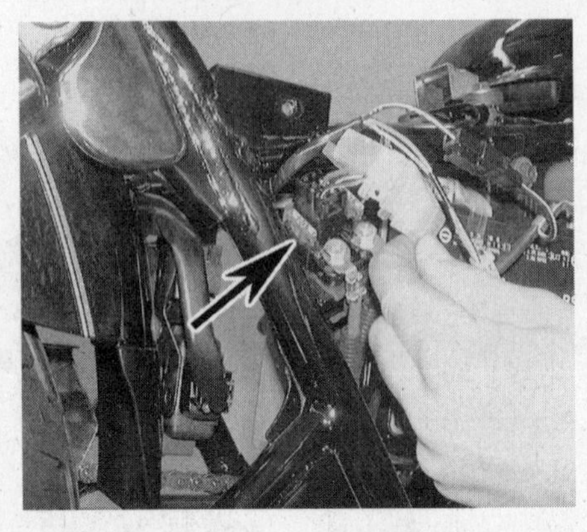

35.2b Fuse holder location – GZ model

36.1a Headlamp subframe uses rubber bushes to isolate vibration – DR model

36.1b Spring loaded adjusters provide beam alignment adjustment – DR model

36.1c Unplug connector and remove dust boot . . .

36.1d . . . then twist and remove retainer ring to release bulb – DR model

type and rating (15A for DR models and 20A for GZ models) should be carried at all times.

2 On the DR model the fuse is contained in a holder, next to the turn signal relay **(see illustration)**. On the GZ model the fuse is mounted on the top of the starter relay, which is located behind the right-hand side cover **(see illustration)**.

36 Headlamp and tail lamp: general

DR model

1 These models are equipped with a new headlamp unit to conform with the general styling alterations of the model. The unit is supported by a steel subframe, which is in turn attached by rubber-bushed bolts to the headlamp brackets on the steering head **(see illustration)**. The unit itself is rubber-mounted, its position within the subframe, and thus headlamp aim, being controlled by vertical and horizontal alignment screws **(see illustration)**. To gain access to the unit for bulb removal purposes, the moulded plastic cowling should be detached. This is retained by three screws **(see illustrations)**.

GZ model

2 On GZ models the headlamp is mounted on a bracket which is bolted to the lower yoke. The headlamp shell is fixed to this bracket by two bolts **(see illustration)**. To access the headlamp bulb, remove the two retaining screws from the base of the shell and manoeuvre the rim free **(see illustration)**. Carefully remove the wiring plug at the rear of

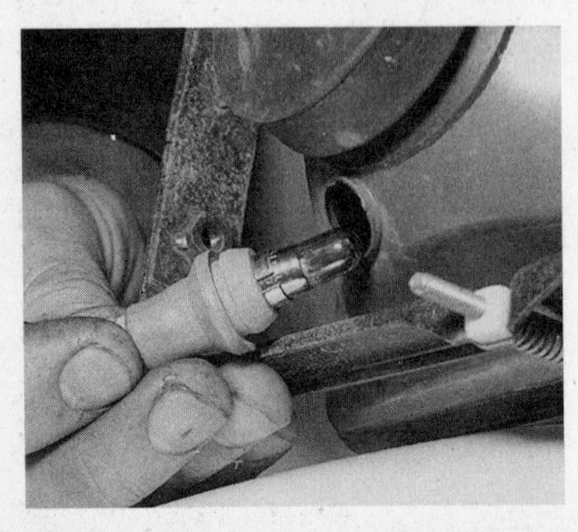

36.1e Parking bulb is a push fit in reflector – DR model

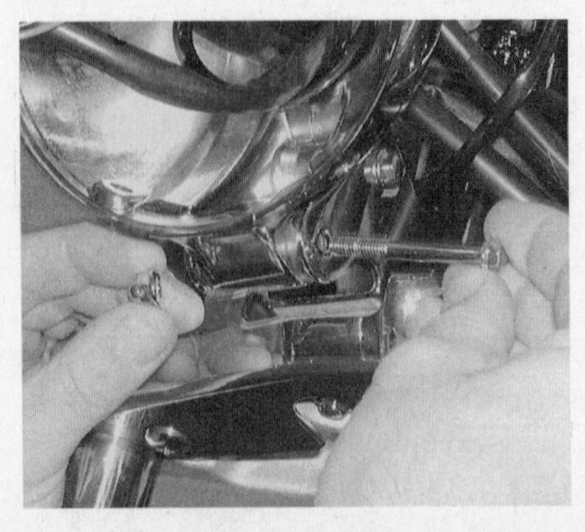

36.2a Remove the two headlamp mounting bolts from the mounting bracket – GZ model

36.2b Remove the lens by removing the retaining screws – GZ model

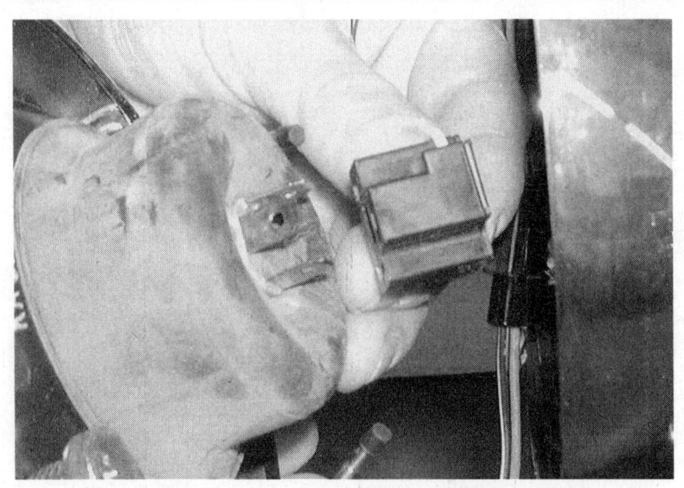
36.2c Unplug the wiring from the rear of the headlamp – GZ model

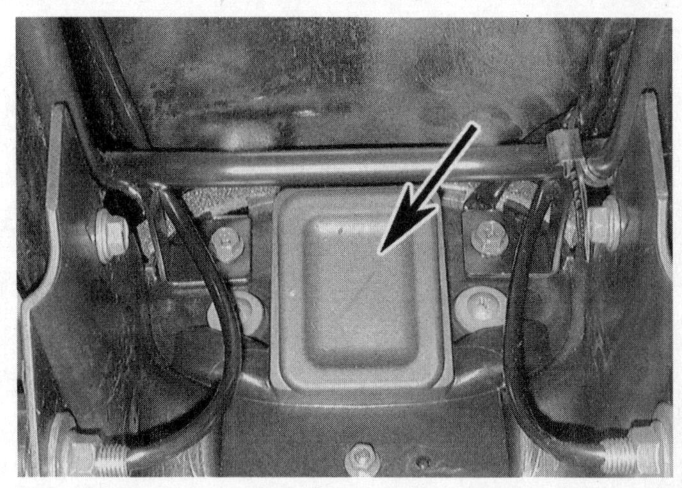
36.4a To gain access to the bulb, remove the rubber cover (arrowed) – GZ model

the headlamp and also remove the parking lamp bulbholder **(see illustration)**.

3 Peel back the rubber cover to access the headlamp bulb. Unhook the spring clip and lift the bulb out. **Note:** *When installing a new bulb, do not touch the glass with your fingers – hold the bulb by its wire connector tangs only.*

4 The tail lamp bulb can be accessed by prising out the rubber cover from inside the mudguard **(see illustration)**. Push in on the bulb and turn it anti-clockwise to remove it. To remove the tail lamp unit, remove the three bolts from inside the mudguard **(see illustrations)**.

GN model

5 Refer to the information for the GS125 in Chapter 6.

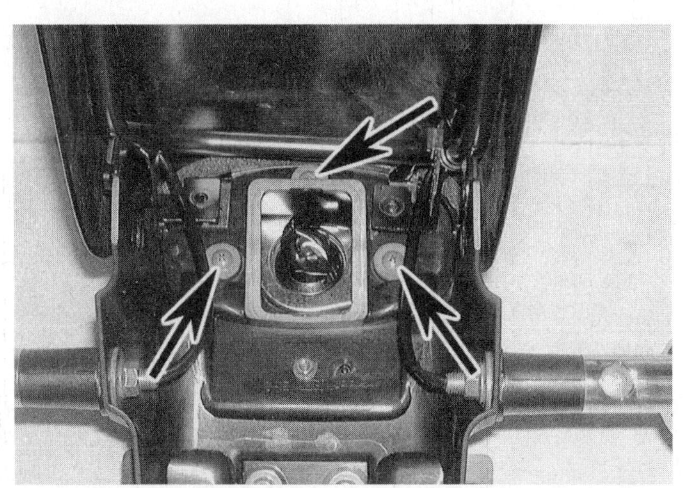
36.4b Remove the three screws . . .

36.4c . . . to free the tail lamp – GZ model

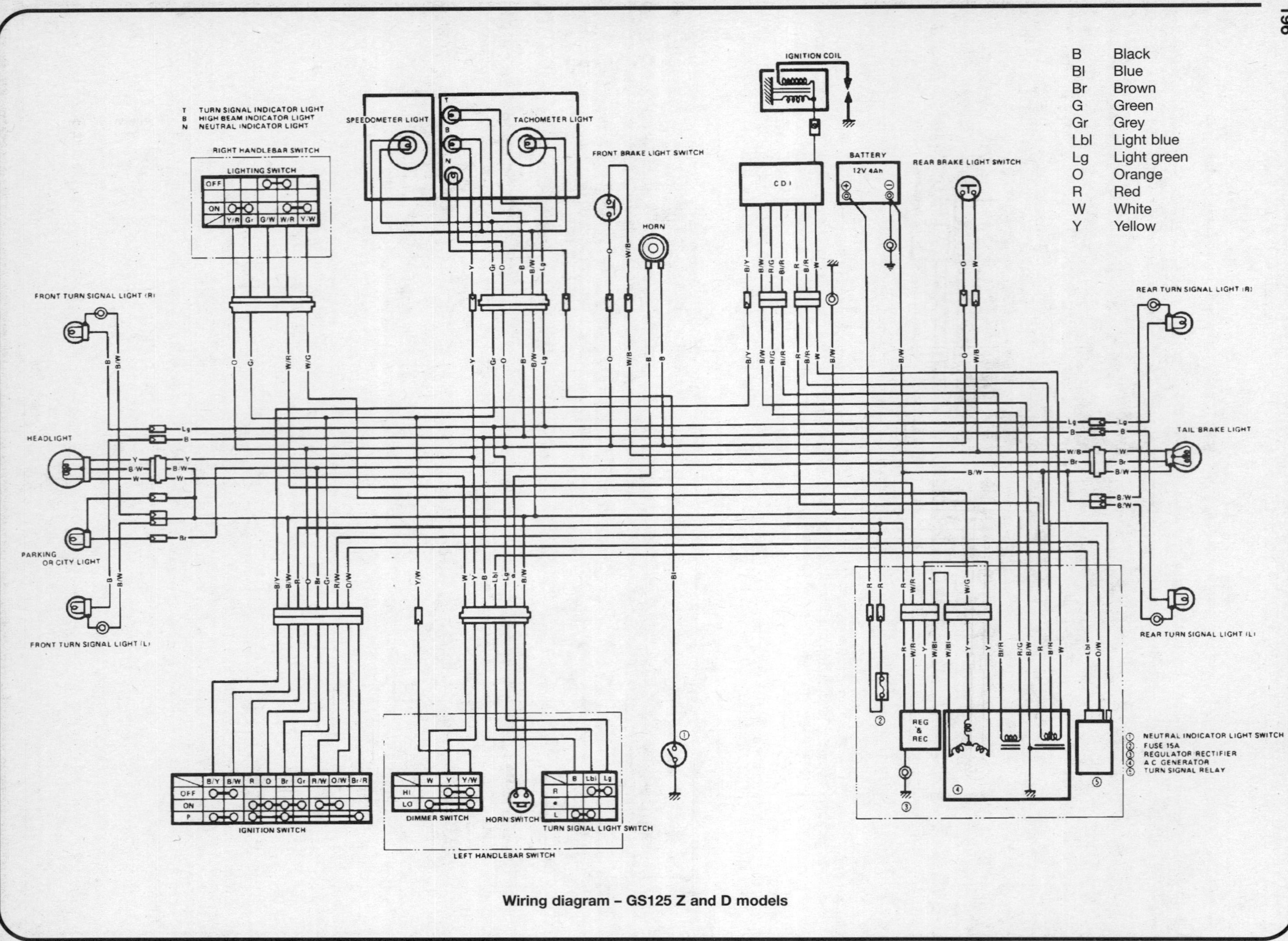

Wiring diagram – GS125 Z and D models

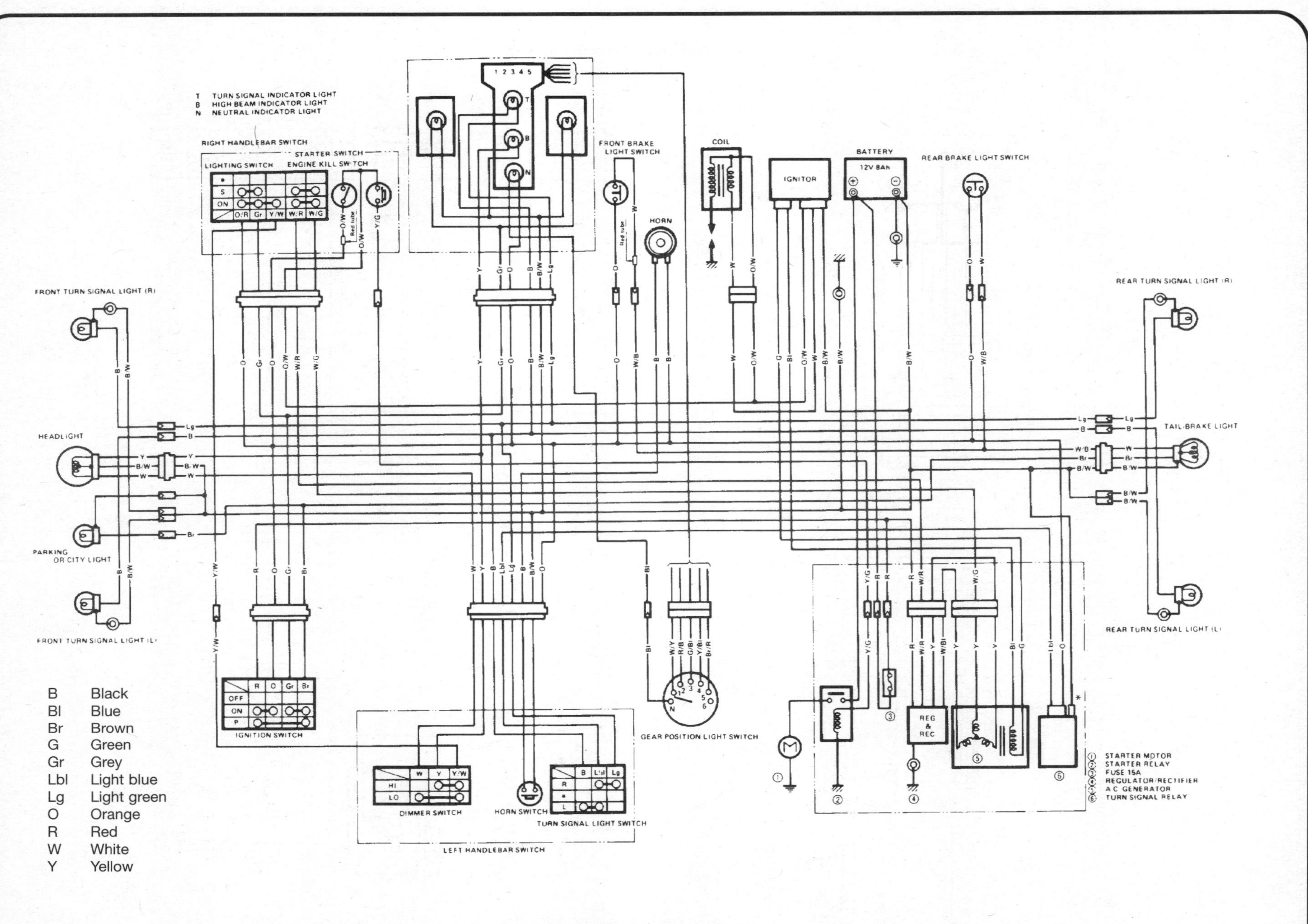

B	Black
Bl	Blue
Br	Brown
G	Green
Gr	Grey
Lbl	Light blue
Lg	Light green
O	Orange
R	Red
W	White
Y	Yellow

Wiring diagram – GS125 ES models

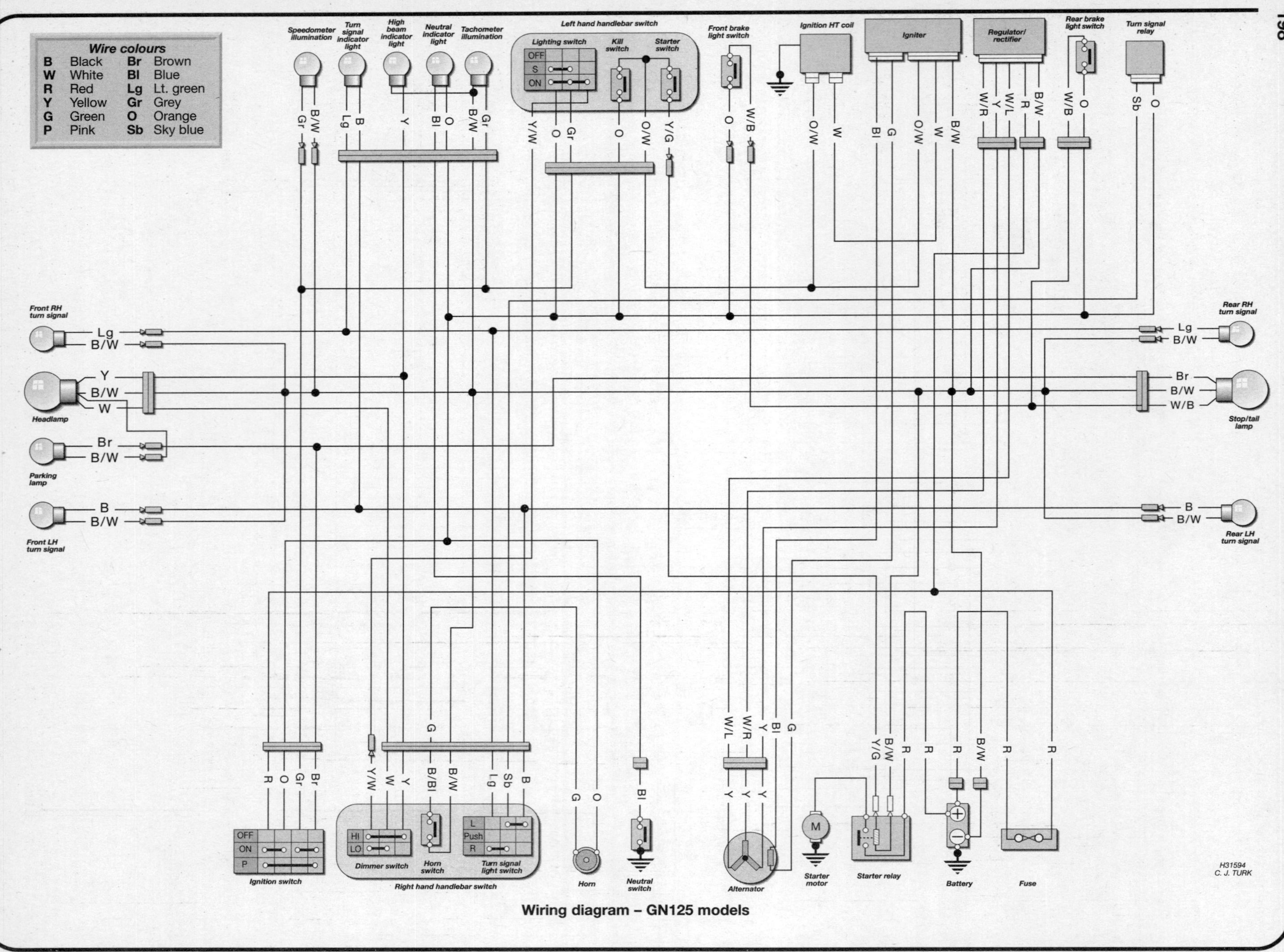

Wiring diagram – GN125 models

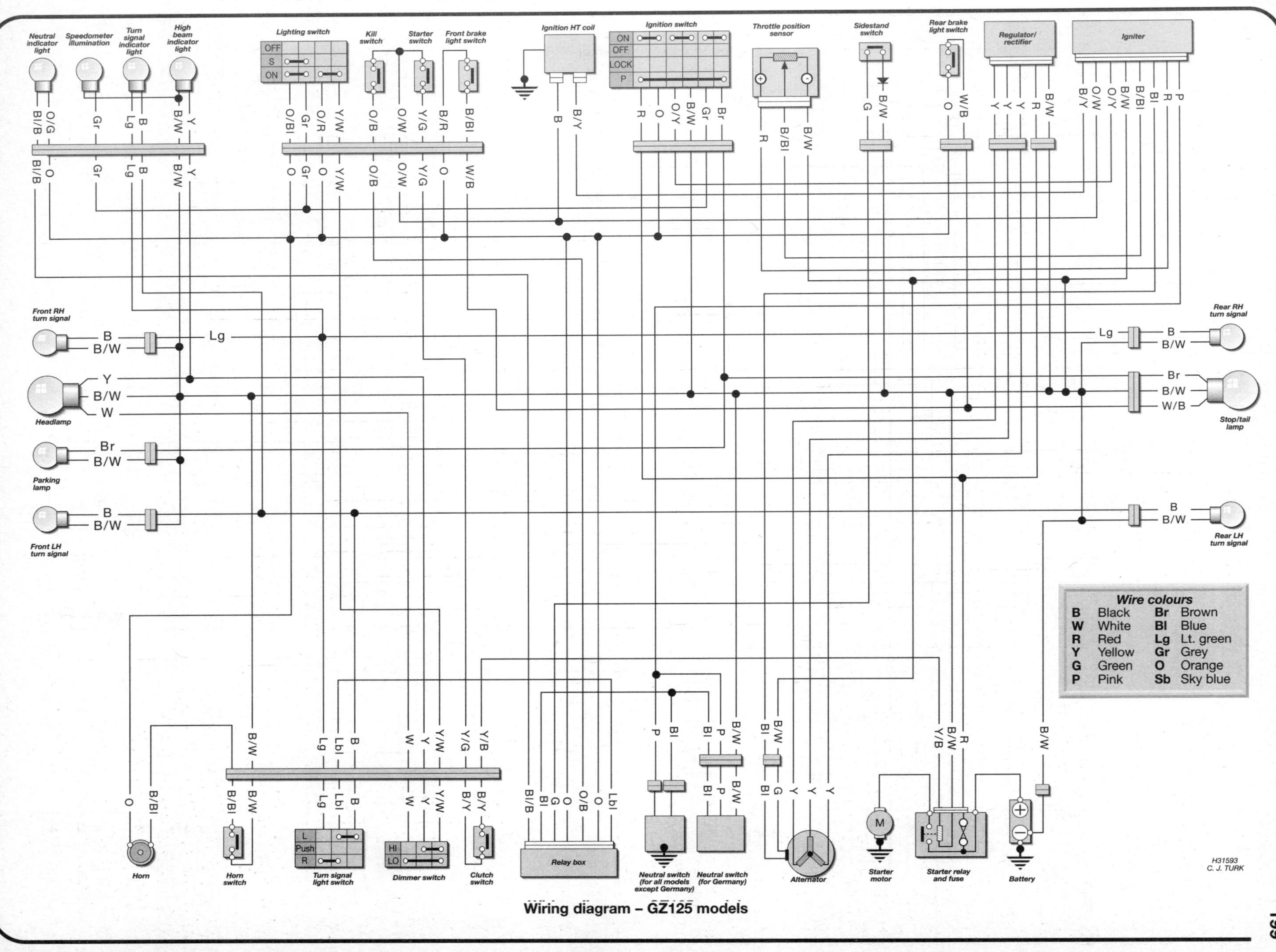

Wiring diagram – GZ125 models

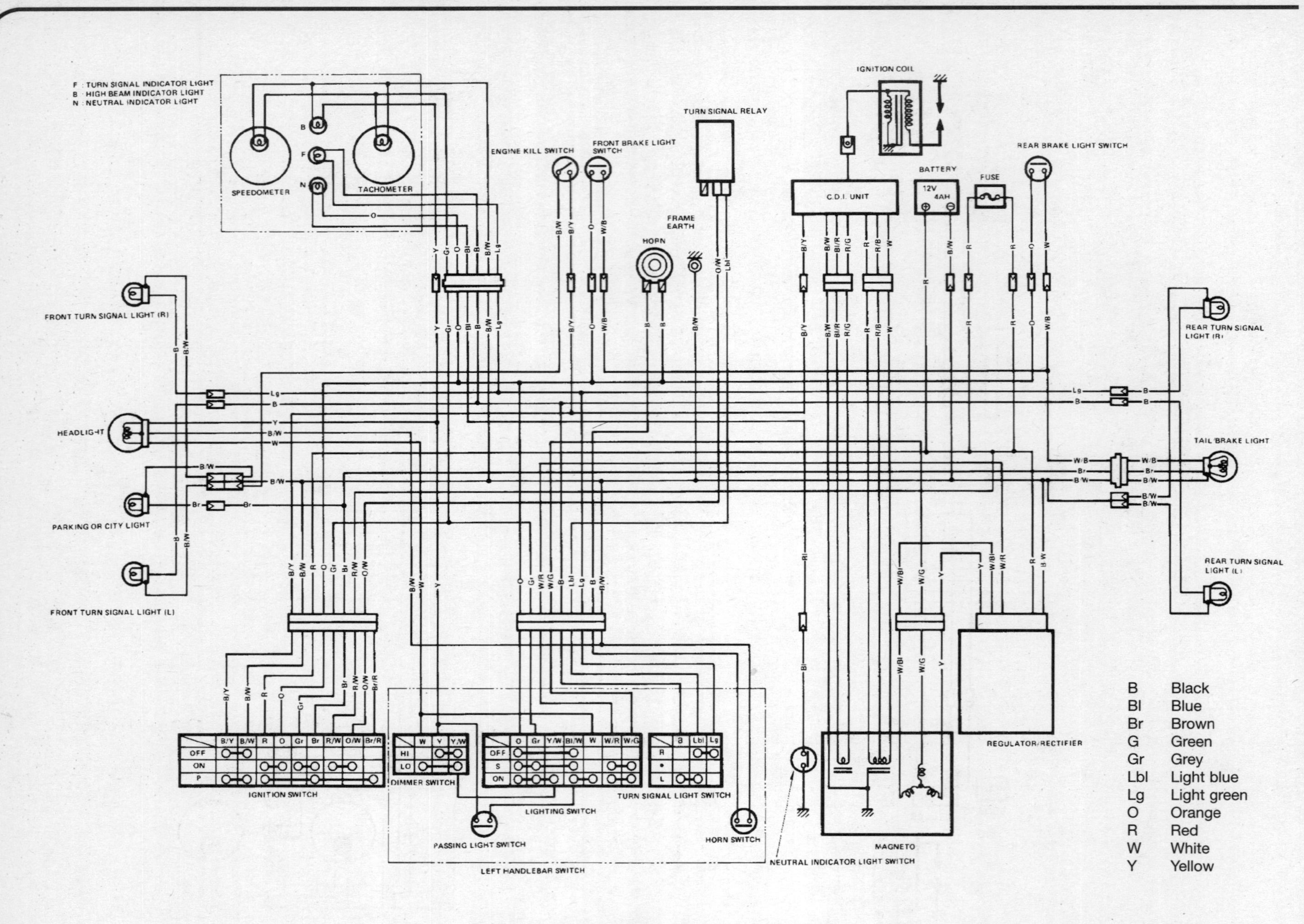

Wiring diagram - DR125 SZ and SD models

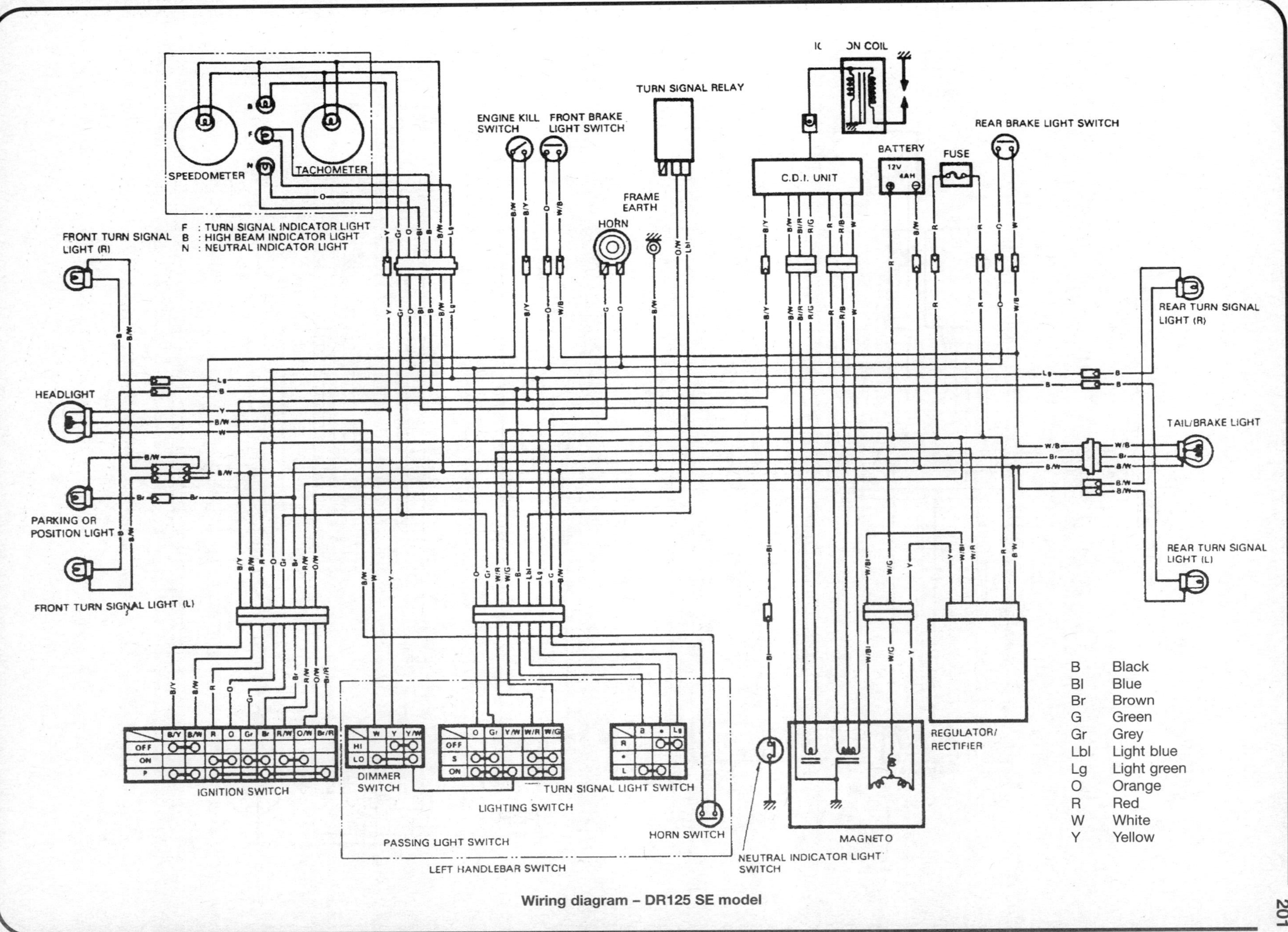

Wiring diagram – DR125 SE model

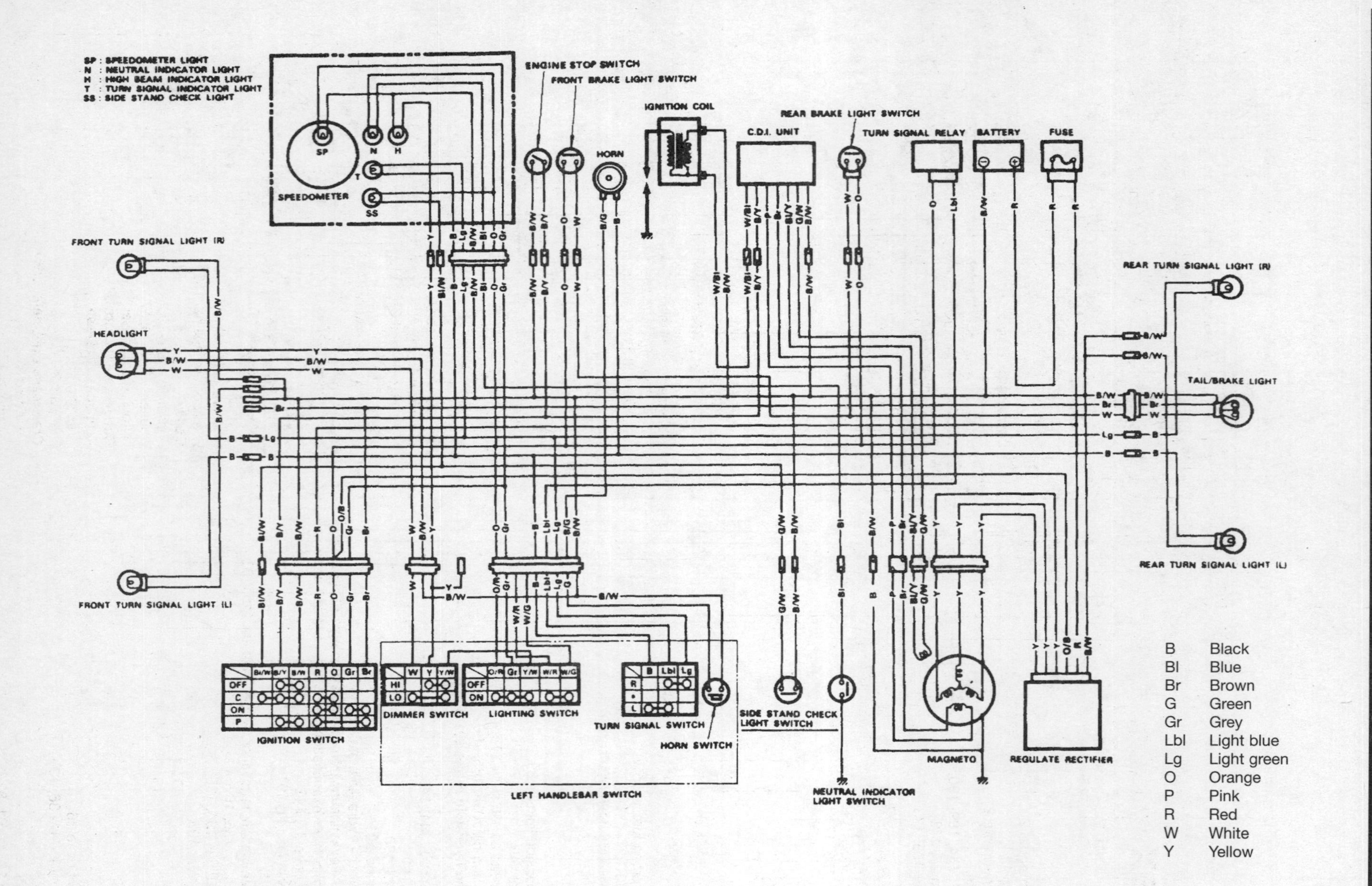

Wiring diagram – DR125 SF, SH and SJ 'Raider' models

Index

D

E

F

G

H

I

K

L

M

N

O

P

R

S

T

V

W

Haynes Motorcycle Manuals – The Complete List

Title	Book No
BMW	
BMW 2-valve Twins (70 - 96)	0249
BMW K100 & 75 2-valve Models (83 - 96)	1373
BMW R850 & R1100 4-valve Twins (93 - 97)	3466
BSA	
BSA Bantam (48 - 71)	0117
BSA Unit Singles (58 - 72)	0127
BSA Pre-unit Singles (54 - 61)	0326
BSA A7 & A10 Twins (47 - 62)	0121
BSA A50 & A65 Twins (62 - 73)	0155
DUCATI	
Ducati 600, 750 & 900 2-valve V-Twins (91 - 96)	3290
HARLEY-DAVIDSON	
Harley-Davidson Sportsters (70 - 99)	0702
Harley-Davidson Big Twins (70 - 99)	0703
HONDA	
Honda NB, ND, NP & NS50 Melody (81 - 85)	◇ 0622
Honda NE/NB50 Vision & SA50 Vision Met-in (85 - 95)	◇ 1278
Honda MB, MBX, MT & MTX50 (80 - 93)	0731
Honda C50, C70 & C90 (67 - 99)	0324
Honda CR80R & CR125R (86 - 97)	2220
Honda XR80R & XR100R (85 - 96)	2218
Honda XL/XR 80, 100, 125, 185 & 200 2-valve Models (78 - 87)	0566
Honda H100 & H100S Singles (80 - 92)	◇ 0734
Honda CB/CD125T & CM125C Twins (77 - 88)	◇ 0571
Honda CG125 (76 - 94)	◇ 0433
Honda NS125 (86 - 93)	◇ 3056
Honda MBX/MTX125 & MTX200 (83 - 93)	◇ 1132
Honda CD/CM185 200T & CM250C 2-valve Twins (77 - 85)	0572
Honda XL/XR 250 & 500 (78 - 84)	0567
Honda XR250L, XR250R & XR400R (86 - 97)	2219
Honda CB250 & CB400N Super Dreams (78 - 84)	◇ 0540
Honda CR250R & CR500R (86 - 97)	2222
Honda Elsinore 250 (73 - 75)	0217
Honda CBR400RR Fours (88 - 99)	3552
Honda VFR400 (NC30) & RVF400 (NC35) V-Fours (89 - 98)	3496
Honda CB400 & CB550 Fours (73 - 77)	0262
Honda CX/GL500 & 650 V-Twins (78 - 86)	0442
Honda CBX550 Four (82 - 86)	◇ 0940
Honda XL600R & XR600R (83 - 96)	2183
Honda CBR600F1 & 1000F Fours (87 - 96)	1730
Honda CBR600F2 & F3 Fours (91 - 98)	2070
Honda CB650 sohc Fours (78 - 84)	0665
Honda NTV600 & 650 V-Twins (88 - 96)	3243
Honda Shadow VT600 & 750 (USA) (88 - 99)	2312
Honda CB750 sohc Four (69 - 79)	0131
Honda V45/65 Sabre & Magna (82 - 88)	0820
Honda VFR750 & 700 V-Fours (86 - 97)	2101
Honda VFR800 V-Fours (97 - 99)	3703
Honda CB750 & CB900 dohc Fours (78 - 84)	0535
Honda CBR900RR FireBlade (92 - 99)	2161
Honda ST1100 Pan European V-Fours (90 - 97)	3384
Honda Shadow VT1100 (USA) (85 - 98)	2313
Honda GL1000 Gold Wing (75 - 79)	0309
Honda GL1100 Gold Wing (79 - 81)	0669
Honda Gold Wing 1200 (USA) (84 - 87)	2199
Honda Gold Wing 1500 (USA) (88 - 98)	2225
KAWASAKI	
Kawasaki AE/AR 50 & 80 (81 - 95)	1007
Kawasaki KC, KE & KH100 (75 - 99)	1371
Kawasaki KMX125 & 200 (86 - 96)	◇ 3046
Kawasaki 250, 350 & 400 Triples (72 - 79)	0134
Kawasaki 400 & 440 Twins (74 - 81)	0281
Kawasaki 400, 500 & 550 Fours (79 - 91)	0910
Kawasaki EN450 & 500 Twins (Ltd/Vulcan) (85 - 93)	2053
Kawasaki EX & ER500 (GPZ500S & ER-5) Twins (87 - 99)	2052
Kawasaki ZX600 (Ninja ZX-6, ZZ-R600) Fours (90 - 97)	2146
Kawasaki ZX-6R Ninja Fours (95 - 98)	3541
Kawasaki ZX600 (GPZ600R, GPX600R, Ninja 600R & RX) & ZX750 (GPX750R, Ninja 750R) Fours (85 - 97)	1780
Kawasaki 650 Four (76 - 78)	0373
Kawasaki 750 Air-cooled Fours (80 - 91)	0574
Kawasaki ZR550 & 750 Zephyr Fours (90 - 97)	3382
Kawasaki ZX750 (Ninja ZX-7 & ZXR750) Fours (89 - 96)	2054
Kawasaki ZX750P (Ninja ZX-7R) & ZX900B/C/E (Ninja ZX-9R) (94 - 00)	3721
Kawasaki 900 & 1000 Fours (73 - 77)	0222
Kawasaki ZX900, 1000 & 1100 Liquid-cooled Fours (83 - 97)	1681
MOTO GUZZI	
Moto Guzzi 750, 850 & 1000 V-Twins (74 - 78)	0339
MZ	
MZ ETZ Models (81 - 95)	◇ 1680
NORTON	
Norton 500, 600, 650 & 750 Twins (57 - 70)	0187
Norton Commando (68 - 77)	0125
PIAGGIO	
Piaggio (Vespa) Scooters (91 - 98)	3492
SUZUKI	
Suzuki GT, ZR & TS50 (77 - 90)	◇ 0799
Suzuki TS50X (84 - 00)	◇ 1599
Suzuki 100, 125, 185 & 250 Air-cooled Trail bikes (79 - 89)	0797
Suzuki GP100 & 125 Singles (78 - 93)	◇ 0576
Suzuki GS, GN, GZ & DR125 Singles (82 - 99)	◇ 0888
Suzuki GT250X7, GT200X5 & SB200 Twins (78 - 83)	◇ 0469
Suzuki GS/GSX250, 400 & 450 Twins (79 - 85)	0736
Suzuki GS500E Twin (89 - 97)	3238
Suzuki GS550 (77 - 82) & GS750 Fours (76 - 79)	0363
Suzuki GS/GSX550 4-valve Fours (83 - 88)	1133
Suzuki GSX-R600 & 750 (96 - 99)	3553
Suzuki GSF600 & 1200 Bandit Fours (95 - 97)	3367
Suzuki GS850 Fours (78 - 88)	0536
Suzuki GS1000 Four (77 - 79)	0484
Suzuki GSX-R750, GSX-R1100 (85 - 92), GSX600F, GSX750F, GSX1100F (Katana) Fours (88 - 96)	2055
Suzuki GS/GSX1000, 1100 & 1150 4-valve Fours (79 - 88)	0737
TRIUMPH	
Triumph 350 & 500 Unit Twins (58 - 73)	0137
Triumph Pre-Unit Twins (47 - 62)	0251
Triumph 650 & 750 2-valve Unit Twins (63 - 83)	0122
Triumph Trident & BSA Rocket 3 (69 - 75)	0136
Triumph Triples & Fours (carburettor engines) (91 - 99)	2162
VESPA	
Vespa P/PX125, 150 & 200 Scooters (78 - 95)	0707
Vespa Scooters (59 - 78)	0126
YAMAHA	
Yamaha DT50 & 80 Trail Bikes (78 - 95)	◇ 0800
Yamaha T50 & 80 Townmate (83 - 95)	◇ 1247
Yamaha YB100 Singles (73 - 91)	◇ 0474
Yamaha RS/RXS100 & 125 Singles (74 - 95)	0331
Yamaha RD & DT125LC (82 - 87)	◇ 0887
Yamaha TZR125 (87 - 93) & DT125R (88 - 95)	◇ 1655
Yamaha TY50, 80, 125 & 175 (74 - 84)	◇ 0464
Yamaha XT & SR125 (82 - 96)	1021
Yamaha 250 & 350 Twins (70 - 79)	0040
Yamaha XS250, 360 & 400 sohc Twins (75 - 84)	0378
Yamaha RD250 & 350LC Twins (80 - 82)	0803
Yamaha RD350 YPVS Twins (83 - 95)	1158
Yamaha RD400 Twin (75 - 79)	0333
Yamaha XT, TT & SR500 Singles (75 - 83)	0342
Yamaha XZ550 Vision V-Twins (82 - 85)	0821
Yamaha FJ, FZ, XJ & YX600 Radian (84 - 92)	2100
Yamaha XJ600S (Diversion, Seca II) & XJ600N Fours (92 - 99)	2145
Yamaha YZF600R Thundercat & FZS600 Fazer (96 - 00)	3702
Yamaha 650 Twins (70 - 83)	0341
Yamaha XJ650 & 750 Fours (80 - 84)	0738
Yamaha XS750 & 850 Triples (76 - 85)	0340
Yamaha TDM850, TRX850 & XTZ750 (89 - 99)	3540
Yamaha YZF750R & YZF1000R Thunderace (93 - 00)	3720
Yamaha FZR600, 750 & 1000 Fours (87 - 96)	2056
Yamaha XV V-Twins (81 - 96)	0802
Yamaha XJ900F Fours (83 - 94)	3239
Yamaha XJ900 S Diversion (94 - 00)	3739
Yamaha FJ1100 & 1200 Fours (84 - 96)	2057
ATVS	
Honda ATC70, 90, 110, 185 & 200 (71 - 85)	0565
Honda TRX300 Shaft Drive ATVs (88 - 95)	2125
Honda TRX300EX & TRX400EX ATVs (93 - 99)	2318
Polaris ATVs (85 to 97)	2302
Yamaha YT, YFM, YTM & YTZ ATVs (80 - 85)	1154
Yamaha YFS200 Blaster ATV (88 - 98)	2317
Yamaha YFB250 Timberwolf ATV (92 - 96)	2217
Yamaha YFM350 Big Bear and ER ATVs (87 - 95)	2126
Yamaha Warrior and Banshee ATVs (87 - 99)	2314
ATV Basics	10450
TECHNICAL TITLES	
Motorcycle Basics Manual	1083
MOTORCYCLE TECHBOOKS	
Motorcycle Basics TechBook	3515
Motorcycle Electrical TechBook (3rd Edition)	3471
Motorcycle Fuel Systems TechBook	3514
Motorcycle Workshop Practice TechBook (2nd Edition)	3470

◇ = *not available in the USA* **Bold type** = *Superbike*

The manuals on this page are available through good motorcyle dealers and accessory shops.
In case of difficulty, contact: **Haynes Publishing**
(UK) **+44 1963 440635** (USA) **+1 805 4986703**
(FR) **+33 1 47 78 50 50** (SV) **+46 18 124016**
(Australia/New Zealand) **+61 3 9763 8100**